SYMPOSIUM ON THE FOUNDATIONS OF MODERN PHYSICS 1987

SYMPOSIUM ON THE FOUNDATIONS OF MODERN PHYSICS 1987

The Copenhagen Interpretation 60 years after the Como Lecture

Joensuu, Finland

6-8 August 1987

Pekka Lahti

University of Turku

Peter Mittelstaedt

University of Cologne

World Scientific

Singapore • New Jersey • Hong Kong

Published by

World Scientific Publishing Co. Pte. Ltd.
P.O. Box 128, Farrer Road, Singapore 9128

U. S. A. office: World Scientific Publishing Co., Inc.
687 Hartwell Street, Teaneck NJ 07666, USA

Library of Congress Cataloging-in-Publication data is available.

SYMPOSIUM ON THE FOUNDATIONS OF MODERN PHYSICS

ISBN 9971-50-382-4
9971-50-460-X (pbk)

Printed in Singapore by Utopia Press.

FOREWORD

This volume contains the proceedings of the "SYMPOSIUM ON THE FOUNDATIONS OF MODERN PHYSICS: The Copenhagen Interpretation 60 years after the Como Lecture" which was held in Joensuu, Finland, at and under the aegis of the Department of Physics of the University of Joensuu during August 6 - 8, 1987.

The aim of the Symposium was to review the main aspects of the Copenhagen interpretation of quantum mechanics, to discuss the relevance of this interpretation for our present understanding of quantum mechanics, to present some systematic interpretations of the theory, and to discuss the role of the observer in quantum physics.

The programme of the Symposium consisted of 12 invited lectures, 11 contributed lectures, a poster session, an experimental demonstration, and three panel discussions. The first panel was chaired by Professor C.F. von Weizsäcker and it was pondering "the essentials of the Copenhagen interpretation". The second panel was chaired by Professor Asher Peres, and it was dealing with "the problem of realism in the sense of a detached observer". The third panel discussed "the philosophical implications of quantum mechanics in a broader sense", and it was chaired by Professor Henry J. Folse. This programme emerged from a close collaboration between the Organizing Committee and the Scientific Advisory Committee. The programme was designed to fit best the aims of the Symposium. In particular, the contributed lectures were selected from the thirty-five submitted papers on the basis of the recommendations of the Scientific Advisory Committee.

The Organizing Committee consisted of Kalervo Laurikainen (chairman, Helsinki), Rauno Hämäläinen (Joensuu), Pekka Lahti (Turku), Pertti Lipas (Jyväskylä), Simo Liukkonen (Otaniemi), Kauko Mansikka (Turku), Jussi Rastas (Otaniemi), and Jouni Viiri (secretary, Joensuu). The members of the Scientific Advisory Committee were Konrad Bleuler (University of Bonn, West Germany), Charles P. Enz (University of Geneva, Switzerland), Peter Mittelstaedt (University of Cologne, West Germany), and Hans Primas (ETH, Zürich, Switzerland). The Joensuu Symposium was host to approximately 80 participants, whose names are listed at the end of this volume.

These proceedings contain the invited and the contributed lectures of the Symposium. This material has been organized into three parts. Part I contains the scientific programme of the Opening Session of the Symposium. Part II collects the invited lectures, whereas Part III contains the contributed lectures of the Symposium. The papers presented at the poster session as well as the experimental demonstration are mentioned at the end of Part III.

The Panel Discussions were recorded and they will be published as a separate booklet by the end of this year in the Report Series of the Department of Physical Sciences of the University of Turku.

Due to the special character of the Symposium, it was thought advisable to include in these Proceedings a published version of the Como lecture of Bohr, "The Quantum Postulate and the Recent Development of Atomic Theory". This paper was originally published in Nature 121 (1928) 580 - 590.

On behalf of the organizers we wish to thank the following organizations for their valuable support to the Symposium: City of Joensuu, Committee of Particle Physics of the Academy of Finland, Finnish Ministry of Education,

Neste Oy Foundation, Research Institute for Theoretical Physics in Helsinki, Turku University Foundation, British Council, French Ministry of Foreigns Affairs, NORDITA.

Cologne, October 1987

Pekka J. Lahti
Peter Mittelstaedt

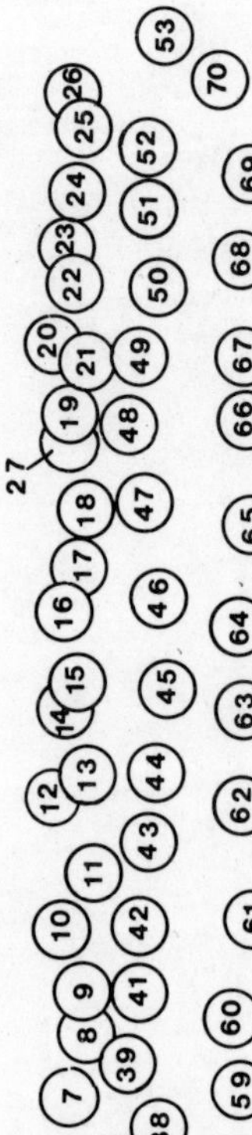
1
2
3
4
5
6
7
8
9
10
11
12
13
14
15
16
17
18
19
20
21
22
23
24
25
26
27
28
29
30
31
32
33
34
35
36
37
38
39
40
41
42
43
44
45
46
47
48
49
50
51
52
53
54
55
56
57
58
59
60
61
62
63
64
65
66
67
68
69
70

SYMPOSIUM ON THE FOUNDATIONS OF MODERN PHYSICS

1. K. Pusa
2. V.-M. Hokkanen
3. L. Smulter
4. H. Rainio
5. G. Wikman
6. P. Pylkkänen
7. C. Garola
8. G. Cattaneo
9. L. Solombrino
10. I. Niiniluoto
11. S. Sandelin
12. K. Rahikainen
13. M. Bitbol
14. P. Alanen
15. B. d'Espagnat
16. H. Primas
17. L. Ryff
18. A. Peres
19. W. Daniel
20. P. Christiansen
21. M. Grabowski
22. T. Hyvönen
23. A. Zeilinger
24. C.P. Enz
25. J. Porra
26. A. Raczynski
27. H. Jämsä
28. P. Lahti
29. P. Sibelius
30. L. Szabo
31. L. Holmberg
32. J. Rastas
33. S. Rügh
34. S. Liukkonen
35. A. Prieur
36. K. Ylinen
37. J. Scheer
38. R. Piirainen
39. E. Huhmar
40. R. Hämäläinen
41. P. Kruszynski
42. J. Fennema
43. D.J. Hooton
44. P. Gregorius
45. Th. Görnitz
46. I. Ohba
47. J. Farina
48. M. Lensu
49. P. Busch
50. P. Burt
51. P. Noyes
52. A. Bach
53. B. Nagel
54. W. Andersson
55. J. Wignall
56. J. Viiri
57. J. Rayski
58. W. de Muynck
59. T. Seppälä
60. M. Jones
61. M. Golshani
62. H. Folse
63. D. Bohm
64. P. Mittelstaedt
65. K.V. Laurikainen
66. J. Mehra
67. G. Szamosi
68. C.F. von Weizsäcker
69. C. Westerlund
70. D. Finkelstein

CONTENTS

Fig. 1: The panelists (from left) C.F. von Weizsäcker, Henry Folse David Finkelstein, Asher Peres (chairman), Peter Mittelstaedt, Paul Busch, and Bernard d'Espagnat are discussing "the problem of realism in the sense of a detached observer".

Fig. 3: Annssi Mälkki is demonstrating the photon split beam experiment.

Fig. 2: The panelists (from left) Jerzy, Rayski, C.F. von Weizsäcker Anton Zeilinger, Henry Folse (chairman), David Bohm, Charles Enz, Kalervo Laurikainen are discussing "the philosophical implications of quantum mechanics in a broader sense".

Fig. 4: Mrs. Liisa Matveinen is playing the kantele at the farewell party.

The Quantum Postulate and the Recent Development of Atomic Theory.[1]

By Prof. N. BOHR, For. Mem. R.S.

IN connexion with the discussion of the physical interpretation of the quantum theoretical methods developed during recent years, I should like to make the following general remarks regarding the principles underlying the description of atomic phenomena, which I hope may help to harmonise the different views, apparently so divergent, concerning this subject.

1. QUANTUM POSTULATE AND CAUSALITY.

The quantum theory is characterised by the acknowledgment of a fundamental limitation in the classical physical ideas when applied to atomic phenomena. The situation thus created is of a peculiar nature, since our interpretation of the experimental material rests essentially upon the classical concepts. Notwithstanding the difficulties which hence are involved in the formulation of the quantum theory, it seems, as we shall see, that its essence may be expressed in the so-called quantum postulate, which attributes to any atomic process an essential discontinuity, or rather individuality, completely foreign to the classical theories and symbolised by Planck's quantum of action.

This postulate implies a renunciation as regards the causal space-time co-ordination of atomic processes. Indeed, our usual description of physical phenomena is based entirely on the idea that the phenomena concerned may be observed without disturbing them appreciably. This appears, for example, clearly in the theory of relativity, which has been so fruitful for the elucidation of the classical theories. As emphasised by Einstein, every observation or measurement ultimately rests on the coincidence of two independent events at the same space-time point. Just these coincidences will not be affected by any differences which the space-time co-ordination of different observers otherwise may exhibit. Now the quantum postulate implies that any observation of atomic phenomena will involve an interaction with the agency of observation not to be neglected. Accordingly, an independent reality in the ordinary physical sense can neither be ascribed to the phenomena nor to the agencies of observation. After all, the concept of observation is in so far arbitrary as it depends upon which objects are included in the system to be observed. Ultimately every observation can of course be reduced to our sense perceptions. The circumstance, however, that in interpreting observations use has always to be made of theoretical notions, entails that for every particular case it is a question of convenience at what point the concept of observation involving the quantum postulate with its inherent 'irrationality' is brought in.

This situation has far-reaching consequences. On one hand, the definition of the state of a physical system, as ordinarily understood, claims the elimination of all external disturbances. But in that case, according to the quantum postulate, any observation will be impossible, and, above all, the concepts of space and time lose their immediate sense. On the other hand, if in order to make observation possible we permit certain interactions with suitable agencies of measurement, not belonging to the system, an unambiguous definition of the state of the system is naturally no longer possible, and there can be no question of causality in the ordinary sense of the word. The very nature of the quantum theory thus forces us to regard the space-time co-ordination and the claim of causality, the union of which characterises the classical theories, as complementary but exclusive features of the description, symbolising the idealisation of observation and definition respectively. Just as the relativity theory has taught us that the convenience of distinguishing sharply between space and time rests solely on the smallness of the velocities ordinarily met with compared to the velocity of light, we learn from the quantum theory that the appropriateness of our usual causal space-time description depends entirely upon the small value of the quantum of action as compared to the actions involved in ordinary sense perceptions. Indeed, in the description of atomic phenomena, the quantum postulate presents us with the task of developing a 'complementarity' theory the consistency of which can be judged only by weighing the possibilities of definition and observation.

This view is already clearly brought out by the much-discussed question of the nature of light and the ultimate constituents of matter. As regards light, its propagation in space and time is adequately expressed by the electromagnetic theory. Especially the interference phenomena *in vacuo* and the optical properties of material media are completely governed by the wave theory superposition principle. Nevertheless, the conservation of energy and momentum during the interaction between radiation and matter, as evident in the photoelectric and Compton effect, finds its adequate expression just in the light quantum idea put forward by Einstein. As is well known, the doubts regarding the validity of the superposition principle on one hand and of the conservation laws on the other, which were suggested by this apparent contradiction, have been definitely disproved through direct experiments. This situation would seem clearly to indicate the impossibility of a causal space-time description of the light phenomena. On one hand, in attempting to trace

[1] The content of this paper is essentially the same as that of a lecture on the present state of the quantum theory delivered on Sept. 16, 1927, at the Volta celebration in Como. For a summary of the theory just previous to the development of the new methods the reader is referred to a lecture of the author, "Atomic Theory and Mechanics," published in this periodical (NATURE, 116, 809; 1925). The rapid development which has taken place since has given rise to a considerable number of publications. The present paper is confined to a few references to recent articles which have a special bearing on the subject now under discussion.

the laws of the time-spatial propagation of light according to the quantum postulate, we are confined to statistical considerations. On the other hand, the fulfilment of the claim of causality for the individual light processes, characterised by the quantum of action, entails a renunciation as regards the space-time description. Of course, there can be no question of a quite independent application of the ideas of space and time and of causality. The two views of the nature of light are rather to be considered as different attempts at an interpretation of experimental evidence in which the limitation of the classical concepts is expressed in complementary ways.

The problem of the nature of the constituents of matter presents us with an analogous situation. The individuality of the elementary electrical corpuscles is forced upon us by general evidence. Nevertheless, recent experience, above all the discovery of the selective reflection of electrons from metal crystals, requires the use of the wave theory superposition principle in accordance with the original ideas of L. de Broglie. Just as in the case of light, we have consequently in the question of the nature of matter, so far as we adhere to classical concepts, to face an inevitable dilemma, which has to be regarded as the very expression of experimental evidence. In fact, here again we are not dealing with contradictory but with complementary pictures of the phenomena, which only together offer a natural generalisation of the classical mode of description. In the discussion of these questions, it must be kept in mind that, according to the view taken above, radiation in free space as well as isolated material particles are abstractions, their properties on the quantum theory being definable and observable only through their interaction with other systems. Nevertheless, these abstractions are, as we shall see, indispensable for a description of experience in connexion with our ordinary space-time view.

The difficulties with which a causal space-time description is confronted in the quantum theory, and which have been the subject of repeated discussions, are now placed into the foreground by the recent development of the symbolic methods. An important contribution to the problem of a consistent application of these methods has been made lately by Heisenberg (*Zeitschr. f. Phys.*, **43**, 172; 1927). In particular, he has stressed the peculiar reciprocal uncertainty which affects all measurements of atomic quantities. Before we enter upon his results it will be advantageous to show how the complementary nature of the description appearing in this uncertainty is unavoidable already in an analysis of the most elementary concepts employed in interpreting experience.

2. Quantum of Action and Kinematics.

The fundamental contrast between the quantum of action and the classical concepts is immediately apparent from the simple formulæ which form the common foundation of the theory of light quanta and of the wave theory of material particles. If Planck's constant be denoted by h, as is well known,

$$E\tau = I\lambda = h, \qquad (1)$$

where E and I are energy and momentum respectively, τ and λ the corresponding period of vibration and wave-length. In these formulæ the two notions of light and also of matter enter in sharp contrast. While energy and momentum are associated with the concept of particles, and hence may be characterised according to the classical point of view by definite space-time co-ordinates, the period of vibration and wave-length refer to a plane harmonic wave train of unlimited extent in space and time. Only with the aid of the superposition principle does it become possible to attain a connexion with the ordinary mode of description. Indeed, a limitation of the extent of the wave-fields in space and time can always be regarded as resulting from the interference of a group of elementary harmonic waves. As shown by de Broglie (Thèse, Paris, 1924), the translational velocity of the individuals associated with the waves can be represented by just the so-called group-velocity. Let us denote a plane elementary wave by

$$A \cos 2\pi(\nu t - x\sigma_x - y\sigma_y - z\sigma_z + \delta),$$

where A and δ are constants determining respectively the amplitude and the phase. The quantity $\nu = 1/\tau$ is the frequency, σ_x, σ_y, σ_z the wave numbers in the direction of the co-ordinate axes, which may be regarded as vector components of the wave number $\sigma = 1/\lambda$ in the direction of propagation. While the wave or phase velocity is given by ν/σ, the group-velocity is defined by $d\nu/d\sigma$. Now according to the relativity theory we have for a particle with the velocity v:

$$I = \frac{v}{c^2}E \text{ and } vdI = dE,$$

where c denotes the velocity of light. Hence by equation (1) the phase velocity is c^2/v and the group-velocity v. The circumstance that the former is in general greater than the velocity of light emphasises the symbolic character of these considerations. At the same time, the possibility of identifying the velocity of the particle with the group-velocity indicates the field of application of space-time pictures in the quantum theory. Here the complementary character of the description appears, since the use of wave-groups is necessarily accompanied by a lack of sharpness in the definition of period and wave-length, and hence also in the definition of the corresponding energy and momentum as given by relation (1).

Rigorously speaking, a limited wave-field can only be obtained by the superposition of a manifold of elementary waves corresponding to all values of ν and σ_x, σ_y, σ_z. But the order of magnitude of the mean difference between these values for two elementary waves in the group is given in the most favourable case by the condition

$$\Delta t \Delta\nu = \Delta x \Delta\sigma_x = \Delta y \Delta\sigma_y = \Delta z \Delta\sigma_z = 1,$$

where Δt, Δx, Δy, Δz denote the extension of the wave-field in time and in the directions of space corresponding to the co-ordinate axes. These relations—well known from the theory of optical instruments, especially from Rayleigh's investigation of the resolving power of spectral apparatus —express the condition that the wave-trains extinguish each other by interference at the space-time boundary of the wave-field. They may be regarded also as signifying that the group as a whole has no phase in the same sense as the elementary waves. From equation (1) we find thus:

$$\Delta t \Delta E = \Delta x \Delta I_x = \Delta y \Delta I_y = \Delta z \Delta I_z = h \quad . \quad (2)$$

as determining the highest possible accuracy in the definition of the energy and momentum of the individuals associated with the wave-field. In general, the conditions for attributing an energy and a momentum value to a wave-field by means of formula (1) are much less favourable. Even if the composition of the wave-group corresponds in the beginning to the relations (2), it will in the course of time be subject to such changes that it becomes less and less suitable for representing an individual. It is this very circumstance which gives rise to the paradoxical character of the problem of the nature of light and of material particles. The limitation in the classical concepts expressed through relation (2) is, besides, closely connected with the limited validity of classical mechanics, which in the wave theory of matter corresponds to the geometrical optics, in which the propagation of waves is depicted through 'rays.' Only in this limit can energy and momentum be unambiguously defined on the basis of space-time pictures. For a general definition of these concepts we are confined to the conservation laws, the rational formulation of which has been a fundamental problem for the symbolical methods to be mentioned below.

In the language of the relativity theory, the content of the relations (2) may be summarised in the statement that according to the quantum theory a general reciprocal relation exists between the maximum sharpness of definition of the space-time and energy-momentum vectors associated with the individuals. This circumstance may be regarded as a simple symbolical expression for the complementary nature of the space-time description and the claims of causality. At the same time, however, the general character of this relation makes it possible to a certain extent to reconcile the conservation laws with the space-time coordination of observations, the idea of a coincidence of well-defined events in a space-time point being replaced by that of unsharply defined individuals within finite space-time regions.

This circumstance permits us to avoid the well-known paradoxes which are encountered in attempting to describe the scattering of radiation by free electrical particles as well as the collision of two such particles. According to the classical concepts, the description of the scattering requires a finite extent of the radiation in space and time, while in the change of the motion of the electron demanded by the quantum postulate one seemingly is dealing with an instantaneous effect taking place at a definite point in space. Just as in the case of radiation, however, it is impossible to define momentum and energy for an electron without considering a finite space-time region. Furthermore, an application of the conservation laws to the process implies that the accuracy of definition of the energy momentum vector is the same for the radiation and the electron. In consequence, according to relation (2), the associated space-time regions can be given the same size for both individuals in interaction.

A similar remark applies to the collision between two material particles, although the significance of the quantum postulate for this phenomenon was disregarded before the necessity of the wave concept was realised. Here this postulate does indeed represent the idea of the individuality of the particles which, transcending the space-time description, meets the claim of causality. While the physical content of the light quantum idea is wholly connected with the conservation theorems for energy and momentum, in the case of the electrical particles the electric charge has to be taken into account in this connexion. It is scarcely necessary to mention that for a more detailed description of the interaction between individuals we cannot restrict ourselves to the facts expressed by formulæ (1) and (2), but must resort to a procedure which allows us to take into account the coupling of the individuals, characterising the interaction in question, where just the importance of the electric charge appears. As we shall see, such a procedure necessitates a further departure from visualisation in the usual sense.

3. Measurements in the Quantum Theory.

In his investigations already mentioned on the consistency of the quantum theoretical methods, Heisenberg has given the relation (2) as an expression for the maximum precision with which the space-time co-ordinates and momentum-energy components of a particle can be measured simultaneously. His view was based on the following consideration: On one hand, the coordinates of a particle can be measured with any desired degree of accuracy by using, for example, an optical instrument, provided radiation of sufficiently short wave-length is used for illumination. According to the quantum theory, however, the scattering of radiation from the object is always connected with a finite change in momentum, which is the larger the smaller the wave-length of the radiation used. The momentum of a particle, on the other hand, can be determined with any desired degree of accuracy by measuring, for example, the Doppler effect of the scattered radiation, provided the wave-length of the radiation is so large that the effect of recoil can be neglected, but then the determination of the space co-ordinates of the particle becomes correspondingly less accurate.

The essence of this consideration is the inevitability of the quantum postulate in the estimation of the possibilities of measurement. A closer investigation of the possibilities of definition would still seem necessary in order to bring out the general complementary character of the description. Indeed, a discontinuous change of energy and momentum during observation could not prevent us from ascribing accurate values to the space-time co-ordinates, as well as to the momentum-energy components before and after the process. The reciprocal uncertainty which always affects the values of these quantities is, as will be clear from the preceding analysis, essentially an outcome of the limited accuracy with which changes in energy and momentum can be defined, when the wave-fields used for the determination of the space-time co-ordinates of the particle are sufficiently small.

In using an optical instrument for determinations of position, it is necessary to remember that the formation of the image always requires a convergent beam of light. Denoting by λ the wave-length of the radiation used, and by ϵ the so-called numerical aperture, that is, the sine of half the angle of convergence, the resolving power of a microscope is given by the well-known expression $\lambda/2\epsilon$. Even if the object is illuminated by parallel light, so that the momentum h/λ of the incident light quantum is known both as regards magnitude and direction, the finite value of the aperture will prevent an exact knowledge of the recoil accompanying the scattering. Also, even if the momentum of the particle were accurately known before the scattering process, our knowledge of the component of momentum parallel to the focal plane after the observation would be affected by an uncertainty amounting to $2\epsilon h/\lambda$. The product of the least inaccuracies with which the positional co-ordinate and the component of momentum in a definite direction can be ascertained is therefore just given by formula (2). One might perhaps expect that in estimating the accuracy of determining the position, not only the convergence but also the length of the wave-train has to be taken into account, because the particle could change its place during the finite time of illumination. Due to the fact, however, that the exact knowledge of the wave-length is immaterial for the above estimate, it will be realised that for any value of the aperture the wave-train can always be taken so short that a change of position of the particle during the time of observation may be neglected in comparison to the lack of sharpness inherent in the determination of position due to the finite resolving power of the microscope.

In measuring momentum with the aid of the Doppler effect—with due regard to the Compton effect—one will employ a parallel wave-train. For the accuracy, however, with which the change in wave-length of the scattered radiation can be measured the extent of the wave-train in the direction of propagation is essential. If we assume that the directions of the incident and scattered radiation are parallel and opposite respectively to the direction of the position co-ordinate and momentum component to be measured, then $c\lambda/2l$ can be taken as a measure of the accuracy in the determination of the velocity, where l denotes the length of the wave-train. For simplicity, we here have regarded the velocity of light as large compared to the velocity of the particle. If m represents the mass of the particle, then the uncertainty attached to the value of the momentum after observation is $cm\lambda/2l$. In this case the magnitude of the recoil, $2h/\lambda$, is sufficiently well defined in order not to give rise to an appreciable uncertainty in the value of the momentum of the particle after observation. Indeed, the general theory of the Compton effect allows us to compute the momentum components in the direction of the radiation before and after the recoil from the wave-lengths of the incident and scattered radiation. Even if the positional co-ordinates of the particle were accurately known in the beginning, our knowledge of the position after observation nevertheless will be affected by an uncertainty. Indeed, on account of the impossibility of attributing a definite instant to the recoil, we know the mean velocity in the direction of observation during the scattering process only with an accuracy $2h/m\lambda$. The uncertainty in the position after observation hence is $2hl/mc\lambda$. Here, too, the product of the inaccuracies in the measurement of position and momentum is thus given by the general formula (2).

Just as in the case of the determination of position, the time of the process of observation for the determination of momentum may be made as short as is desired if only the wave-length of the radiation used is sufficiently small. The fact that the recoil then gets larger does not, as we have seen, affect the accuracy of measurement. It should further be mentioned, that in referring to the velocity of a particle as we have here done repeatedly, the purpose has only been to obtain a connexion with the ordinary space-time description convenient in this case. As it appears already from the considerations of de Broglie mentioned above, the concept of velocity must always in the quantum theory be handled with caution. It will also be seen that an unambiguous definition of this concept is excluded by the quantum postulate. This is particularly to be remembered when comparing the results of successive observations. Indeed, the position of an individual at two given moments can be measured with any desired degree of accuracy; but if, from such measurements, we would calculate the velocity of the individual in the ordinary way, it must be clearly realised that we are dealing with an abstraction, from which no unambiguous information concerning the previous or future behaviour of the individual can be obtained.

According to the above considerations regarding the possibilities of definition of the properties of individuals, it will obviously make no difference in the discussion of the accuracy of measurements of position and momentum of a particle if collisions with other material particles are considered instead of scattering of radiation. In both cases we see that the uncertainty in question equally affects the description of the agency of measurement and of the object. In fact, this uncertainty cannot be avoided in a description of the behaviour of individuals with respect to a co-ordinate system fixed in the ordinary way by means of solid bodies and unperturbable clocks. The experimental devices—opening and closing of apertures, etc.—are seen to permit only conclusions regarding the space-time extension of the associated wave-fields.

In tracing observations back to our sensations, once more regard has to be taken to the quantum postulate in connexion with the perception of the agency of observation, be it through its direct action upon the eye or by means of suitable auxiliaries such as photographic plates, Wilson clouds, etc. It is easily seen, however, that the resulting additional statistical element will not influence the uncertainty in the description of the object. It might even be conjectured that the arbitrariness in what is regarded as object and what as agency of observation would open up a possibility of avoiding this uncertainty altogether. In connexion with the measurement of the position of a particle, one might, for example, ask whether the momentum transmitted by the scattering could not be determined by means of the conservation theorem from a measurement of the change of momentum of the microscope—including light source and photographic plate—during the process of observation. A closer investigation shows, however, that such a measurement is impossible, if at the same time one wants to know the position of the microscope with sufficient accuracy. In fact, it follows from the experiences which have found expression in the wave theory of matter, that the position of the centre of gravity of a body and its total momentum can only be defined within the limits of reciprocal accuracy given by relation (2).

Strictly speaking, the idea of observation belongs to the causal space-time way of description. Due to the general character of relation (2), however, this idea can be consistently utilised also in the quantum theory, if only the uncertainty expressed through this relation is taken into account. As remarked by Heisenberg, one may even obtain an instructive illustration to the quantum theoretical description of atomic (microscopic) phenomena by comparing this uncertainty with the uncertainty, due to imperfect measurements, inherently contained in any observation as considered in the ordinary description of natural phenomena. He remarks on that occasion that even in the case of macroscopic phenomena we may say, in a certain sense, that they are created by repeated observations. It must not be forgotten, however, that in the classical theories any succeeding observation permits a prediction of future events with ever-increasing accuracy, because it improves our knowledge of the initial state of the system. According to the quantum theory, just the impossibility of neglecting the interaction with the agency of measurement means that every observation introduces a new uncontrollable element. Indeed, it follows from the above considerations that the measurement of the positional coordinates of a particle is accompanied not only by a finite change in the dynamical variables, but also the fixation of its position means a complete rupture in the causal description of its dynamical behaviour, while the determination of its momentum always implies a gap in the knowledge of its spatial propagation. Just this situation brings out most strikingly the complementary character of the description of atomic phenomena which appears as an inevitable consequence of the contrast between the quantum postulate and the distinction between object and agency of measurement, inherent in our very idea of observation.

4. Correspondence Principle and Matrix Theory.

Hitherto we have only regarded certain general features of the quantum problem. The situation implies, however, that the main stress has to be laid on the formulation of the laws governing the interaction between the objects which we symbolise by the abstractions of isolated particles and radiation. Points of attack for this formulation are presented in the first place by the problem of atomic constitution. As is well known, it has been possible here, by means of an elementary use of classical concepts and in harmony with the quantum postulate, to throw light on essential aspects of experience. For example, the experiments regarding the excitation of spectra by electronic impacts and by radiation are adequately accounted for on the assumption of discrete stationary states and individual transition processes. This is primarily due to the circumstance that in these questions no closer description of the space-time behaviour of the processes is required.

Here the contrast with the ordinary way of description appears strikingly in the circumstance that spectral lines, which on the classical view would be ascribed to the same state of the atom, will, according to the quantum postulate, correspond to separate transition processes, between which the excited atom has a choice. Notwithstanding this contrast, however, a formal connexion with the classical ideas could be obtained in the limit, where the relative difference in the properties of neighbouring stationary states vanishes asymptotically and where in statistical applications the discontinuities may be disregarded. Through this connexion it was possible to a large extent to interpret the regularities of spectra on the basis of our ideas about the structure of the atom.

The aim of regarding the quantum theory as a rational generalisation of the classical theories led to the formulation of the so-called correspondence principle. The utilisation of this principle for the interpretation of spectroscopic results was based on a symbolical application of classical electrodynamics, in which the individual transition processes were each associated with a harmonic in the motion of the atomic particles to be expected according to ordinary mechanics. Except in the limit mentioned, where the relative difference between adjacent stationary states may be neglected, such a fragmentary application of the classical theories could only in certain cases lead to a strictly quantitative description of the phenomena. Especially the connexion developed by

Ladenburg and Kramers between the classical treatment of dispersion and the statistical laws governing the radiative transition processes formulated by Einstein should be mentioned here. Although it was just Kramers' treatment of dispersion that gave important hints for the rational development of correspondence considerations, it is only through the quantum theoretical methods created in the last few years that the general aims laid down in the principle mentioned have obtained an adequate formulation.

As is known, the new development was commenced in a fundamental paper by Heisenberg, where he succeeded in emancipating himself completely from the classical concept of motion by replacing from the very start the ordinary kinematical and mechanical quantities by symbols, which refer directly to the individual processes demanded by the quantum postulate. This was accomplished by substituting for the Fourier development of a classical mechanical quantity a matrix scheme, the elements of which symbolise purely harmonic vibrations and are associated with the possible transitions between stationary states. By requiring that the frequencies ascribed to the elements must always obey the combination principle for spectral lines, Heisenberg could introduce simple rules of calculation for the symbols, which permit a direct quantum theoretical transcription of the fundamental equations of classical mechanics. This ingenious attack on the dynamical problem of atomic theory proved itself from the beginning to be an exceedingly powerful and fertile method for interpreting quantitatively the experimental results. Through the work of Born and Jordan as well as of Dirac, the theory was given a formulation which can compete with classical mechanics as regards generality and consistency. Especially the element characteristic of the quantum theory, Planck's constant, appears explicitly only in the algorithms to which the symbols, the so-called matrices, are subjected. In fact, matrices, which represent canonically conjugated variables in the sense of the Hamiltonian equations, do not obey the commutative law of multiplication, but two such quantities, q and p, have to fulfil the exchange rule

$$pq - qp = \sqrt{-1}\frac{h}{2\pi}. \quad . \quad . \quad . \quad (3)$$

Indeed, this exchange relation expresses strikingly the symbolical character of the matrix formulation of the quantum theory. The matrix theory has often been called a calculus with directly observable quantities. It must be remembered, however, that the procedure described is limited just to those problems, in which in applying the quantum postulate the space-time description may largely be disregarded, and the question of observation in the proper sense therefore placed in the background.

In pursuing further the correspondence of the quantum laws with classical mechanics, the stress placed on the statistical character of the quantum theoretical description, which is brought in by the quantum postulate, has been of fundamental importance. Here the generalisation of the symbolical method made by Dirac and Jordan represented a great progress by making possible the operation with matrices, which are not arranged according to the stationary states, but where the possible values of any set of variables may appear as indices of the matrix elements. In analogy to the interpretation considered in the original form of the theory of the 'diagonal elements' connected only with a single stationary state, as time averages of the quantity to be represented, the general transformation theory of matrices permits the representation of such averages of a mechanical quantity, in the calculation of which any set of variables characterising the 'state' of the system have given values, while the canonically conjugated variables are allowed to take all possible values. On the basis of the procedure developed by these authors and in close connexion with ideas of Born and Pauli, Heisenberg has in the paper already cited above attempted a closer analysis of the physical content of the quantum theory, especially in view of the apparently paradoxical character of the exchange relation (3). In this connexion he has formulated the relation

$$\Delta q \Delta p \sim h \quad . \quad . \quad . \quad . \quad (4)$$

as the general expression for the maximum accuracy with which two canonically conjugated variables can simultaneously be observed. In this way Heisenberg has been able to elucidate many paradoxes appearing in the application of the quantum postulate, and to a large extent to demonstrate the consistency of the symbolic method. In connexion with the complementary nature of the quantum theoretical description, we must, as already mentioned, constantly keep the possibilities of definition as well as of observation before the mind. For the discussion of just this question the method of wave mechanics developed by Schrödinger has, as we shall see, proved of great help. It permits a general application of the principle of superposition also in the problem of interaction, thus offering an immediate connexion with the above considerations concerning radiation and free particles. Below we shall return to the relation of wave mechanics to the general formulation of the quantum laws by means of the transformation theory of matrices.

5. Wave Mechanics and Quantum Postulate.

Already in his first considerations concerning the wave theory of material particles, de Broglie pointed out that the stationary states of an atom may be visualised as an interference effect of the phase wave associated with a bound electron. It is true that this point of view at first did not, as regards quantitative results, lead beyond the earlier methods of quantum theory, to the development of which Sommerfeld has contributed so essentially. Schrödinger, however, succeeded in developing a wave-theoretical method which has opened up new aspects, and has proved to be of decisive importance for the great progress in atomic physics during the last years. Indeed, the proper vibrations of the Schrödinger wave equation have been found to furnish a representation of the stationary states of an atom meeting all requirements. The energy of each state is connected with the corresponding period of vibration according to the general quantum relation (1). Furthermore, the number of nodes in the various characteristic vibrations gives a simple interpretation to the concept of quantum number which was already known from the older methods, but at first did not seem to appear in the matrix formulation. In addition, Schrödinger could associate with the solutions of the wave equation a continuous distribution of charge and current, which, if applied to a characteristic vibration, represents the electrostatic and magnetic properties of an atom in the corresponding stationary state. Similarly, the superposition of two characteristic solutions corresponds to a continuous vibrating distribution of electrical charge, which on classical electrodynamics would give rise to an emission of radiation, illustrating instructively the consequences of the quantum postulate and the correspondence requirement regarding the transition process between two stationary states formulated in matrix mechanics. Another application of the method of Schrödinger, important for the further development, has been made by Born in his investigation of the problem of collisions between atoms and free electric particles. In this connexion he succeeded in obtaining a statistical interpretation of the wave functions, allowing a calculation of the probability of the individual transition processes required by the quantum postulate. This includes a wave-mechanical formulation of the adiabatic principle of Ehrenfest, the fertility of which appears strikingly in the promising investigations of Hund on the problem of formation of molecules.

In view of these results, Schrödinger has expressed the hope that the development of the wave theory will eventually remove the irrational element expressed by the quantum postulate and open the way for a complete description of atomic phenomena along the line of the classical theories. In support of this view, Schrödinger, in a recent paper (*Ann. d. Phys.*, **83**, p. 956; 1927), emphasises the fact that the discontinuous exchange of energy between atoms required by the quantum postulate, from the point of view of the wave theory, is replaced by a simple resonance phenomenon. In particular, the idea of individual stationary states would be an illusion and its applicability only an illustration of the resonance mentioned. It must be kept in mind, however, that just in the resonance problem mentioned we are concerned with a closed system which, according to the view presented here, is not accessible to observation. In fact, wave mechanics just as the matrix theory on this view represents a symbolic transcription of the problem of motion of classical mechanics adapted to the requirements of quantum theory and only to be interpreted by an explicit use of the quantum postulate. Indeed, the two formulations of the interaction problem might be said to be complementary in the same sense as the wave and particle idea in the description of the free individuals. The apparent contrast in the utilisation of the energy concept in the two theories is just connected with this difference in the starting-point.

The fundamental difficulties opposing a space-time description of a system of particles in interaction appear at once from the inevitability of the superposition principle in the description of the behaviour of individual particles. Already for a free particle the knowledge of energy and momentum excludes, as we have seen, the exact knowledge of its space-time co-ordinates. This implies that an immediate utilisation of the concept of energy in connexion with the classical idea of the potential energy of the system is excluded. In the Schrödinger wave equation these difficulties are avoided by replacing the classical expression of the Hamiltonian by a differential operator by means of the relation

$$p = \sqrt{-1}\frac{h}{2\pi}\frac{\delta}{\delta q}, \quad . \quad . \quad . \quad (5)$$

where p denotes a generalised component of momentum and q the canonically conjugated variable. Hereby the negative value of the energy is regarded as conjugated to the time. So far, in the wave equation, time and space as well as energy and momentum are utilised in a purely formal way.

The symbolical character of Schrödinger's method appears not only from the circumstance that its simplicity, similarly to that of the matrix theory, depends essentially upon the use of imaginary arithmetic quantities. But above all there can be no question of an immediate connexion with our ordinary conceptions because the 'geometrical' problem represented by the wave equation is associated with the so-called co-ordinate space, the number of dimensions of which is equal to the number of degrees of freedom of the system, and hence in general greater than the number of dimensions of ordinary space. Further, Schrödinger's formulation of the interaction problem, just as the formulation offered by matrix theory, involves a neglect of the finite velocity of propagation of the forces claimed by relativity theory.

On the whole, it would scarcely seem justifiable, in the case of the interaction problem, to demand a visualisation by means of ordinary space-time pictures. In fact, all our knowledge concerning the internal properties of atoms is derived from experiments on their radiation or collision reactions, such that the interpretation of experimental facts ultimately depends on the abstractions of radiation in free space, and free material particles. Hence, our whole space-time view of physical phenomena, as well as the definition of energy and momentum, depends ultimately upon these abstractions. In judging the applications of these auxiliary ideas we should only demand inner consistency, in which connexion special regard has to be paid to the possibilities of definition and observation.

In the characteristic vibrations of Schrödinger's wave equation we have, as mentioned, an adequate representation of the stationary states of an atom allowing an unambiguous definition of the energy of the system by means of the general quantum relation (1). This entails, however, that in the interpretation of observations, a fundamental renunciation regarding the space-time description is unavoidable. In fact, the consistent application of the concept of stationary states excludes, as we shall see, any specification regarding the behaviour of the separate particles in the atom. In problems where a description of this behaviour is essential, we are bound to use the general solution of the wave equation which is obtained by superposition of characteristic solutions. We meet here with a complementarity of the possibilities of definition quite analogous to that which we have considered earlier in connexion with the properties of light and free material particles. Thus, while the definition of energy and momentum of individuals is attached to the idea of a harmonic elementary wave, every space-time feature of the description of phenomena is, as we have seen, based on a consideration of the interferences taking place inside a group of such elementary waves. Also in the present case the agreement between the possibilities of observation and those of definition can be directly shown.

According to the quantum postulate any observation regarding the behaviour of the electron in the atom will be accompanied by a change in the state of the atom. As stressed by Heisenberg, this change will, in the case of atoms in stationary states of low quantum number, consist in general in the ejection of the electron from the atom. A description of the 'orbit' of the electron in the atom with the aid of subsequent observations is hence impossible in such a case. This is connected with the circumstance that from characteristic vibrations with only a few nodes no wave packages can be built up which would even approximately represent the 'motion' of a particle. The complementary nature of the description, however, appears particularly in that the use of observations concerning the behaviour of particles in the atom rests on the possibility of neglecting, during the process of observation, the interaction between the particles, thus regarding them as free. This requires, however, that the duration of the process is short compared with the natural periods of the atom, which again means that the uncertainty in the knowledge of the energy transferred in the process is large compared to the energy differences between neighbouring stationary states.

In judging the possibilities of observation it must, on the whole, be kept in mind that the wave mechanical solutions can be visualised only in so far as they can be described with the aid of the concept of free particles. Here the difference between classical mechanics and the quantum theoretical treatment of the problem of interaction appears most strikingly. In the former such a restriction is unnecessary, because the 'particles' are here endowed with an immediate 'reality,' independently of their being free or bound. This situation is particularly important in connexion with the consistent utilisation of Schrödinger's electric density as a measure of the probability for electrons being present within given space regions of the atom. Remembering the restriction mentioned, this interpretation is seen to be a simple consequence of the assumption that the probability of the presence of a free electron is expressed by the electric density associated with the wave-field in a similar way to that by which the probability of the presence of a light quantum is given by the energy density of the radiation.

As already mentioned, the means for a general consistent utilisation of the classical concepts in the quantum theory have been created through the transformation theory of Dirac and Jordan, by the aid of which Heisenberg has formulated his general uncertainty relation (4). In this theory also the Schrödinger wave equation has obtained an instructive application. In fact, the characteristic solutions of this equation appear as auxiliary functions which define a transformation from matrices with indices representing the energy values of the system to other matrices, the indices of which are the possible values of the space coordinates. It is also of interest in this connexion to mention that Jordan and Klein (*Zeitsch. f. Phys.*, **45**, 751; 1927) have recently arrived at the formulation of the problem of interaction expressed by the Schrödinger wave equation, taking as starting-point the wave representation of individual particles and applying a symbolic method closely related to the deep-going treatment of the radiation problem developed by Dirac from the point of view of the matrix theory, to which we shall return below.

6. Reality of Stationary States.

In the conception of stationary states we are, as mentioned, concerned with a characteristic application of the quantum postulate. By its very nature this conception means a complete renunciation as regards a time description. From the point of view taken here, just this renunciation forms the necessary condition for an unambiguous definition of the energy of the atom. Moreover, the conception of a stationary state involves, strictly speaking, the exclusion of all interactions with individuals not belonging to the system. The fact that such a closed system is associated with a particular energy value may be considered as an immediate expression for the claim of causality contained in the theorem of conservation of energy. This circumstance justifies the assumption of the supra-mechanical stability of the stationary states, according to which the atom, before as well as after an external influence, always will be found in a well-defined state, and which forms the basis for the use of the quantum postulate in problems concerning atomic structure.

In a judgment of the well-known paradoxes which this assumption entails for the description of collision and radiation reactions, it is essential to consider the limitations of the possibilities of definition of the reacting free individuals, which is expressed by relation (2). In fact, if the definition of the energy of the reacting individuals is to be accurate to such a degree as to entitle us to speak of conservation of energy during the reaction, it is necessary, according to this relation, to co-ordinate to the reaction a time interval long compared to the vibration period associated with the transition process, and connected with the energy difference between the stationary states according to relation (1). This is particularly to be remembered when considering the passage of swiftly moving particles through an atom. According to the ordinary kinematics, the effective duration of such a passage would be very small as compared with the natural periods of the atom, and it seemed impossible to reconcile the principle of conservation of energy with the assumption of the stability of stationary states (cf. *Zeits. f. Phys.*, **34**, 142 ; 1925). In the wave representation, however, the time of reaction is immediately connected with the accuracy of the knowledge of the energy of the colliding particle, and hence there can never be the possibility of a contradiction with the law of conservation. In connexion with the discussion of paradoxes of the kind mentioned, Campbell (*Phil. Mag.*, i. 1106 ; 1926) suggested the view that the conception of time itself may be essentially statistical in nature. From the view advanced here, according to which the foundation of space-time description is offered by the abstraction of free individuals, a fundamental distinction between time and space, however, would seem to be excluded by the relativity requirement. The singular position of the time in problems concerned with stationary states is, as we have seen, due to the special nature of such problems.

The application of the conception of stationary states demands that in any observation, say by means of collision or radiation reactions, permitting a distinction between different stationary states, we are entitled to disregard the previous history of the atom. The fact that the symbolical quantum theory methods ascribe a particular phase to each stationary state the value of which depends upon the previous history of the atom, would for the first moment seem to contradict the very idea of stationary states. As soon as we are really concerned with a time problem, however, the consideration of a strictly closed system is excluded. The use of simply harmonic proper vibrations in the interpretation of observations means, therefore, only a suitable idealisation which in a more rigorous discussion must always be replaced by a group of harmonic vibrations, distributed over a finite frequency interval. Now, as already mentioned, it is a general consequence of the superposition principle that it has no sense to co-ordinate a phase value to the group as a whole, in the same manner as may be done for each elementary wave constituting the group.

This inobservability of the phase, well known from the theory of optical instruments, is brought out in a particularly simple manner in a discussion of the Stern-Gerlach experiment, so important for the investigation of the properties of single atoms. As pointed out by Heisenberg, atoms with different orientation in the field may only be separated if the deviation of the beam is larger than the diffraction at the slit of the de Broglie waves representing the translational motion of the atoms. This condition means, as a simple calculation shows, that the product of the time of passage of the atom through the field, and the uncertainty due to the finite width of the beam of its energy in the field, is at least equal to the quantum of action. This result was considered by Heisenberg as a support of relation (2) as regards the reciprocal uncertainties of energy and time values. It would seem, however, that here we are not simply dealing with a measurement of the energy of the atom at a given time. But since the period of the proper vibrations of the atom in the field is connected with the total energy by relation (1), we realise that the condition for separability mentioned just means the loss of the phase. This circumstance removes also the apparent contradictions, arising in certain problems concerning the coherence of resonance radiation, which have been discussed frequently, and were also considered by Heisenberg.

To consider an atom as a closed system, as we have done above, means to neglect the spontaneous emission of radiation which even in the absence of external influences puts an upper limit to the lifetime of the stationary states. The fact that this neglect is justified in many applications is connected with the circumstance that the coupling between the atom and the radiation field, which is to be expected on classical electrodynamics, is in general very small compared to the coupling between the particles in the atom. It is, in fact, possible in a description of the state of an atom to a considerable extent to neglect the reaction of radiation, thus disregarding the unsharpness in the energy values connected with the lifetime of the stationary states according to relation (2) (cf. *Proc. Camb. Phil. Soc.*, 1924 (Supplement), or *Zeits. f. Phys.*, **13**, 117 ; 1923). This is the reason why it is possible to draw conclusions concerning the properties of radiation by using classical electrodynamics.

The treatment of the radiation problem by the new quantum theoretical methods meant to begin with just a quantitative formulation of this correspondence consideration. This was the very starting-point of the original considerations of Heisenberg. It may also be mentioned that an instructive analysis of Schrödinger's treatment of the radiation phenomena from the point of view of the correspondence principle has been recently given by Klein (*Zeits. f. Phys.*, **41**, 707 ; 1927). In the more rigorous form of the theory developed by Dirac (*Proc. Roy. Soc.*, A, vol. 114, p. 243 ; 1927) the radiation field itself is included in the closed system under consideration. Thus it became possible in a rational way to take account of the individual character of radiation demanded by the quantum theory and to build up a dispersion theory, in which the final width of the spectral lines is taken into consideration.

The renunciation regarding space-time pictures characterising this treatment would seem to offer a striking indication of the complementary character of the quantum theory. This is particularly to be borne in mind in judging the radical departure from the causal description of Nature met with in radiation phenomena, to which we have referred above in connexion with the excitation of spectra.

In view of the asymptotic connexion of atomic properties with classical electrodynamics, demanded by the correspondence principle, the reciprocal exclusion of the conception of stationary states and the description of the behaviour of individual particles in the atom might be regarded as a difficulty. In fact, the connexion in question means that in the limit of large quantum numbers where the relative difference between adjacent stationary states vanishes asymptotically, mechanical pictures of electronic motion may be rationally utilised. It must be emphasised, however, that this connexion cannot be regarded as a gradual transition towards classical theory in the sense that the quantum postulate would lose its significance for high quantum numbers. On the contrary, the conclusions obtained from the correspondence principle with the aid of classical pictures depend just upon the assumptions that the conception of stationary states and of individual transition processes are maintained even in this limit.

This question offers a particularly instructive example for the application of the new methods. As shown by Schrödinger (*Naturwiss.*, **14**, 664; 1926), it is possible, in the limit mentioned, by superposition of proper vibrations to construct wave groups small in comparison to the 'size' of the atom, the propagation of which indefinitely approaches the classical picture of moving material particles, if the quantum numbers are chosen sufficiently large. In the special case of a simple harmonic vibrator, he was able to show that such wave groups will keep together even for any length of time, and will oscillate to and fro in a manner corresponding to the classical picture of the motion. This circumstance Schrödinger has regarded as a support of his hope of constructing a pure wave theory without referring to the quantum postulate. As emphasised by Heisenberg, the simplicity of the case of the oscillator, however, is exceptional and intimately connected with the harmonic nature of the corresponding classical motion. Nor is there in this example any possibility for an asymptotical approach towards the problem of free particles. In general, the wave group will gradually spread over the whole region of the atom, and the 'motion' of a bound electron can only be followed during a number of periods, which is of the order of magnitude of the quantum numbers associated with the proper vibrations. This question has been more closely investigated in a recent paper by Darwin (*Proc. Roy. Soc.*, A, vol. 117, 258; 1927), which contains a number of instructive examples of the behaviour of wave groups. From the viewpoint of the matrix theory a treatment of analogous problems has been carried out by Kennard (*Zeits. f. Phys.*, **47**, 326; 1927).

Here again we meet with the contrast between the wave theory superposition principle and the assumption of the individuality of particles with which we have been concerned already in the case of free particles. At the same time the asymptotical connexion with the classical theory, to which a distinction between free and bound particles is unknown, offers the possibility of a particularly simple illustration of the above considerations regarding the consistent utilisation of the concept of stationary states. As we have seen, the identification of a stationary state by means of collision or radiation reactions implies a gap in the time description, which is at least of the order of magnitude of the periods associated with transitions between stationary states. Now, in the limit of high quantum numbers these periods may be interpreted as periods of revolution. Thus we see at once that no causal connexion can be obtained between observations leading to the fixation of a stationary state and earlier observations on the behaviour of the separate particles in the atom.

Summarising, it might be said that the concepts of stationary states and individual transition processes within their proper field of application possess just as much or as little 'reality' as the very idea of individual particles. In both cases we are concerned with a demand of causality complementary to the space-time description, the adequate application of which is limited only by the restricted possibilities of definition and of observation.

7. The Problem of the Elementary Particles.

When due regard is taken of the complementary feature required by the quantum postulate, it seems, in fact, possible with the aid of the symbolic methods to build up a consistent theory of atomic phenomena, which may be considered as a rational generalisation of the causal space-time description of classical physics. This view does not mean, however, that classical electron theory may be regarded simply as the limiting case of a vanishing quantum of action. Indeed, the connexion of the latter theory with experience is based on assumptions which can scarcely be separated from the group of problems of the quantum theory. A hint in this direction was already given by the well-known difficulties met with in the attempts to account for the individuality of ultimate electrical particles on general mechanical and electrodynamical principles. In this respect also the general relativity theory of gravitation has not fulfilled expectations. A satisfactory solution of the problems touched upon would seem to be possible only by means of a rational quantum-theoretical transcription of the general field theory, in which the ultimate quantum of electricity has found its natural position as an expression of the feature of individuality characterising the quantum theory. Recently Klein (*Zeits. f. Phys.*, **46**, 188; 1927) has directed attention to the possibility of connecting this problem with the five-dimensional unified representation of electromagnetism and gravitation proposed by Kaluza. In fact, the conservation of electricity appears in this theory as an analogue to the conservation theorems for energy and momentum. Just as these concepts are complementary to the space-time description, the appropriateness of the ordinary four-dimensional description as well as its symbolical utilisation in the quantum theory would, as Klein emphasises, seem to depend essentially on the circumstance that in this description electricity always appears in well-defined units, the conjugated fifth dimension being as a consequence not open to observation.

Quite apart from these unsolved deep-going problems, the classical electron theory up to the present time has been the guide for a further development of the correspondence description in connexion with the idea first advanced by Compton that the ultimate electrical particles, besides their mass and charge, are endowed with a magnetic moment due to an angular momentum determined by the quantum of action. This assumption, introduced with striking success by Goudsmit and "hlenbeck into the discussion of the origin of the ıomalous Zeeman effect, has proved most fruitl in connexion with the new methods, as own especially by Heisenberg and Jordan. One ight say, indeed, that the hypothesis of the agnetic electron, together with the resonance oblem elucidated by Heisenberg (*Zeits. f. Phys.*, , 239; 1927), which occurs in the quantumeoretical description of the behaviour of atoms th several electrons, have brought the correondence interpretation of the spectral laws and e periodic system to a certain degree of complen. The principles underlying this attack have en made it possible to draw conclusions regardg the properties of atomic nuclei. Thus Dennison *roc. Roy. Soc.*, A, vol. 115, 483; 1927), in nnexion with ideas of Heisenberg and Hund, s succeeded recently in a very interesting way showing how the explanation of the specific at of hydrogen, hitherto beset with diffilties; can be harmonised with the assumption at the proton is endowed with a moment of omentum of the same magnitude as that of the ctron. Due to its larger mass, however, a ıgnetic moment much smaller than that of the ctron must be associated with the proton.

The insufficiency of the methods hitherto deloped as concerns the problem of the elementary rticles appears in the questions just mentioned m the fact that they do not allow of an unıbiguous explanation of the difference in the behaviour of the electric elementary particles and the 'individuals' symbolised through the conception of light quanta expressed in the so-called exclusion principle formulated by Pauli. In fact, we meet in this principle, so important for the problem of atomic structure as well as for the recent development of statistical theories, with one among several possibilities, each of which fulfils the correspondence requirement. Moreover, the difficulty of satisfying the relativity requirement in quantum theory appears in a particularly striking light in connexion with the problem of the magnetic electron. Indeed, it seemed not possible to bring the promising attempts made by Darwin and Pauli in generalising the new methods to cover this problem naturally, in connexion with the relativity kinematical consideration of Thomas so fundamental for the interpretation of experimental results. Quite recently, however, Dirac (*Proc. of the Roy. Soc.*, A, 117, 610; 1928) has been able successfully to attack the problem of the magnetic electron through a new ingenious extension of the symbolical method and so to satisfy the relativity requirement without abandoning the agreement with spectral evidence. In this attack not only the imaginary complex quantities appearing in the earlier procedures are involved, but his fundamental equations themselves contain quantities of a still higher degree of complexity, that are represented by matrices.

Already the formulation of the relativity argument implies essentially the union of the space-time co-ordination and the demand of causality characterising the classical theories. In the adaptation of the relativity requirement to the quantum postulate we must therefore be prepared to meet with a renunciation as to visualisation in the ordinary sense going still further than in the formulation of the quantum laws considered here. Indeed, we find ourselves here on the very path taken by Einstein of adapting our modes of perception borrowed from the sensations to the gradually deepening knowledge of the laws of Nature. The hindrances met with on this path originate above all in the fact that, so to say, every word in the language refers to our ordinary perception. In the quantum theory we meet this difficulty at once in the question of the inevitability of the feature of irrationality characterising the quantum postulate. I hope, however, that the idea of complementarity is suited to characterise the situation, which bears a deep-going analogy to the general difficulty in the formation of human ideas, inherent in the distinction between subject and object.

Printed in Great Britain by R. & R. Clark, Limited, Edinburgh.

Part I: OPENING SESSION

OPENING ADDRESS

Prof.emer. K.V.Laurikainen

On behalf of the organizers of this Symposium I wish to welcome all of you to the third North-Karelian Symposium on the Foundations of Modern Physics. I hope that this series of meetings contributes a little to the clarification of the basic philosophy of quantum theory. There is still much to do in this field.

The fact that meetings of this kind have been arranged in Northern Karelia is due to the positive interest which the City of Joensuu has shown in these conferences. Let me say quite frankly that this city has an exceptional Mayor, Mr. Aaro Heikkilä, who also has an interest in philosophy, understanding the philosophical importance of the Copenhagen interpretation even better than many physicists. Mayor Heikkilä has to a great extent influenced the cultural activity in Joensuu where now various meetings and festivals take place regularly. We know some drawbacks in the situation, because it has been really difficult to accomodate all participants in an appropriate way.

We decided rather late to arrange this Symposium, after learning that the City of Joensuu had reserved funds for such a purpose. It was uncertain how much financial support we should get from other sources, and therefore we decided to arrange a rather small meeting. We also thought that discussions between a few suitable persons would now be more appropriate than a big meeting where people give their talks without much contact with one another. Plans were mainly based on the support which could be expected from the City of Joensuu. So the organizers have special reason to express their gratitude to the City of Joensuu and especially to its Mayor for making this meeting possible.

Later we obtained additional funds from other sources mentioned in the papers which have been distributed to all participants and to the press. I wish to express our sincere thanks also to the other sponsors, and I have special reason to mention in this sense the British Council, the Frech Ministry of Foreign Affairs, and the Finnish oil refinement company Neste; their support shows that the importance of this Symposium is understood also outside those bodies which regularly grant funds for physics meetings in Finland.

The number of participants has grown larger than what we originally planned. There has been much interest in this meeting although we announced it rather late. In order to select the contributions to be presented orally in this Symposium, a scientific advisory committee was formed to help organizers in this difficult selection. I wish to thank the members of our advisory committee, Professors Konrad Bleuler, Charles Enz, Peter Mittelstaedt, and Hans Primas, for their valuable assistance.

I wish still point out one rather personal aspect concerning our discussions. I hope that Wolfgang Pauli's philosophical thought would be one of the main themes in this Symposium. The new perspective which Pauli asserts has not been seriously discussed in any physics meeting as far as I know, and few people seem to know anything about his philosophical thought. Pauli's view differs essentially from the general trend of the research on the foundations of quantum theory today. If his remarks are taken seriously, many questions have to be re-evaluated, both in physics and in Western culture in general.

Pauli's philosophy connects physics with much more general problems. Characteristic of Pauli's thought was his generalization of the concept of complementarity to a quaternarian philosophy. This presupposes that in addition to the normal (+-)-opposites there are also complementary opposites. This generalized complementarity is connected with the medieval concept of conjunction or joining of opposites. In Taoism it is symbolized with the aid of the Yin-Yang figure which Niels Bohr used in his coat-of-arms in the Danish Elephant Order. It presents two opposite qualities which together form a whole, continuously giving room for each other. This is a good symbol for complementarity, but Pauli used it in a much more general sense than Bohr.

According to Pauli, this idea of "conjunction" is intensively present within the "collective unconscious" of our time, i.e., it is now "in the air" and wishes to become accepted. One symptom of the presence of thought parallel to Pauli's view was the Venice Declaration published in the Final Report of the Venice Symposium arranged by Unesco last year. It was a meeting of 17 invited representatives of different fields from mathematics to cultural anthropology and scientific novelists from different parts of the world. Among the five invited physicists I'll mention the Nobel Prize winner Abdus Salam, whom most of us know.

The participants of the Venice Symposium expressed their conviction that it is necessary now to start discussions concerning the relation of modern science to the great religous traditions. The theme of this Symposium was "Science and the Boundaries of Knowledge: the Prologue of our Cultural Past". Copies of the Venice Declaration will be available from the secretaries.

I hope that we, in our discussions during these three intensive days, will also consider such more general implications of modern physics. Pauli's philosophy opens such perspectives.

This Symposium is arranged in remembrance of the birth of the Copenhagen interpretation. The precise 60th anniversary would perhaps be about one month later, as Bohr gave his Como lecture on the 16th of September. The anniversary is not, however, quite exact, as Bohr continued working on the final text until the middle of October before he sent it for publication. After the lecture Bohr was with Pauli for one week for the final formulation of the text. At any rate, at this time 60 years ago Bohr was busy writing and rewriting and again rewriting the text for this article which still inspires quite conflicting philosophical thoughts, especially with respect to its ontological consequences.

The English text of the Como lecture was published in Nature in February 1928 and was entitled "The Quantum Postulate and the Recent Development of Atomic Theory". In this article Bohr introduced his concept of complementarity and presented all the main ideas of the so-called Copenhagen interpretation. What this interpretation really means and what its philosophical implications are in their full depth still arises discussions.

We are happy to have very well-known experts representing different attitudes in this meeting. Perhaps we will succeed in shedding a little additional light on the mysteries of the atomic world.

Niels Bohr's Discussions with Albert Einstein, Werner Heisenberg, and Erwin Schrödinger: The Origins of the Principles of Uncertainty and Complementarity

Jagdish Mehra[1,2]

In this paper, the main outlines of the discussions between Niels Bohr with Albert Einstein, Werner Heisenberg, and Erwin Schrödinger during 1920–1927 are treated. From the formulation of quantum mechanics in 1925–1926 and wave mechanics in 1926, there emerged Born's statistical interpretation of the wave function in summer 1926, and on the basis of the quantum mechanical transformation theory—formulated in fall 1926 by Dirac, London, and Jordan—Heisenberg formulated the uncertainty principle in early 1927. At the Volta Conference in Como in September 1927 and at the fifth Solvay Conference in Brussels the following month, Bohr publicly enunciated his complementarity principle, which had been developing in his mind for several years. The Bohr–Einstein discussions about the consistency and completeness of quantum mechanics and of physical theory as such—formally begun in October 1927 at the fifth Solvay Conference and carried on at the sixth Solvay Conference in October 1930—were continued during the next decades. All these aspects are briefly summarized.

DEDICATION

During over two decades of our friendship and association, Ilya Prigogine has often talked to me about the principal architects of quantum theory—Max Planck, Albert Einstein, Niels Bohr, Werner Heisenberg, Max Born, Paul Dirac, and Pascual Jordan—and asked me questions

[1] Instituts Internationaux de Physique et de Chimie (Solvay), U.L.B. Campus Plaine, Code Postal 231, 1050 Brussels, Belgium.
[2] Address in U.S.A.: 7830 Candle Lane, Houston, Texas 77071.

about their respective ideas and attitudes. He is profoundly interested in all the historical and philosophical aspects of physical theories. I take great pleasure in dedicating this article—with affection, admiration, respect, and gratitude—to Ilya Prigogine in honor of his 70th birthday. It also marks the 60th anniversary of Werner Heisenberg's enunciation of the principle of uncertainty and Niels Bohr's principle of complementarity.

1. INTRODUCTION

From 1913 onwards, Niels Bohr engaged in scientific and philosophical discussions about the problems of atomic and quantum theory with various discussion partners. During the decade of 1920–1930, Bohr carried on intense discussions with Hendrik Kramers, Paul Ehrenfest, Albert Einstein, Wolfgang Pauli, Werner Heisenberg, Max Born, Erwin Schrödinger, Oskar Klein, and others. In 1913 Bohr had first given expression to the principle of analogy, which later became the correspondence principle, and which was applied to obtain the solution of atomic problems from 1913 to 1925[(1)]. Bohr's 1913 papers are reprinted in Ref. 1. In the latter half of 1925 and the first half of 1926, the quantum mechanics of Heisenberg[(2)] and Born, Heisenberg, Jordan,[(3)] and Dirac,[(4)] and the wave mechanics of Schrödinger,[(5)] were formulated; their formal mathematical equivalence was established by Schrödinger,[(6)] Pauli (letter of Pauli to Jordan, April 12, 1926; see Ref. 7), and Carl Eckart.[(8)] In the summer of 1926 Born[(9)] gave the statistical interpretation of the wave function, and in the fall of 1926 the transformation theory of Fritz London,[(10)] P. A. M. Dirac,[(11)] and Pascal Jordan[(12)] was formulated. Early in 1927, Heisenberg employed the transformation theory to derive his indeterminacy relations and formulate the uncertainty principle.[(13)] In the fall of 1926 Bohr engaged in deep discussions with Schrödinger,[(14)] during the latter's visit to Copenhagen, and in the early spring of 1927 he carried on discussions with Heisenberg. (See the citations to W. Heisenberg and K. Stolzenburg in Ref. 14; these discussions are dealt with in detail in Vol. 6 of Mehra and Rechenberg.[(15)] The earlier discussions of Bohr with Heisenberg are dealt with in the citations in Ref. 14, in Vol. 2 of Mehra and Rechenberg,[(15)] and in Mehra.[(15)]) At the Como Conference in September 1927, on the occasion of the centenary of Alessandro Volta's death, Bohr first enunciated his principle of complementarity.[(16)] (See also Ref. 72, pp. 345–361, and footnote 78 on p. 349.) Quantum mechanics, in it various formulations, complete with its physical interpretation, was presented by its various protagonists at the fifth Solvay Conference in late October 1927.[(17)] On that occasion, Niels Bohr and Albert Einstein engaged in discussions about the consistency and completeness of quantum mechanics with several other participants—including Ehrenfest, Born, Heisenberg, Dirac, Schrödinger, and Louis de Broglie—present. Bohr and Einstein had encountered each other previously in Berlin in April 1920 and in Leyden in December 1925, when other topics of discussion between them had been at the forefront.[18] The discussions between Bohr and Einstein about classical determinism and statistical causality, and whether the quantum mechanical description was "consistent" and "complete," would be resumed in October 1930 in Brussels (Ref. 19; see also Ref. 17), and be continued in various forms until Einstein's death in April 1955 (See Ref. 17, Chapter 6 and Appendix).

In this article, I shall review the essential aspects of the ideas that were expressed in Bohr's discussions with Einstein, Heisenberg, and Schrödinger. Some of these discussions directly led to Bohr's formulation of the principle of complementarity, which had actually been evolving in Bohr's mind for some time.

2. NIELS BOHR AND ALBERT EINSTEIN: FIRST ENCOUNTER AND EARLY DISCUSSIONS[20,21]

In Albert Einstein's work of 1916–1917 on the treatment of emission and absorption of radiation according to quantum theory,[22] it was found that probability laws governed atomic phenomena and a "statistical residue" definitely remained. Einstein had this problem in mind when, on January 27, 1920, he wrote to Max Born: "I myself do not believe that the solution to the quanta has to be found by giving up the continuum. Similarly, it could be assumed that one could arrive at general relativity by giving up the coordinate system. I believe now, as before, that one has to look for redundancy in determination by using differential equations so that the *solutions* themselves no longer have the character of continuum. But how?"[23] (See letter from Einstein to Born, January 27, 1920, Ref. 23, p. 21.) Later on, in the same letter, Einstein remarked: "That business about causality causes me a lot of trouble, too. Can the quantum absorption and emission of light ever be understood in terms of the complete causality requirement, or would a statistical residue remain? I must admit that there I lack the courage of my convictions. But I would be very unhappy to renounce *complete* causality ··· (The question whether strict causality exists or not has a definite meaning, even though there can probably never be a definite answer to it.)" (See letter from Einstein to Born, January 27, 1920, Ref. 23, p. 22.)

These considerations formed the early basis for Einstein's position concerning quantum mechanics several years later. Einstein believed unconditionally that a continuum theory, based on differential equations, had to be retained; he believed that quantum phenomena could be obtained by redundancy in determination, i.e., more equations than unknowns. The latter had been his hope in his formulation of the equations of general relativity theory; there the equations had seemed to be overdetermined, and Einstein had hoped that he could bring in the quantum phenomena (discontinuities) through this overdetermination.[(24)] On March 3, 1920, Einstein wrote to Born: "In my spare time I always brood about the problem of quanta from the point of view of relativity. I do not think the theory can work without the continuum. But I do not seem to be able to give tangible form to my pet idea which is to understand the structure of the quanta by redundancy in determination, using differential equations."[27] (See letter from Einstein to Born, January 27, 1920, Ref. 23, p. 26.) This indeed was the reason why Einstein would become so enthusiastic about Schrödinger's work on wave mechanics in early 1926.

In his essay on "Does Field Theory Provide Possibilities for Solving the Problem of Quanta?"[(25)] in 1923, Einstein would approach the problem of incorporating quanta into a general field theory, based on the principles of causality and continuity. Soon thereafter, he would write to his friend Michele Besso: "The idea I am toiling with in order to reach full understanding of quantum phenomena refers to an overdetermination of the laws by having more differential equations than field variables. For in this way the arbitrariness of the initial conditions could be overcome without renouncing field theory. Although this approach may well turn out to be a failure, it has to be attempted for, after all, it is logically possible.... The mathematics is exceedingly complicated and the relation to experience is even more indirect. But it remains a logical possibility, to do justice to reality, without any *sacrificium intellectus*." (See letter from Einstein to Besso, January 5, 1924, Ref. 26.)

Albert Einstein entertained such thoughts before he ever met Niels Bohr, and several years before the quantum mechanics of Born, Heisenberg, Jordan, and Dirac, and the wave mechanics of Schrödinger, were formulated, and long before the Bohr–Einstein discussions on classical determinism versus statistical causality took place in Brussels and were continued in subsequent years. (For an account of the Bohr–Einstein discussions, see Ref. 27.)

Niels Bohr believed that although classical physics and quantum theory were connected asymptotically through his correspondence principle, they were irreconcilable. Einstein found this view repugnant. He wrote to Max Born on June 4, 1919: "The quantum theory gives me a feeling very much like yours. One really ought to be ashamed of its success, because it has been obtained in accordance with the Jesuit maxim: 'Let not thy left hand know what thy right hand doeth.' (See letter from Einstein to Born, June 4, 1919, Ref. 23, pp. 10, 11.) Einstein had already shown in 1909 that Maxwell's equations might yield pointlike singular solutions in addition to waves.[(28)] In 1927 Einstein would successfully apply this idea to the field equations of general relativity; he firmly believed in a unified causal theory of all physical phenomena, and this idea would prompt him to support Louis de Broglie's theory of pilot waves at the fifth Solvay Conference. (See de Broglie in Mehra, Ref. 17.)

In the spring of 1920 Niels Bohr visited Berlin, where he gave a lecture before the Berlin Physical Society on April 27, 1920 about "the present state of the theory of spectra and possibilities of its development in the near future." This subject was closely related to the theory of light quanta, but in his talk Bohr referred only once to the conception of "radiation quanta," which also he probably did out of respect for Einstein who was in the audience. Immediately, however, Bohr added: "I shall not discuss here the familiar difficulties to which the 'hypothesis of light quanta' leads in connection with the phenomena of interference, for the explanation of which the classical theory of radiation has shown itself to be so remarkably suited."[(29)] In Berlin, Niels Bohr met Albert Einstein, Max Planck, and James Franck; with Franck he immediately established a strong rapport.

In the discussions between Bohr and Einstein in Berlin, Einstein expressed his fundamental belief that a complete theory of light must somehow incorporate both corpuscular and undulatory features, whereas Bohr defended the classical wave theory of light. Bohr insisted that the frequency ν appearing in the energy $h\nu$ of the quantum is defined by experiments on interference phenomena "which apparently demand for their interpretations a wave constitution of light" and "the light-quantum theory thus makes nonsense of its own basic equations." (quoted in Ref. 21, p. 121). In the spring of 1920, the roles of Bohr and Einstein were quite opposite of what they became thereafter. However, their characteristic views could be recognized: Bohr sought a complete break with the ideas of classical mechanics, while Einstein endorsed the wave-particle duality of light, and was convinced that these two aspects could be causally related (Ref. 21, pp. 121, 122).

Einstein was deeply impressed by Bohr's personality, in spite of the diversity of their views. Soon after Bohr's return to Copenhagen from Berlin, Einstein wrote to him: "Not often in life was I so delightfully impressed already by the mere presence of somebody as by yours. Now I understand why Ehrenfest is so fond of you." (letter from Einstein to Bohr, May 2, 1920; quoted in Ref. 21, p. 123, and Ref. 20, Vol. 1, Part 1, p. 257). Bohr described his visit to Einstein as "one of the greatest events in my life." (letter from Bohr to Einstein, June 24, 1920, quoted in Ref. 21, p. 123).

And Einstein wrote to Ehrenfest: "Bohr was here and I am just as keen on him as you are. He is a very sensitive fellow and goes about this world as if hypnotized." (letter from Einstein to Ehrenfest, Ref. 21, p. 123). In a letter to Arnold Sommerfeld, Einstein expressed his admiration of Bohr's intuition.[30]

The conflict between Bohr and Einstein reached its first peak after the discovery of the Compton effect in October 1922.[31] This effect was immediately explained by Arthur Holly Compton[32] and, independently, by Peter Debye,[33] as the directed scattering of individual light quanta or photons by electrons, with resultant recoil of the electron, thus conserving momentum and energy in individual atomic processes.

The Compton effect was proof positive of the existence of light quanta, which had been doubted by many serious physicists (including Max Planck) ever since Einstein introduced them in 1905 and explained the photoelectric effect.[34] Bohr himself had used the emission and absorption of light quanta in his theory of the hydrogen spectrum merely as a heuristic device, without ever believing in their existence.[35] Like Planck, Bohr believed that a merely "corpuscular theory of light" would lead to enormous difficulties in explaining electrostatic fields, and one would have to sacrifice some of the proudest achievements of Maxwell's electrodynamics. Bohr did not see how the correspondence limit or analogy between the light quantum and classical wave radiation could be established, and he had declared: "Even if Einstein sends me a cable announcing the proof of the light quantum, the message cannot reach me because it has to be propagated by electromagnetic waves." (W. Heisenberg, in conversations with J. Mehra, Geneva, July 1962).

Yet Bohr was extremely bothered by the problem of explaining the Compton effect *without the light quantum.* He was therefore very glad when, toward the end of 1923, John Slater, the young American from Harvard, brought to Copenhagen the idea of the "virtual oscillator" by means of which Slater attempted to reconcile the discrete theory of light quanta with the continuous wave theory of the electromagnetic field.[36] On the basis of this idea, Bohr developed, in 1924, with Kramers and Slater, the outline of a new "Quantum Theory of Radiation,"[38] which Kramers applied to the theory of dispersion.[39] The Bohr–Kramers–Slater theory of radiation completely abandoned Einstein's conception of a quantum structure of radiation, and replaced it by an entirely probabilistic approach based only on a statistical conservation of energy and momentum in atomic processes. When this notion was first proposed in early 1924, men like Einstein and Pauli, who believed in strict energy-momentum conservation as the divine plan of an orderly universe, regarded Bohr's idea as being completely heretical. (See Ref. 20, Vol. 1, Part 2, Chapter V.)

On 29 April 1924, Einstein wrote to Max Born: "Bohr's opinion about radiation is of great interest. But I should not be forced into abandoning strict causality without defending it more strongly than I have so far. I find the idea quite intolerable that an electron exposed to radiation should *choose of its own free will* [*aus freiem Entschluss*], not only its moment to jump off, but also its direction. In that case, I would rather be a cobbler, or even an employee in a gambling house, than a physicist. It is true, my attempts to give tangible form to the quanta have foundered again and again, but I am far from giving up hope [for a long time yet]." (letter from Einstein to Born, Ref. 23, p. 82). In his letter to Paul Ehrenfest, dated May 1, 1924, Einstein listed a number of suggestions why he rejected Bohr's suggestions, the principal reason being that "a final abandonment of strict causality is very hard for me to tolerate." (letter from Einstein to Ehrenfest, May 1, 1924, quoted in Ref. 21, p. 124).

In April 1925, Walther Bothe and Hans Geiger obtained the results of their *coincidence experiment*, showing that the secondary Compton radiation indeed emerged after scattering by a *single* electron.(39) This simple result represented the demise of the radiation theory of Bohr, Kramers, and Slater, and the triumph of strict energy-momentum conservation in individual atomic processes and its vehicle, the light-quantum. Einstein was convinced that *it had to be so*, and was glad that *it was so*. Bohr wrote a touching letter to Rutherford about the terrible difficulties of physics, or of physics as he had conceived it to be, and told him how miserable he was (letter from Bohr to Rutherford, April 18, 1925, *Bohr Archives*).

In December 1925, Bohr and Einstein met again, this time on the occasion of the celebration of Hendrik Lorentz' fiftieth anniversary of his doctorate on December 11. Paul Ehrenfest had been in Leyden since 1912, where he had become Lorentz' successor. He had enjoyed friendly relations with Einstein ever since he visited him in Prague in 1912; he also greatly admired Bohr, with whom he had frequent contact since May 1918. Hendrik Kramers, Bohr's long-time collaborator, was a student of Ehrenfest. (For an account of the beginning of H. A. Kramers' association with N. Bohr, see Ref. 20, Vol. 1, Part 1, Section III. 4.) In the fall of 1925, Ehrenfest's students George Uhlenbeck and Samuel Goudsmit had introduced the hypothesis of electron spin, based on Pauli's assignation of four quantum numbers to the electron; their hypothesis was published in a short note in *Naturwissenschaften* in November 1925.(40) Their note initiated a quick response: on November 21, 1925, just the day after its publication, Werner Heisenberg from Göttingen wrote to Goudsmit —whom he knew quite well—stating his essential agreement with the idea of the rotating electron but asked how he had got rid of the factor 2 in the doublet formula. In fact, they did not know how to proceed with the calculation. Fortunately, a little afterwards Albert Einstein came to Leyden and provided the necessary hint. He suggested that the calculation should be made in the coordinate system in which the electron was at rest. By performing the calculation Uhlenbeck found that there was indeed a difficulty about the factor 2 in the doublet formula.

The negative result was, however, soon balanced by the response which the Uhlenbeck and Goudsmit hypothesis received from Einstein and Bohr in Leyden. Before arriving in Leyden for the Lorentz jubilee celebration, Bohr had passed through Hamburg and met Pauli, who had warned him against accepting the hypothesis of the rotating electron during his visit to Holland. But then Bohr was completely won over, as he wrote several months later to Ralph Kronig: "When I came to Leyden to the Lorentz festivals, Einstein asked the very first moment I saw him what I believed about the spinning electron. Upon my question about the cause of the necessity of the mutual coupling between spin axis and the orbital motion, he explained that this coupling was an immediate consequence of the theory of relativity. This remark acted as a complete revelation to me, and I have never faltered in my conviction that we at last were at the end of our sorrows." (Letter from Niels Bohr to Ralph Kronig, March 26, 1926. For an account of the history of electron spin, see Ref. 20, Vol. 1, Part 2, Section VI. 4). Thus Bohr, who had shown only little interest in the magnetic electron before, became "completely like a prophet for the electron-magnet gospel." (letter from N. Bohr to P. Ehrenfest, December 22, 1925, quoted in Ref. 20, Vol. 1, Part 2, p. 703).

Very little is known about the discussions between Bohr and Einstein at the Lorentz festival in Leyden. In the meanwhile, Bohr had accepted Einstein's theory of the light-quantum, but he was deeply concerned about the difficulties of applying the notions of classical physics to quantum theory. In a letter to Einstein, dated April 13, 1927, Bohr referred to their encounter in Leyden, which had given him "great pleasure," and—as if continuing the discussion at Leyden—he again emphasized that the concepts of classical physics "give us only the choice between Scylla and Charybdis, depending on whether we direct our attention to the continuous or the discontinuous features of the description." (letter from N. Bohr to A. Einstein, April 13, 1927, quoted in Ref. 21, p. 125).

In his letter to Einstein, Bohr enclosed a copy of the proofs of Heisenberg's paper on uncertainty relations. He sought to connect Heisenberg's indeterminacy principle with his discussion with Einstein in Leyden; he now wrote that, as Heisenberg had shown in his paper, inconsistencies could be avoided only because of the fact that "the limitations of our concepts coincide with the limitations of our capacities of observation" (Ref. 21, p. 125). This indicates that "Bohr had already envisaged his complementarity interpretation in April 1927." (Ref. 21, p. 125). Turning to the problem of light quanta, Bohr wrote: "In view of this new formulation [Heisenberg's indeterminacy relations] it becomes possible to reconcile the requirement of energy conservation with the implications of the wave theory of light, since according to the character of the description the different aspects of the problem never manifest themselves simultaneously." (letter from N. Bohr to A. Einstein, April 13, 1927, quoted in Ref. 21, pp. 125, 126).

3. NIELS BOHR AND WERNER HEISENBERG: EARLY DISCUSSIONS AND THE BIRTH OF QUANTUM MECHANICS

The first time Werner Heisenberg encountered Niels Bohr was on the occasion of the lectures on atomic structure, which Bohr delivered in Göttingen from June 12–22, 1922 at the invitation of the Wolfskehl Commission.[3] Arnold Sommerfeld had taken two of his brightest pupils, Wolfgang Pauli and Werner Heisenberg, to attend Bohr's lectures. Heisenberg was impressed with the pleasant, warm, and glowing personality of Niels Bohr and enjoyed his lectures. Although he had already learned the content of what Bohr had to say in Sommerfeld's courses, he noticed that, "it all sounded quite different from Bohr's own lips. We could clearly sense that he had reached his results not so much by calculation and demonstration as by intuition and inspiration, and that he found it difficult to justify his findings before Göttingen's famous school of mathematics."[(41)] There were discussions after each lecture, in which—especially after the third lecture—Heisenberg participated. Bohr had talked about the calculations of his collaborator Hendrik Kramers on the Stark effect of the hydrogen atom, in particular when the strength of the electric field was very weak and the electric splitting of the components was of the same order as the fine-structure splitting.[(42)] Bohr had concluded by saying: "No experiments have yet been performed on the transition of the fine structure to the usual Stark effect by the gradual increase of an electric field. The quantum theory yields very many details of the phenomenon to be expected. Even if we really should not be unprepared to find that the quantum theory is false, it would surprise us very much if such a detailed picture obtained from the quantum theory should not be valid; for our belief in the formal reality of the quantum conditions is so strong that we should won-

[3] The members of the Wolfskehl Commission were the mathematicians Ernst Ehlers, David Hilbert, Felix Klein, Hermann Minkowski, and Carl Runge. They had invited Bohr to lecture at Gottingen in spring 1921 on the problems of atomic theory. Illness prevented him from doing so in 1921, but he delivered his lectures in June 1922.

der very much if experiments were to give a different answer than what is demanded by the theory."[43] Now Heisenberg, who knew Kramers' paper quite well because he had reviewed it earlier in Sommerfeld's Seminar, dared to dissent from this opinion and this gave rise to his first discussion with Niels Bohr. Heisenberg raised a serious objection because Kramers' result did not agree with any of the classical frequencies of the atom. On the other hand, the phenomenon of the quadratic Stark effect could be related to the dispersion of light of small frequencies by bound electrons in an atom; moreover, in the existing description of dispersion only the classical frequency of the electron's motion always appeared. Heisenberg had put Bohr on the spot with the question concerning the validity of the correspondence principle in treating the quadratic Stark effect. Bohr was not prepared to deal with Heisenberg's objection. (For a discussion of this problem, see Ref. 20, Vol. 2, Section III. 1, and Vol. 1, Part 1, Section III. 4.) "Bohr answered that one should take here into account the reaction of the radiation on the atom, but he was obviously worried by this objection. When the discussion was over, Bohr came to me and suggested that we should go for a walk together on the Hainberg outside Göttingen. Of course, I was very willing."[44]

Heisenberg was very happy that Bohr took the time to talk to him, and he was very impressed by the manner in which these private discussions went on. During the walk, which took Bohr and Heisenberg on one of the trails—passing the *Café 'Zum Rohns'*—to the top of the Hainberg, from where one had an excellent view of Göttingen and its surroundings, Heisenberg learned more about Bohr's ideas than from his previous study of the papers. As Heisenberg recalled: "That discussion which took us back and forth over Hainberg's wooded heights was the first thorough discussion I can remember on the fundamental physical and philosophical problems of modern atomic theory, and it has certainly had a decisive influence on my later career. For the first time I understood that Bohr's view of atomic theory was much more skeptical than that of many other physicists—e.g., Sommerfeld—at that time, and that his insight into the structure of the theory was not a result of a mathematical analysis of the basic assumptions, but rather of an intense occupation with the actual phenomena, such that it was possible for him to sense the relationships intuitively rather than derive them formally." (Ref. 44, p. 95, and Ref. 20, Vol. 2, p. 130).

The discussion about atomic physics between Bohr and Heisenberg on the Hainberg in Göttingen ended by Bohr inquiring about the young man's background and plans. Bohr invited him to come to Copenhagen for a few weeks the following spring, and perhaps later, possibly on a scholarship, to work there for a longer period. Heisenberg was extremely flattered by Niels

Bohr's personal interest in his future; to be invited by Bohr meant a great honor. He knew that his friend Wolfgang Pauli was about to go to Copenhagen in fall 1922; he would be able to follow him very soon. However, it took some time before Heisenberg was able to go to Denmark. In the meantime he remained in contact with Bohr's Institute, mainly through Pauli, discussing in letters the progress of his work on the anomalous Zeeman effects and the helium atom.

In a letter to Heisenberg, dated January 31, 1924, Bohr inquired "whether it would suit you to come to Copenhagen for a few weeks." Bohr added that he would be able to pay the expenses for Heisenberg's travel and stay, and concluded: "I often remember with great joy our meetings in Göttingen, and I very much hope that we shall be able to collaborate here in Copenhagen for a longer period. I shall be grateful if you would write to me about your future plans. Now, however, I hope first that you will be able to accept my invitation for a shorter visit in the near future." (letter from Bohr to Heisenberg, January 31, 1924, quoted in Ref. 20, Vol. 2, p. 131).

Heisenberg went for his first visit to Copenhagen at Easter 1924. (For an account of Heisenberg's first visit to Copenhagen, see Ref. 35 and Ref. 20, Vol. 2, Chapter III.) He had looked forward to criticizing Bohr's methods and results in atomic theory. Before he had had the chance, however, Bohr took him on a walking tour of Denmark, showing him the sights and talking to him about history and philosophy, and finally physics. Heisenberg was charmed. He found Bohr to be friendly, inspiring, kind, and one who had thought about the problems of atomic physics like no one else. Heisenberg had gone to Copenhagen to battle against the correspondence principle with the prophet himself; instead he became its evangelist.

Heisenberg returned to Copenhagen for six months in the fall of 1924. He worked with Bohr and Kramers on specific problems of atomic theory, in which he sought to formulate the content of the correspondence principle in terms of equations from which new physical results could be derived. For instance, he treated the problem of the polarization of resonance fluorescence light emitted by atoms.[45] Together with Kramers, Heisenberg extended Kramers' dispersion formula to the incoherent scattering of light by atoms.[46] The success thus obtained by what he called the *sharpening (Verschärfung)* of the correspondence principle increased Heisenberg's confidence in the Copenhagen approach, and he hoped, as he recalled later, that "Perhaps it would be possible one day, simply by clever guessing, to achieve the passage to a complete mathematical scheme of quantum mechanics."[47] In April 1925 Heisenberg returned to Göttingen to take up his duties as *Privatdozent* during the summer semester.

In Göttingen, Heisenberg sought to guess the intensities of the hydrogen lines on the basis of a "sharpened" correspondence principle, but in this specific problem he failed. He concluded that the difficulties arising from the rules of quantization were of a more general nature and had to be treated first. These difficulties were due, not so much to a departure from classical mechanics, but rather to a breakdown of the kinematics underlying this mechanics. Heisenberg employed a completely new idea: he assumed that the classical equation of an electron could be retained but the kinematical interpretation of the quantity $x(t)$ as a position depending on time had to be rejected. He expressed $x(t)$ as a Fourier series in terms of Fourier coefficients and frequencies that corresponded to the transition from a state n to a state $n-\alpha$. He motivated the introduction of the transition amplitude, $a(n, n-\alpha)$, by saying that the intensities and, therefore, the probabilities proportional to $|a(n, n-\alpha)|^2$, are *observable*, in contrast to the function $x(t)$.

The importance of the idea of employing only observable quantities in physical theories had been discussed often in Göttingen ever since Mach,[(48)] Einstein,[(49)] and Minkowski[(50)] had introduced it. Born, Pauli, Jordan, and Heisenberg had all discussed it at length in the context of quantum theory.[4] But Heisenberg made the idea of employing only observable quantities as the guiding philosophical principle of his quantum-theoretical reinterpretation *(Umdeutung)* of kinematical variables.[(2)] Heisenberg quantum-theoretically reinterpreted the classical combination law of frequencies in terms of frequencies depending upon the transitions between two states. He showed that in the product of two quantum variables, $x(t)\,y(t)$, expressed in terms of Fourier series and reinterpreted quantum-theoretically according to his prescription, the coefficient should obey a noncommutative product rule, i.e., in quantum theory $x(t)\,y(t) \neq y(t)\,x(t)$; the result of the product depends on the order in which the product is taken. By this reinterpretation the correspondence principle was incorporated into the very foundations of the theory.

Heisenberg noticed that this reinterpretation introduced a great new difficulty: whereas in classical theory $x(t)\,y(t)$ is always equal to $y(t)\,x(t)$, this is not necessarily the case in quantum theory. Therefore, he concluded that, in general, it was not clear how to formulate a product of two dynamical variables in quantum theory. (See Ref. 35, p. 22, and Ref. 20, Vol. 2, Chapter IV, Section 4.) Heisenberg looked for an example in which he could employ his quantum-theoretical reinterpretation of classical mechanical quantities by avoiding the new difficulty concerning the product. He chose the example of the anharmonic oscillator, with λ as the

[4] For a full account of the background of employing only observable quantities as a guiding principle, see Ref. 20, Vol. 2, Section V. 2.

perturbation parameter and λx^3 as the perturbation term: he wrote its classical equation, in which he reinterpreted the position $x(t)$ and the frequency $\omega(t)$ by means of his quantum-theoretical scheme and he obtained the necessary equations in the perturbation parameter, $\lambda = 0$ (the harmonic oscillator solution) and the first order $\lambda = 1$. Heisenberg found that the transition amplitudes, $a(n, n-\alpha)$, were determined only up to a constant, and he did not know what to do with this constant. This was the beginning of June 1925, and his program was stuck. (See footnote 3 and Section IV. 5 of Vol. 2 of Ref. 20.)

With the coming of spring in 1925, Heisenberg had developed a case of severe hay fever, and to cure it he decided to take a week or ten days off in June 1925 at the rocky island of Helgoland in the North Sea. (See Ref. 35, pp. 23, 24, and Ref. 20, Vol. 2, Section IV. 4.) At Helgoland, Heisenberg sought to give his vague ideas a more definite shape. There he solved two important problems: First, he reformulated quantum-theoretically the quantum condition in one dimension, $\int m\dot{x}\, dx = J = nh$, thereby obtaining an equation which was equivalent to the Thomas–Kuhn–Reiche sum rule. Since for the ground state no transition is possible, he put $a(n, n-\alpha) = 0$, if n is the quantum number of the ground state. The derivation of the quantum condition, and the subsequent determination of the transition amplitudes, was thus the first problem that was solved. (See Ref. 35, p. 25, and Ref. 20, Vol. 2, Sections IV. 4 and IV. 5.) The second problem was whether energy conservation would hold in the new scheme, a question which had become important after the Bothe–Geiger experiment. He knew that, by suitably writing the classical equation for the anharmonic oscillator, it could be shown that dH/dt, the time derivative of its energy or Hamiltonian H, was equal to zero and the energy was conserved. He reinterpreted the expression for the Hamiltonian of the anharmonic oscillator, and went through the calculation of the terms up to the second order in λ, making errors along the way and rechecking them. He found that no time-dependent terms remained in the kinematically reinterpreted Hamiltonian. The example of the anharmonic oscillator showed him that a dynamical problem in quantum theory could be solved with the help of his scheme. Heisenberg was very excited and elated by this result (see Ref. 41, p. 61).

On his way back from Helgoland to Göttingen, Heisenberg stopped to see Pauli in Hamburg. Pauli was his critical genius, and he had learned to respect Pauli's critical faculties since their first encounter in Sommerfeld's Seminar in the fall of 1920. Pauli encouraged him to go on. During the next couple of weeks Heisenberg exchanged several letters with him, and on July 9, 1925 sent him the manuscript of the finished paper. Pauli's opinion of it was favorable.

Having received Pauli's favorable verdict, Heisenberg gave the paper,[2] around the middle of July, to Max Born and asked him to do with it what he thought was right. When Born read Heisenberg's paper, he was just "fascinated," and "I... was soon involved in it that I thought the whole day and could hardly sleep at night... In the morning I suddenly saw the light: Heisenberg's symbolic multiplication was nothing but the matrix calculus, well known to me since my student days from the lectures of Rosanes in Breslau." (See Ref. 20, Vol. 3, Chapter I, and Ref. 35, pp. 27, 28.) Soon after receiving it, Born sent Heisenberg's paper to *Zeitschrift für Physik* for publication.

Born put Heisenberg's quantum condition into the matrix notation, and determined that the two matrix products **pq** and **qp** (of momenta **p** and coordinates **q**) were not identical. Born guessed that the nondiagonal elements of the matrix **pq** − **qp** were zero, and the quantum condition could be written, in general, as

$$\mathbf{pq} - \mathbf{qp} = \frac{h}{2\pi i}\mathbf{1} \qquad (1)$$

where h is Planck's constant and **1** the unit matrix, but it was only a guess and he could not prove it. The proof was given independently by Pascual Jordan and P. A. M. Dirac.

On July 19, 1925, Born travelled to Hanover to attend a meeting of the German Physical Society, where Pauli also came from Hamburg. At the railway station, Born told Pauli about the matrices and his difficulty in finding the value of the nondiagonal elements. Born invited Pauli to collaborate with him, to which Pauli gave a sarcastic refusal. (See Ref. 20, Vol. 3, Chapter I, and Ref. 35, pp. 27, 28.)

On his return from Hanover, Born immediately persuaded Jordan to help him in his work, which led to Born and Jordan's formulation of quantum mechanics, being completed on September 27, 1925.[51] This paper contained a résumé of matrix methods, the interpretation of Heisenberg's symbolic multiplication, the proof of the formula for the product difference of **pq** and **qp**, Eq. (1), proof of energy conservation, and the proof of Bohr's frequency condition. It already contained an attempt, made entirely by Jordan, at the quantization of the electromagnetic field by regarding its components as matrices.

Further development toward the completion of quantum mechanics began immediately afterwards in the collaboration of Born, Heisenberg, and Jordan. This collaboration began when Jordan wrote a letter to Heisenberg early in September 1925—Heisenberg was in Copenhagen for a few weeks before he returned to Göttingen for the winter semester—with Heisenberg and Born and Jordan all contributing their bits. The general editing of the paper was done by Jordan, and the leading introduction was written by Heisenberg. This paper by Born, Heisenberg, and Jordan was thus the third paper in the series after Heisenberg's discovery,[(2)] and it gave a logically consistent exposition of matrix mechanics. It was completed by the end of October 1925, and is usually called the *"Drei-Männer-Arbeit."*[(52)] It was really a learned paper bringing in all the mathematical learning of Göttingen, against which Pauli had severely complained personally to Max Born and in his letters to collegues [letter from Pauli to Kronig, October 9, 1925 (Ref. 7); most of Pauli's carly letters to Heisenberg were destroyed during World War II]: eigenvalues and eigenvectors, canonical transformations, principal axis transformation, Hilbert's quadratic form in an infinite number of variables, general commutation relations, and physical applications—including the quantization of the electromagnetic field and the calculation of fluctuations in this field by Jordan. This paper contained essentially the entire apparatus of modern matrix mechanics.

Wolfgang Pauli took up the problem of the hydrogen atom and solved it within the next few weeks by means of matrix methods, employing all the formal mathematical learning against which he had complained earlier. He made an ingenious application of the integration method which Wilhelm Lenz had used earlier for determining the effect of crossed electric and magnetic fields on the energy states of the hydrogen atom in the Bohr–Sommerfeld theory.[(53)] With the help of the "Lenz vector" Pauli obtained the Balmer formula and showed hos the situation with respect to the forbidden orbits could now be understood naturally.[(54)]It was exactly two years since Pauli had first seriously doubted Bohr's theory of the hydrogen atom, and now one had come around to full circle. This was indeed a triumphant moment for the new quantum mechanics, and Niels Bohr celebrated it by writing another letter to Rutherford, informing him that the reasons for his misery in the previous spring had now disappeared (letter from Bohr to Rutherford, January 27, 1926, *Bohr Archives*).

Soon after the publication of the papers of Heisenberg[(2)] and Born and Jordan,[(51)] Cornelius Lanczos, then at the University of Frankfurt, conceived of a "field-like representation of quantum mechanics."[(55)] Also, from November 14, 1925 to January 22, 1926, Max Born gave a series of lectures at the Masschusetts Institute of Technology, Cambridge, Massachusetts, which were later published under the title *Problems of Atomic Dynamics.*[(56)] At M.I.T., Born collaborated with the mathematician Norbert Wiener, and together they developed "a generalization of matrix mechanics into a kind of operator mechanics."[(57,58)]

However, just before the Born–Heisenberg–Jordan paper was published in *Zeitschrift für Physik* in January 1926, another paper, containing the complete scheme of quantum mechanics, made its appearance in the *Proceedings of the Royal Society.*[4] Its author was Paul Adrien Maurice Dirac and, briefly, it came about as follows. In July 1925, after giving the manuscript of his paper on quantum-theoretical kinematics to Max Born, Heisenberg left for Leyden and Cambridge. In Leyden he stayed as a guest of Paul Ehrenfest, and discussed spectroscopic questions with Ehrenfest, Uhlenbeck, and Goudsmit. Uhlenbeck and Goudsmit would soon propose the hypothesis of electron spin.[40] From Leyden, Heisenberg went to Cambridge, where he stayed as a guest of R. H. Fowler—with whom he had become acquainted in Copenhagen—and gave a talk on "Term Zoology and Zeeman Botany" at the Kapitza Club on July 28, 1925 (W. Heisenberg, Talk on "term zoology and Zeeman botany," *Minute Book* of the Kapitza Club; Heisenberg spoke at the 94th meeting of the Club on July 28, 1925). Privately, he mentioned his new ideas on the quantum-theoretical reformulation of kinematical quantities to Fowler. Fowler requested Heisenberg to send the proof sheets of his paper as soon as they became available, which he did in the beginning of September 1925. Fowler found Heisenberg's paper interesting, and wanted to know what Dirac's reaction would be. At that time, Dirac was too much enamored with the Hamiltonian formalism as the basis of atomic physics, and thought that anything not connected with it would not be much good. At first, Dirac thought there was not much in Heisenberg's paper, and he put it aside for a week or so. When Dirac went back to it, it suddenly became clear to him that Heisenberg's idea had provided the key to the "whole mystery." (See Ref. 35, p. 33; Ref. 20, Vol. 4, Part 1, Chapter IV; and Ref. 59.) During the following weeks Dirac tried to connect Heisenberg's quantum-theoretical reinterpretation of kinematical quantities with the action-angle variables of the Hamilton–Jacobi theory. During a long walk on a Sunday in September 1925, it occurred to Dirac that the commutator might be the analogue of the Poisson bracket. He verified this the next day and found that they fit. (See Ref. 35, p. 34; Ref. 20, Vol. 2, Part 1, Chapter IV; and Ref. 59.)

From the quantum conditions expressed in angular variables Dirac found the correspondence between Heisenberg's commutation brackets and the classical Poisson brackets for the variables X and Y,

$$XY - YX = \frac{ih}{2\pi}\sum_r \left\{\frac{\partial X}{\partial q_r}\frac{\partial Y}{\partial p_r} - \frac{\partial Y}{\partial q_r}\frac{\partial X}{\partial p_r}\right\} \tag{2}$$

where q_r and p_r can be regarded as the action-angle variables (w_r and J_r).

Dirac was now safely back on Hamiltonian ground. He showed his new results to Fowler, who fully appreciated their importance. Fowler knew what was going on in Copenhagen and Göttingen, and he realized that there would be competition from these places. He thought that the results obtained in England in this field had to be published at once, and urged the *Proceedings of the Royal Society* to give immediate priority to the publication of Dirac's paper on "The Fundamental Equations of Quantum Mechanics."[4] Sir James Jeans, who was then editor of the *Proceedings* and Secretary of the Royal Society, was ready and willing to oblige. All of Dirac's papers from 1925 to 1933 were thus published very fast.

In his fundamental paper,[4] Dirac first gave a summary of Heisenberg's ideas, simplifying the mathematics and making it at once more elegant. He anticipated all the essential results of the papers of Born and Jordan[51] and Born, Heisenberg, and Jordan.[3,52] He developed a quantum algebra, derived Heisenberg's quantization rules, and obtained the canonical equations of motion for quantum systems. In the same paper, Dirac introduced an early form of creation and annihilation operators, pointing out their analogs in classical theory.

Dirac quickly followed this paper by another a few weeks later.[60,61] In it he developed the algebra of q-numbers, that is, the dynamical variables which satisfy all the rules of normal numbers except that their product is not necessarily commutative. He gave detailed theorems on the operations with q-numbers, and applied the rules he had obtained to multiply periodic systems in close analogy with the old quantum rules.

Dirac's aim was to apply his scheme to the hydrogen atom. He wrote its Hamiltonian by simply replacing position and momentum variables in the classical Hamiltonian by q-numbers, and proceeded to obtain the Balmer formula in order to show that this abstract scheme could give results closely related to the experiments. Dirac, however, did not go into the details of this calculation as Pauli[54] (in his paper published during the same month, March 1926) had already shown that this could be done, and Dirac mentioned it in a footnote. (Dirac referred to Pauli's "not yet published paper" on the hydrogen atom in a footnote in Ref. 60, p. 570.) He then went on to calculate the various features of the splitting and intensities of spectral lines in a magnetic field (including the Zeeman effect) in agreement with the experiments.

With all this work on the principles of quantum mechanics Dirac was awarded the Ph. D. degree in May 1926 at Cambridge.[62]

4. THE CREATION OF WAVE MECHANICS

Since 1921 Erwin Schrödinger had been at the University of Zurich, where he occupied the chair of theoretical physics, which Albert Einstein,

Peter Debye, and Max von Laue had held previously (Ref. 20, Vol. 5, Part 1, Chapter II). Schrödinger was a Viennese and a man of vast personal culture that included the study of Greek literature and philosophy in the original and the writing of poetry (Ref. 35, p. 37, and Ref. 20, Vol. 5, Part 1, Chapter I). A distinguished physicist by any measure, Schrödinger traced his scientific lineage to Boltzmann through his teacher Fritz Hasenöhrl, but he had himself not yet set the world aglow although he had done excellent work on problems of Brownian motion, specific heat and quantum mechanics, and of general relativity theory (Ref. 35, p. 37, and Ref. 20, Vol. 5, Part 1, Chapter I). By the summer of 1925 Schrödinger had become tired of his stay in Zurich because, as he wrote to Sommerfeld, "the Swiss are just too uncongenial" *('die Schweizer sind gar zu ungemütlich')* and he wanted to go home to Austria (letter from Schrödinger to Sommerfeld, July 21, 1925, in *Sommerfeld Correspondence*, Deutsches Museum, Munich). He was negotiating for the chair of theoretical physics at Innsbruck, but since the University of Innsbruck sought to dicker about the salary, he let the matter drop in favor of Arthur March. Within eighteen months Schrödinger would be appointed as Max Planck's successor at the University of Berlin. (See Ref. 35, p. 37, and Ref. 20, Vol. 5, Part 2, Chapter IV, Section 5.)

In the fall of 1925 Schrödinger suffered not only from the lack of congeniality of his colleagues in Zurich, but the work of Heisenberg and of Born and Jordan on matrix mechanics added to his discomfort, for he remarked: "... I was discouraged *(abgeschreckt)*, if not repelled *(abgestossen)*, by what appeared to me a rather difficult method of transcendental algebra, defying any visualization *(Anschaulichkeit)*. (See Ref. 6, footnote 2, p. 735; Ref. 20, Vol. 5, Part 2, Chapter IV, Section 5.) He decided to sublimate his social and scientific unhappiness by conceiving and delivering a scheme of atomic mechanics which not only seemed to be a genuine alternative to the matrix or *q*-number mechanics of Heisenberg, Born, Jordan, and Dirac, but helped in completing the edifice of quantum mechanics and in inaugurating the discussions that led to its physical and philosophical interpretation.

In four communications to *Annalen der Physik*, submitted from the end of January to the end of June 1926, Schrödinger developed his theory of wave mechanics, entitled "Quantization as an Eigenvalue Problem." He kept detailed notebooks on his attempts to formulate wave mechanics and its applications to atomic problems; in these attempts, his earlier studies on tensor-analytical mechanics and Hamilton's optical-mechanical analogy—also preserved in his notebooks—came in handy. (Erwin Schrödinger kept almost all of his notebooks in which he had written up his notes on various topics he studied and thought about, for example, notebooks on *Tensoranalytische Mechanik*, *Eigenwertproblem des Atoms*, etc. These notebooks have been made use of in Ref. 20, Vol. 5 on *Erwin Schrödinger and the Rise of Wave Mechanics* to indicate Schrödinger's developing thought processes.) He arrived at his fundamental equation

$$H\left(q, \frac{h}{2\pi i}\frac{\partial}{\partial q}\right)\psi(q) = E\,\psi(q) \tag{3}$$

where H is the Hamiltonian, $\psi(q)$ the wave function, and E the energy eigenvalues, and solved the problem of the spectrum of the hydrogen atom.[63] In the mathematical aspects of some of his work he had invaluable help from Hermann Weyl, then also in Zurich at the *Eidgenössische Technische Hochschule*, and Schrödinger acknowledged it. (Schrödinger thanked Hermann Weyl for help in solving the eigenvalue differential equation for the hydrogen atom. See Ref. 63, footnote 1, p. 363). Weyl's 1908 thesis[64] under Hilbert had dealt with integral equations, eigenvalue problems, orthogonal functions, etc., and it was a fortuitous combination of circumstances that brought Schrödinger and Weyl together.

In his communications, Schrödinger provided the basis of treating all those problems of atomic physics that had been impossible to handle in the Bohr–Sommerfeld theory. In Schrödinger's work the fundamental ideas of Einstein and Louis de Broglie[65] found a natural place. Schrödinger soon recognized that in spite of fundamental disparities the two approaches, his own and Heisenberg and Born's, did not clash but rather complemented each other. In fact, in the early spring of 1926, prior to the publication of his third communication, Schrödinger discovered what he called "a formal, mathematical identity" of wave mechanics and matrix mechanics. (See Ref. 6 and Ref. 20, Vol. 5, Part 2, Chapter IV.) The same formal equivalence was demonstrated, independently, by Carl Eckart (see Ref. 8 and Ref. 20, Vol. 5, Part 2, Chapter IV) in the United States and by Pauli (see Ref. 7 and Ref. 20, Vol. 5, Part 2, Chapter IV) in a letter to Jordan. Many years later, Max Born said in an obituary of Schrödinger: "What is more magnificent in theoretical physics than his first six papers on wave mechanics?"[66]

Soon after the publication of Schrödinger's papers, wave mechanics was successfully applied to a large number of energy-eigenvalue problems of atomic physics. (See Ref. 20, Vol. 5, Part 2, Chapter IV. Heisenberg himself employed Schrödinger's scheme to solve the helium problem; see also Refs. 67 and 68.) It soon became clear that the theory could be extended to deal with types of problems not initially envisaged by Schrödinger.

5. ERWIN SCHRÖDINGER'S LECTURES IN BERLIN AND MUNICH AND VISIT TO COPENHAGEN

From the spring of 1926, Max Planck in Berlin and Wilhelm Wien and Arnold Sommerfeld in Munich had repeatedly invited Erwin Schrödinger to come to Berlin and Munich, respectively, and speak on his new atomic theory. Schrödinger, who was very happy about the appraisal and approval that his work had received in both places, actually fulfilled these pressing and welcome invitations as soon as he could free himself from the obligations of the summer semester in Zurich. He first travelled to Berlin and spoke before the German Physical Society on July 16 on the *"Grundlagen einer Wellenlehre begründeten Atomphysik"* ("Foundations of an Atomic Physics Based on Wave Theory"). (See Ref. 20, Vol. 5, Part 2, Chapter IV, p. 800.) He stayed on in Berlin for a few days, as a personal guest of Max Planck, and then he returned to Zurich via Munich. In Munich he delivered a similar lecture, entitled *"Grundgedanken einer auf Wellenlehre begründeten Atomphysik"* ("Basic Ideas of an Atomic Physics Founded on Wave Theory") to the Bavarian Section *(Gauverein)* of the German Physical Society. (See Ref. 20, Vol. 5, Part 2, Chapter IV, p. 800.)

These two lectures were the first reviews of wave mechanics by its creator. Schrödinger had especially available the results of his fourth communication ("Quantization as an Eigenvalue Problem. Part IV"),[69] in which the theory had received an important generalization and—as it seemed to Schrödinger—a reasonably "visualizable" *("anschauliche")* interpretation also. In these lectures Schrödinger covered all the results he had hitherto obtained in his main communications. (See Ref. 20, Vol. 5, Part 2, Chapter IV, p. 801.) Interestingly enough, he left out a closer discussion of the formal equivalence of the undulatory and matrix mechanics. (See Ref. 20, Vol. 5, Part 2, Chapter IV, p. 801.)

Schrödinger primarily emphasized the progress achieved in his fourth communication,[69] in which he ascribed to the wave function ψ an "electrodynamical significance" in order to account for the fact that a mechanical system can emit electromagnetic waves of a frequency equal to the term difference and to deduce their intensity and polarization. According to Schrödinger, the concept of the electric charge density in the case of many-particle systems could be phrased as follows: "$\psi\bar{\psi}$ [where $\bar{\psi}$ is the Hermitian conjugate of ψ] is a kind of *weight function* in the system's configuration space. The *wave mechanical* configuration is a *superposition* of many, strictly speaking of all, point-mechanical configurations kinematically possible. Thus, each point-mechanical configuration contributes to the true wave mechanical configuration with a certain *weight*, which is given precisely by $\psi\bar{\psi}$." (See Ref. 20, Vol. 5, Part 2, Chapter IV,

p. 797.) While for macroscopic systems and motions the weight function is practically concentrated in small space regions, the varying distribution $\psi\bar{\psi}$ over a region plays a crucial role for microscopic systems. Schrödinger combined his electrodynamic interpretation of the wave function with the idea that particles of corpuscular physics are essentially only wave groups composed of numerous, strictly speaking, infinitely many wave functions.

Schrödinger admitted that the "new interpretation may shock us at first glance", especially since he himself had always insisted on taking the ψ vibrations as something having an easily intelligible physical reality. (See Ref. 20, Vol. 5, Part 2, Chapter IV, p. 797.) Now he found himself forced to modify his earlier statements: the ψ vibrations could still be considered as based on something "tangibly real" *("greifbar Reales")*, he said, namely on "the very real electrodynamically effective fluctuations of the electric space density." (See Ref. 20, Vol. 5, Part 2, Chapter IV, p. 797.) The ψ function, having lost its (primitive) interpretation as directly representing the spatial distribution of the electric charge, still describes the electric fluctuations in a slightly more involved manner. However, if this description is consistent, one has to demand that the $\psi\bar{\psi}$ integral assumes a constant value, preferably unity, for nonconservative systems also, and this constancy had evidently to follow from the time-dependent wave equation. After demonstrating this constancy, Schrödinger proceeded to interpret "the *current-density* of the weight function [$\psi\bar{\psi}$] in the configuration space," which he also called "the *continuity equation* of the weight function, and then derived the *"continuity equation of electricity."* (See Ref. 20, Vol. 5, Part 2, Chapter IX, pp. 789-799, and Ref. 69, pp. 137-138.) Schrödinger remarked: "Since [either zero or stationary current density] occurs in the unperturbed normal state [of the atomic system] at any rate, we may in a certain sense speak of *a return to electrostatic and magnetostatic atomic models.* In this way the lack of radiation in the normal state would, indeed, find a startlingly simple explanation." (See Ref. 69, pp. 138, 139.) This last point, which Schrödinger had already mentioned in his letter to Wilhelm Wien, dated June 18, 1926 (letter from Schrödinger to Wien, June 18, 1926, in the *Wien Collection*, Deutsches Museum, Munich), seemed indeed to support the visualizable *(anschaulische)* interpretation of the new atomic theory, if atoms described by a wave equation need not radiate in what had previously been called the stationary states, i.e., classical electrodynamics then remained valid to a certain extent. As Schrödinger wrote to Wien: "The vibration equation for the ψ function enables us to consider again these static models [for atoms]; although the ψ function oscillates in these models, the distribution of electricity does not change." (See letter from Schrödinger to Wien, June 18, 1926, in the *Wien Collection*, Deutsches Museum, Munich.) Admittedly, the new situation weakened the "reality" of the ψ vibrations, but Schrödinger argued: "This does not matter at all. If one can only control, with their help [i.e., of the ψ-vibrations], distributions and fluctuations of electricity, which are real in the highest sense, then one may be allowed to call them a substitute *(Hilfs)* concept in the same sense as one speaks of the electrodynamic potentials, of which only the derivatives can be observed. (See letter from Schrödinger to Wien, June 18, 1926, in the *Wien Collection*, Deutsches Museum, Munich.)

Schrödinger's communications on wave mechanics were most favorably received by Max Planck, Albert Einstein, and Hendrik Lorentz,[170] as well as by Arnold Sommerfeld and Wilhelm Wien. (See Schrödinger's correspondence with Sommerfeld and Wien, in the *Sommerfeld and Wien Collections*, Deutsches Museum, Munich. It has been amply used in Ref. 20, Vol. 5.) They greatly appreciated the fact that Schrödinger had restored the use of differential equations in quantum theory, with their solutions as various kinds of familiar polynomials, and one did not have to resort to such esoteric devices as matrices and q-numbers. Werner Heisenberg, however, had believed throughout that the solution of the problems of atomic mechanics would lead to *one, unique, general mathematical scheme,* and when he discovered his scheme, well, that was it. With the arrival of Schrödinger's theory, Heisenberg was unhappy, and he believed, indeed hoped, that it was wrong. (See Ref. 35, pp. 39, 40.) When in June and July 1926 Max Born[(9)] applied the Schrödinger method to the collision problems, a work which led to the statistical interpretation of Schrödinger's wave function, Heisenberg reproached him for going over "to the enemy camp." (See Ref. 35, p. 40, and Ref. 67.) Heisenberg attended Schrödinger's lecture at Munich and, even many years later, he vividly recalled Schrödinger's colloquium and the discussion afterwards. Thus, for instance, he recalled in 1968: "In July 1926 Schrödinger was invited to Munich by Wilhelm Wien to report on his theory. The experimental physicists in Munich, headed by Wien, were enthusiastic about the possibility that now perhaps this whole 'quantum mystery of atomic physics' might be dealt with, and one would be able to return to the classical concepts of honest fields, such as one had learned from Maxwell's [electromagnetic] theory. I listened to this lecture by Schrödinger, as I was then staying with my parents in Munich for the [summer] vacation; and I was quite horrified by his interpretation, because I simply could not believe it. I objected (in the discussion) that with such an interpretation one would not even be able to explain Planck's heat radiation law. But general opinion at that time was extremely hostile toward my objection. Wien answered me very harshly in that he could understand how I felt about the fact that the whole quantum jumping, the matrices and all that had become superfluous; anyway, it would be better for me to leave the field to Schrödinger, who would certainly solve all the difficulties in the [near] future. This was not very encouraging; I did not have the slightest chance to get across my point of view in the discussion." (See Ref. 20, Vol. 5, Part 2, Chapter IV. 5, p. 803.) Heisenberg further recalled: "My arguments had clearly failed to impress anyone—even Sommerfeld, who felt more kindly toward me, succumbed to the persuasive force of Schrödinger's mathematics. And so I went home rather sadly. It must have been that same evening that I wrote to Niels Bohr about the unhappy outcome of the discussion. Perhaps it was as a result of this letter that he invited Schrödinger to spend part of September in Copenhagen. Schrödinger agreed, and I, too, sped back to Denmark." (See Ref. 41, pp. 72, 73.)

From the very beginning, Heisenberg had been seriously opposed to the *"anschauliche"* interpretation of wave mechanics. Thus, for instance, he had written to Wolfgang Pauli in June: "The more I ponder on the physical part of Schrödinger's theory, the more detestable I find it. One should imagine the rotating electron, whose charge is distributed over the entire space and which has an axis in a fourth and fifth dimension. What Schrödinger writes on the visualizability of his theory ··· I find rubbish. The great achievement of Schrödinger's theory is the calculation of matrix elements." (See letter from Heisenberg to Pauli, June 8, 1926, in Ref. 71.) Schrödinger's colloquium in Munich had merely confirmed Heisenberg's negative opinion. A few days after the colloquium, he wrote to Pauli: "As nice as Schrödinger is personally, I find his physics so strange: one feels 26 years younger when listening to it. Indeed, Schrödinger throws overboard everything which is 'quantum-theoretical': namely, the photoelectric effect, the Franck[-Hertz] collisions, the Stern–Gerlach effect, etc. It is not then difficult to establish a theory [of the kind Schrödinger has in mind]. However, it does not agree with experience." (See letter from Heisenberg to Pauli, July 28, 1926, in Ref. 71.) But the unfortunate discussion following Schrödinger's Munich lecture persuaded Heisenberg of the necessity of having a really detailed and penetrating discussion with Erwin Schrödinger elsewhere, most profitably in Copenhagen where Niels Bohr especially could participate.

In a letter dated September 11, 1926, Bohr actually invited Schrödinger to come to Copenhagen and deliver a lecture to the Danish Physical Society *(Fysisk Forening)* on wave mechanics. At the same time Bohr expressed the hope "that you will introduce some discussions for the narrower circle of those who work here at the Institute, and in which we can deal more deeply with the open questions of atomic theory." [See Ref. 20, Vol. 1, Part 2, Section V. 2 (for Schrödinger's views on the Bohr–Kramers–Slater theory, see pp. 540, 552, and 553); see also Vol. 5, Part 2, Chapter IV, and K. Stolzenburg (cited in Ref. 14) for an account of

Schrödinger's visit to Copenhagen.] Among this narrower circle belonged, in particular, Werner Heisenberg—from May, 1926 the main *"Assistent"* and *"Lektor"* at the Copenhagen Institute for Theoretical Physics—and Oskar Klein; Paul Dirac was also present at Bohr's Institute at that time. Schrödinger accepted Bohr's invitation quite readily (in a letter to Bohr, dated September 21, 1926) and announced his arrival in Copenhagen on October 1 (Schrödinger to Bohr, telegram of September 27, 1926). There he was received with great eagerness; thus Heisenberg recalled: "Bohr's discussions with Schrödinger began at the railway station and were continued daily from early morning until late at night. Schrödinger stayed at Bohr's house so that nothing would interrupt the conversations." (See Ref. 41, pp. 73-75.)

No contemporary notes about the content of the Copenhagen discussions exist; it does not seem to have occurred to anyone—neither Bohr, nor Schrödinger, nor any of the other participants—to keep such notes. However, Heisenberg later gave many lively accounts of the discussions. (See Ref. 20, Vol. 5, Part 2, Chapter IV, p. 823, and footnote 295.) As Heisenberg recalled, in Copenhagen Schrödinger especially attacked the idea of sudden quantum jumps. Schrödinger believed that the idea of quantum jumps was bound to end in nonsense. He reminded Bohr that "according to his (Bohr's) theory, if an atom is in a stationary state, the atom revolves periodically but does not emit light, when, according to Maxwell's theory it must. Next the electron is said to jump from one orbit to the next and to emit radiation. Is this jump supposed to be gradual or sudden? If it is gradual, the orbital frequency and energy of the electron must change gradually as well. But in that case, how do you explain the persistence of fine spectral lines? On the other hand, if the jump is sudden, Einstein's idea of light-quanta will admittedly lead us to the right wave number, but then we must ask ourselves how precisely the electron behaves during the jump. Why does it not emit a continuous spectrum, as electromagnetic theory demands? And what laws govern its motion during the jump? In other words, the whole idea of quantum jumps is sheer fantasy." (See Ref. 41, pp. 73–75.)

Bohr agreed with Schrödinger's objections, but pointed out that they did not prove that there are no quantum jumps; only we cannot imagine them and the concepts with which we describe events in daily life and experiments in classical physics are inadequate when it comes to describing quantum jumps. "Nor should we be surprised to find it so," Bohr said, "seeing that the processes involved are not the objects of direct experience." (See Ref. 41, pp. 73-75.) Schrödinger countered by saying that "if there are electrons in the atom, and if these are particles—as all of us believe—then they must surely move in some way. Right now I am not concerned with a precise description of this motion, but it ought to be possible to determine the principle of how they behave in the stationary state or during the transition from one state to the next. But from the mathematical form of wave or quantum mechanics alone, it is clear that we cannot expect reasonable answers to these questions. The moment, however, that we change the picture and say that there are no discrete electrons, only electron waves or waves of matter, then everything looks quite different. We no longer wonder about the fine lines. The emission of light is as easily explained as the transmission of radio waves through the aerial of the transmitter, and what seemed to be insoluble contradictions have suddenly disappeared." (See Ref. 41, pp. 73-75.) Bohr disagreed with this and pointed out that the contradictions did not disappear: "You speak of the emission of light by the atom or more generally of the interaction between the atom and the surrounding radiation field, and you think that all the problems are solved once we assume that there are material waves but no quantum jumps. But just take the case of thermodynamic equilibrium between the atom and the radiation field—remember, for instance, the Einsteinian derivation of Planck's radiation law. The derivation demands that the energy of the atom should assume discrete values and change discontinuously from time to time; discrete values for the frequencies cannot help us here. You can't seriously be trying to cast doubt on the whole basis of quantum theory." (See Ref. 41, pp. 73-75.)

Schrödinger conceded that all these relationships had not yet been fully explained, but pointed out that Bohr and his associates had also so far failed to discover a satisfactory physical interpretation of quantum mechanics. Bohr agreed that there were inconsistencies, as for example when one watched sudden flashes of light on a scintillation screen or the sudden rush of an electron through a cloud chamber; one could not just ignore these observations as if they did not exist at all.

On Bohr's defense of the concept of quantum jumps as being essential in describing the behavior of atoms and radiation, Schrödinger became quite despondent, and finally exclaimed: "If all this quantum jumping were here to stay, I should be sorry I ever got involved with quantum theory." To which Bohr replied: "But the rest of us are extremely grateful that you did; your wave mechanics has contributed so much to the mathematical clarity and simplicity that it represents a gigantic advance over all previous forms of quantum mechanics." (See Ref. 41, pp. 73-75.)

Heisenberg also recalled that the continuous, strenuous discussions and conversations with Bohr exhausted Schrödinger. "After a few days Schrödinger fell ill, perhaps as a result of his enormous effort; in any case, he was forced to keep to his bed with a feverish cold. While Mrs. Bohr nursed him and brought in tea and cake, Niels Bohr kept sitting on the edge of the bed talking to Schrödinger: 'But you must surely admit that....'" (See Ref. 41, pp. 73-75.) In other words Bohr—whom Heisenberg described as having been "an almost remorseless fanatic" (Ref. 41, pp. 73-75) in the debate with Schrödinger—still *had to convince* his miserable guest to accept the Copenhagen position. However, "no real understanding could be expected since, at that time, neither side was able to offer a complete and coherent interpretation of quantum mechanics. For all that, we in Copenhagen felt convinced toward the end of Schrödinger's visit that we were on the right track, though we fully realized how difficult it would be to convince even leading physicists that they must abandon all attempts to construct perceptual models of atomic processes." (See Ref. 41, pp. 73-75.)

Schrödinger, who delivered his lecture entitled *"Grundlagen der undulatorischen Mechanik"* ("Foundations of an Undulatory Mechanics") before the Danish Physical Society on October 4, 1926, stayed in Copenhagen only a few days. A couple of weeks later he reported to Wilhelm Wien: "... it was very nice that I was able to become thoroughly acquainted with Bohr—whom I had never known before—in his own surroundings, and to talk with him for hours about these matters which are so very dear to all of us." (See letter from Schrödinger to Wien, October 21, 1926, in the *Wien Collection*, Deutsches Museum, Munich.) In his letter to Wien, Schrödinger briefly sketched the contents of discussions in Copenhagen, especially the points of disagreement. He wrote: "Quite certainly, the point of view of [using] visualizable pictures, which de Broglie and I assume, has not been carried through nearly far enough in order to render an account of the most important facts [of atomic theory]. It is of course probable that here and there a wrong path was taken that must now be abandoned. But that, even if one is Niels Bohr, one could possibly say at this point: the visualizable wave pictures work as little as the visualizable point [-particle] models, there being something in *the results of observation* which cannot be grasped by our erstwhile way of thinking; this I do not believe. I believe it even less since for me the comprehensibility of the external processes in nature is an axiom, say, in the sense: to *grasp* experience means nothing more than establishing the *best possible* organization among the different facts of experience." (See letter from Schrödinger to Wien, October 21, 1926, in the *Wien Collection*, Deutsches Museum, Munich.) In Schrödinger's opinion, the facts of experience could not contradict each other, as Bohr—for many years—had tended to assume in atomic physics; only the "theoretical connections" *("gedankliche Verbindungsglieder")* or the physical interpretations could do so. In particular, it seemed "premature" *("voreilig")* to give the "completely general conceptions of space and time and the connection of the interaction of neighboring space-time points," concepts that had been preserved even in general relativity theory. (See letter from Schrödinger to Wien, October 21, 1926, in the *Wien Collection*, Deutsches Museum, Munich.)

Schrödinger further remarked to Wien: "I can only say that I don't care at all for this whole play of waves, if it should turn out to be nothing more than a comfortable computational device to evaluate matrix elements. (Heisenberg had written to Pauli in his letter of June 8, 1926; for the letters from Heisenberg to Pauli, see Refs. 67, 68, 71; see also letter from Heisenberg to Pauli in Ref. 71; and Ref. 20, Vol. 5, Part 2, p. 741.) In spite of all these theoretical points of dispute, however, the relationship with Bohr, and especially Heisenberg, both of whom behaved toward me in a touchingly kind, nice, caring and attentive manner, was totally, cloudlessly amiable and cordial." (See letter from Schrödinger to Wien, October 21, 1926, in the *Wien Collection*, Deutsches Museum, Munich.)

The reports from Copenhagen on the results of the discussions with Schrödinger sounded a little different. For example, Niels Bohr wrote to Ralph Fowler on October 26, 1926: "... The discussions gradually centered themselves on the problems of physical reality of the postulates of atomic theory. We all agreed that a continuity theory [such as Schrödinger's] leads to expectations fundamentally different from those of the usual discontinuity theory [of Born, Heisenberg, Jordan, and Dirac]. Schrödinger himself continued in the hope that the idea of stationary states and transitions was altogether avoidable, but I think we succeeded at least in convincing him that for the fulfilement of his hope he must be prepared to pay a cost, as regards reformation of fundamental concepts, formidable in comparison with that hitherto contemplated by the supporters of the idea of a continuity theory of atomic phenomena." (N. Bohr to R. H. Fowler, October 26, 1926, *Bohr Archives*, quoted in Ref. 20, Vol. 5, Part 2, Chapter IV, and in K. Stolzenburg, cited in Ref. 14). Schrödinger obviously believed that matrix mechanics implied that single stationary states possessed no physical reality, a point of view which Bohr found to be "a confounding of the means and aims of Heisenberg's theory." (N. Bohr to R. H. Fowler, October 26, 1926, *Bohr Archives*, quoted in Ref. 20, Vol. 5, Part 2, Chapter IV, and in K. Stolzenburg, cited in Ref. 14). On the other hand, Bohr considered wave mechanics "so wonderfully suited to bring out the true correspondence between the quantum theory and the classical ideas." (N. Bohr to R. H. Fowler, October 26, 1926, *Bohr Archives*, quoted in Ref. 20, Vol. 5, Part 2, Chapter IV, and in K. Stolzenburg, cited in Ref. 14). This correspondence was explored by Bohr and his collaborators in the months following Schrödinger's visit to Copenhagen and helped enormously in establishing the physical interpretation of quantum mechanics. The efforts of the Copenhagen physicists on the physical interpretation of quantum mechanics led, in particular, to Heisenberg's uncertainty relations and to Bohr's complementarity view—which became the central parts of the so-called Copenhagen interpretation of quantum mechanics.

6. MAX BORN'S STATISTICAL INTERPRETATION OF THE WAVE FUNCTION

A new interpretation of Schrödinger's wave function was proposed in connection with the quantum mechanical study of atomic scattering processes. In order to account for the quantum mechanical explanation of the process of collision between a free particle—such as an α-particle or an electron—and an atom, Born employed the formalism of Schrödinger's wave mechanics. Born first gave a preliminary account of his quantum mechanical treatment of collision processes.[(9)] In this paper he made the statement that $|\psi_{nm}^{(E)}(\alpha, \beta, \gamma)|^2$ and not $\psi_{nm}^{(E)}$ measures the probability that the electron which approaches the scattering center, say along the z axis, is found scattered in the direction defined by the unit vector whose components are α, β, γ; $E = h^2/2m\lambda^2$ is the energy of the electron, and $p = h/\lambda$ is its momentum, h being Planck's constant and λ the de Broglie wavelength. Born's probabilistic interpretation of the wave function was noted in Einstein's earlier work.[(72)] In the collision experiments, Born took into account corpuscular aspects and sought to associate the wave function with the particles. In this, he was primarily influenced by Einstein's conception of the relation between the light quanta and the field of electromagnetic waves. Einstein had regarded the electromagnetic wave field as a kind of "ghost field" *("Gespensterfeld")* whose waves served to guide the motion of the corpuscular light quanta, and the squared wave amplitudes (intensities) determined the probability of the presence of light quanta. (Letter from Born to Einstein, November 30, 1926, quoted in Ref. 73; Einstein's reply to Born, December 4, 1926, Ref. 23, pp. 90-91.) Born argued that just as the intensity of light waves was a measure of the density of light quanta, "it was almost self-understood to regard $|\psi|^2$ as the probability density of particles." (See Ref. 72, p. 285.) In addition, Born's probability interpretation was also rooted in the Bohr–Kramers–Slater theory of radiation and their conception of the virtual radiation field.[(37)] As Heisenberg, in recalling their treatment of induced and spontaneous emission, remarked: "In the summer of 1926 Born established his theory of collision processes and interpreted correctly the wave in multidimensional configuration space as a probability wave by developing and elaborating an idea previously expressed by Bohr, Kramers, and Slater." (Ref. 47, pp. 40-47).

In two papers, Born developed the theory of collision processes further. (See Ref. 9 and Ref. 72, p. 285; see also Ref. 20, Vol. 5, Part 2, Chapter IV.) After interpreting ψ as a probability wave, Born expanded ψ in terms of a complete orthonormal set of eigenfunctions ψ_n of the Schrödinger equation, $H\psi = E\psi$, Eq. (3), as

$$\psi = \sum c_n \psi_n \tag{4}$$

In accordance with the completeness relation,

$$\int |\psi(q)|^2 \, dq = \sum_n |c_n|^2 \tag{5}$$

How were c_n to be interpreted? Since for a single normalized eigenfunction $\psi(q)$, corresponding to a single particle, the right-hand side of Eq. (5) is unity, Born concluded that $\int |\psi(q)|^2 \, dq$ represents the number of particles and $|c_n|^2$ the statistical frequency of the occurrence of the state characterized by the index n. Born justified this assumption by calculating the "expectation value" of the energy E for ψ and obtained

$$E = \sum_n |c_n|^2 E_n \tag{6}$$

where E_n is the energy eigenvalue of ψ_n. Born's statistical interpretation of the wave function was immediately applied with resounding success to various problems of atomic scattering.[74]

Schrödinger had been confronted with the probability interpretation of the wave function soon after its conception. He had seen the proofs of Max Born's paper on the collision problem in wave mechanics (See Ref. 9, Ref. 72, p. 285, and Ref. 20, Vol. 5, Part 2, Chapter IV.) prior to publication and, in a letter to Wien, he had confided his concern about the probability interpretation of the wave function. (See letter from Schrödinger to Wien, August 23, 1926, in the *Sommerfeld and Wien Collections*, Deutsches Museum, Munich; cited in Ref. 20, Vol. 5, Part 2, p. 827.) In the discussions in Copenhagen, the question of the probability interpretation did play a role. As Oskar Klein recalled after several decades: "Schrödinger wanted to interpret that which Born interpreted as probability density as really the density of the electrons. Then they [i.e., Bohr and Heisenberg] showed that if that was so and if by means of his currents and densities he coupled the thing to the electromagnetic field, then the probability of the spontaneous emission would be proportional to the number of atoms in the upper state multiplied by the number of atoms in the lower state. It would be quite against anything already known." (Oscar Klein, AHQP Interview, July 16, 1963, p. 2). Still, Schrödinger remained unshaken, and shortly afterwards he wrote to Max Born: "I have, however, the impression that you and others, who essentially share your opinion, are too deeply under the spell of those concepts (like stationary states, quantum jumps, etc.), which have obtained civic rights in our thinking in the last dozen years; hence you cannot do full justice to an attempt to break away from this scheme of thought." (See letter from Schrödinger to Born, November 2, 1926, Ref. 20, Vol. 5, Part 2, Chapter IV, p. 829.)

Evidently, Schrödinger counted the probability interpretation among the essential concepts, arising from Bohr's atomic theory of 1913, which he now wanted to abolish. He wrote to Bohr: "What is before my eyes, is only one thesis: one should not, even if a hundred trials fail, give up the hope of arriving at the goal—I do not say by means of classical pictures, but by logically consistent conceptions—of the real structure of space-time processes. It is extremely probable that this is possible." (See letter from Schrödinger to Bohr, Ref. 20, Vol. 5, Part 2, Chapter IV, p. 829.) No, Schrödinger would not easily give up what he considered to be his program of a visualizable *(anschauliches)* understanding of what happens in atoms and molecules. With this program, Schrödinger acted against the entire development which atomic theory had pursued from 1913 under the leadership of Niels Bohr and which had eventually led to the quantum mechanical theory proposed by Heisenberg, Born, Jordan, and Dirac.

7. TRANSFORMATION THEORY OF PAUL DIRAC, PASCUAL JORDAN, AND FRITZ LONDON

With Born's statistical interpretation of the wave function in hand in July 1926 (see Ref. 9, Ref. 72, p. 285, and Ref. 20, Vol. 5, Part 2, Chapter IV), serious and prolonged discussions began about the fundamental physical meaning of quantum mechanics as represented by the two schemes—Born, Heisenberg, Jordan, and Dirac's scheme, on the one hand, and that of Schrödinger, on the other. Their equivalence was established rigorously by the transformation theory of Dirac,[(10)] Jordan,[(12)] and Fritz London[(11)] by late fall 1926, and the question of physical interpretation became paramount. Bohr, Heisenberg, Pauli, and Schrödinger, primarily, took part in these discussions.

The problem of the interpretation of quantum theory had occupied Niels Bohr increasingly since 1923 when the question of the nature of radiation became crucial for the understanding of the Compton effect. For Heisenberg, who had eagerly pressed forward by abandoning the use of classical concepts such as electron orbits in atoms, the problem of the interpretation arose late in 1925 when he thought about the simultaneous existence of the discrete energy spectrum of electrons bound in atoms and the continuous spectrum of free electrons moving along well-defined paths. It now occurred to him that, in some sense which was not yet clear, a space-time description should also be possible for the electrons in atoms.

In the fall of 1926 Heisenberg returned to the question of the space-time description of electron's behavior in the atom. Pauli pointed out to him that Schrödinger's wave function could be considered in momentum space, as $\psi(p)$, just as well as in coordinate space, as $\psi(q)$, to which Heisenberg responded: "The fundamental equivalence of p and q pleases me very much. Thus, in the wave formulation, the equation $pq-qp=h/2\pi i$ always corresponds to the fact that it makes no sense to speak of a monochromatic wave at a definite moment (or in a very small time interval)." (Heisenberg to Pauli, October 28, 1926, Ref. 71). At this place, in the margin of the letter, Pauli noted: "It also makes no sense to speak of a state (energy) in a time interval which is small compared to the period [because the state or the energy can be defined only over the entire period]." In his letter, Heisenberg continued: "If the [spectral] line may be taken as being not too sharp, i.e., the time interval is not too small, that of course makes sense. Analogously, there is no point in talking about the position of a particle of a definite velocity. However, it makes sense if one does not consider the velocity and the position too accurately. It is quite clear that, macroscopically, it is meaningful to talk about the position and velocity of a body." (Heisenberg to Pauli, October 28, 1926, Ref. 71).

Thus far, Heisenberg had only vaguely formulated his ideas about a "coarse" space-time description, reflecting his new understanding based upon wave mechanics. In the fall of 1926 Heisenberg was in Copenhagen, where he had taken up his new duties as a lecturer as the successor of Kramers who had been appointed to a professorship in Utrecht. Bohr, with whom he discussed daily, had been developing his own approach to the problem of the interpretation by emphasizing *the duality of the wave and particle pictures in quantum theory*. Heisenberg preferred to abide by the quantum mechanical scheme, as formulated by Born, Heisenberg, and Jordan, and by Dirac; he believed that the wave features should be brought in only by means of the transformation theory which Dirac[(10)] had worked out in Copenhagen in the fall of 1926.

Dirac had shown conclusively that the matrix S, employed in solving the problem of the principal axis transformation in the case of a Hermitian Hamiltonian function $H(p, q)$, could be identified with Schrödinger's wave function. In other words, for each column vector, there exists the identity

$$S_{q,E}=\psi_E(q) \tag{7}$$

where E is the discrete or continuous eigenvalue of the energy matrix. In order to handle the problem of continuous indices, Dirac introduced the *delta-function*, δ; with its help, the momentum p, conjugate to a continuous position variable q, could be written formally as

$$p(q', q'') = \frac{h}{2\pi i}\,\delta'(q' - q'') = \frac{h}{2\pi i}\frac{\partial}{\partial q'} \tag{8}$$

and the Born–Jordan matrix equation for diagonalizing the Hamiltonian H,

$$H(p, q)\, S_E(q) = E \cdot S_E(q) \tag{9}$$

could thus be transformed into Schrödinger's wave equation,

$$H\left(q, \frac{h}{2\pi i}\frac{\partial}{\partial q}\right)\psi_E q = E\psi_E q \qquad (3), (10)$$

8. NIELS BOHR'S DISCUSSIONS WITH HEISENBERG IN FALL 1926 AND THE FORMULATION OF THE UNCERTAINTY PRINCIPLE

The discussions [Ref. 20, Vol. 2 (1982) and Vol. 6 (1988); Stolzenburg, cited in Ref. 14], sometimes stormy, between Bohr and Heisenberg about the interpretation of quantum mechanics had begun soon after Schrödinger's visit to Copenhagen, and continued in the following winter months. Pauli was kept informed by Heisenberg about the course of these discussions. Heisenberg noted: "During these months I spoke with Bohr almost daily about the fundamental problems of quantum theory. Bohr sought to make the duality between the wave picture and the corpuscular picture as the starting point of the physical interpretation, while I tried to derive my conclusions—without the help of wave mechanics—by appealing only to quantum mechanics and Dirac's transformation theory." (See Ref. 47, p. 45; and O. Klein, AHQP Interview, February 28, 1963, p. 10.) These two different starting points led Heisenberg to the indeterminacy relations and Bohr to the principle of complementarity. As Heisenberg recalled: "Bohr and I tried from different angles and therefore it was difficult to agree. Whenever Bohr could give an example in which I couldn't find the answer, then it was clear that we had not understood what the actual situation was ··· Shortly after Christmas, we both were in a kind of despair. In some way we couldn't agree and so we were a bit angry about it. So about mid-February 1927 Bohr left for a skiing vacation in Norway. Earlier he had thought about taking me with him, but then he decided against it. He wanted to be alone and think, while I stayed on in Copenhagen." (See W. Heisenberg, AHQP Interview, February 25, 1963, p. 16; conversations with J. Mehra, Geneva, July 1962.) Heisenberg made an effort to bring some order into his thoughts and results of the past few months. Bohr spent his winter vacation in Norway from about the middle of February to mid-March 1927. (See letter from O. Klein to M. Saha on behalf of N. Bohr, February 18, 1927, cited in K. Stolzenburg, Ref. 14.)

On February 23, 1927, Heisenberg wrote a long letter to Pauli, in which he dealt with the problem of observing simultaneously the position and momentum of atomic systems. He stated that the "commutation relation," $pq - qp = h/2\pi i$, has the following physical interpretation: Given the exact momentum p of an electron in an atom, its position is then completely undetermined, and *vice versa* (Heisenberg to Pauli, February 23, 1927, in Ref. 71). To support this point of view, and to render it more visual, Heisenberg discussed the thought experiment *(Gedankenexperiment)* for the observation of an electron by means of a γ-ray microscope, an analogy which occurred to him from his doctoral oral examination under Wilhelm Wien several years before (Ref. 20, Vol. 2, Chapter I. 8; also Ref. 35, p. 10) Then he turned to the exact calculation of the accuracy involved in the observation of p and q.

The probability amplitude of the position of an object, which lies within the space interval $q_0 - q_1 < q < q_0 + q_1$, is given by

$$S(q) = \text{const} \cdot \exp\left[-\frac{(q-q_0)^2}{2q_1^2} - \frac{2\pi i p_0(q-q_0)}{h} \right] \tag{11}$$

where the first term represents a Gaussian distribution and the second, the general wave function. From $S(q)$, he determined $S(p)$ with the help of the transformation equation

$$\begin{aligned} S(p) &= \int S(q)\, e^{2\pi i pq/h}\, dq \\ &= \text{const} \cdot \exp\left[-2\pi^2 q_1^2 (p - p_0^2)^2 + \frac{2\pi i}{h}(p - p_0)\, q_0 \right] \end{aligned} \tag{12}$$

Hence, for a given uncertainty $\delta q = q_1$ in the position, the probability distribution, $|S(p)|^2$, of the momentum p is nonzero in the region $p_0 - p_1 < p < p_0 + p_1$, such that

$$\frac{4\pi^2 q_1^2 p_1^2}{h^2} \approx 1 \tag{13}$$

The simultaneous measurement of the position and momentum of an electron is thus limited by the *uncertainty relation*,

$$\delta p \cdot \delta q \approx \frac{h}{2\pi} \tag{14}$$

Heisenberg asked Pauli for his severe criticism [*"unnachsichtige Kritik"*] (Heisenberg to Pauli, February 23, 1927, in Ref. 71) As he recalled: "I wanted to get Pauli's reaction before Bohr was back because I felt that when Bohr comes back he will be angry with my interpretation. So I first wanted to have some support, and see whether somebody else liked it. Now Pauli's reaction was extremely enthusiastic. He said something like *"Morgenröte der Neuzeit"*—the light of day in quantum theory." (See Heisenberg, AHQP Interview, February 25, 1963, pp. 16, 17.) Pauli at once approved Heisenberg's ideas on the uncertainty principle, and thought that this interpretation endowed quantum mechanics with a coherent physical meaning. Heisenberg informed Jordan that he had derived a physical interpretation of Dirac's(10) and Jordan's(12) mathematical work on transformation theory. (See Heisenberg to Jordan, March 7, 1927, quoted in Stolzenburg, Ref. 14, p. 128). After receiving Pauli's favorable verdict, Heisenberg informed Bohr: "I myself have worked very vigorously during recent weeks in order to carry through the program (with respect to the Dirac–Jordan quantum mechanics [i.e., transformation theory]), about which we talked before your departure. I believe that I have fully succeeded. The case, in which *p as well as q* are given with a certain accuracy, can be formulated without going beyond the Dirac–Jordan mathematics.... Further, one finds that the transition from "micro- to macromechanics"(75) can be understood very easily: the classical mechanics is entirely a part of quantum mechanics. As for the old question concerning 'statistical or causal law,' the situation is this: one *cannot* say that quantum mechanics is statistical. However, one can obtain only statistical results, if one wants to calculate 'future events' from the 'present,' since one cannot take into account all the initial conditions of the present." (Heisenberg to Bohr, March 10, 1927, cited in Stolzenburg, Ref. 14, p. 128).

Heisenberg's paper on the uncertainty principle in quantum mechanics was received by *Zeitschrift für Physik* on March 23, 1927.(13) He began his considerations by making the following statement of his conviction: "We regard a physical theory to be perceptual [*anschaulich*] [only] if we can think of the experimental consequences of this theory qualitatively in all simple cases, and [when] we have recognized that the application of this theory never leads to inner contradictions." (Ref. 13, p. 172). The visualizable interpretation of quantum mechanics had thus far been full of contradictions; for this reason Heisenberg believed that this interpretation could not be consistent with the usual kinematic and mechanical concepts. "Quantum mechanics arose exactly out of the attempt to break away from those habitual kinematic concepts, and substitute in their place concrete, experimentally given magnitudes. Since this seems to have been achieved, the mathematical scheme of quantum mechanics would therefore not require a revision." (Ref. 13, p. 172) Heisenberg investigated the conditions under which the concepts of position, orbit, velocity, and energy of classical physics could be taken over in quantum theory. He concluded: "All concepts, which are needed to describe a mechanical system in classical physics, can also be defined—analogously to classical notions—exactly for atomic processes. The experiments, which serve to define them, also lead to an inherent indeterminacy, if we demand from them the simultaneous determination of two canonically conjugate quantities.... In this case, the quantum theory can be closely compared to the relativity theory." (Ref. 13, p. 172). The speed of light enters into the definition of simultaneity in relativity theory. The experiments always yield a finite light velocity in agreement with the postulate of a constant velocity of light; hence this postulate does not contradict the obvious notions of position, velocity, and time in relativity theory.

In his special relativity theory (1905), Einstein[49] had emphasized the fundamental importance of employing "observable magnitudes" only in the construction of a physical theory. This conception of Einstein had guided Heisenberg in his discovery of a quantum-theoretical kinematics.[2] But when in April 1926 Heisenberg met Einstein in Berlin, the latter had told him: "... it may be heuristically useful to keep in mind what one has actually observed. But on principle, it is quite wrong to try founding a theory on observable magnitudes alone. In reality the very opposite happens. It is the theory which decides what we can observe." (Ref. 35, p. 46; Ref. 41, p. 63).

With this conceptual framework in mind, Heisenberg applied the Dirac–Jordan transformation theory to obtain Eq. (14) for the indeterminacy in the simultaneous measurement of p and q, about which he had written to Pauli (Heisenberg to Pauli, February 23, 1927, in Ref. 71). He concluded: "The more accurately is the position determined, the more uncertain is the momentum and *vice versa*; in this we find a directly visualizable explanation of the relation

$$pq - qp = \frac{h}{2\pi i}. \qquad (1), (15)$$

(Ref. 13, p. 172.) Heisenberg obtained the relation corresponding to Eq. (14) for the simultaneous determination of the energy E and time t

from an analysis of the Stern–Gerlach experiment. Also, in the experiment on resonance fluorescence light, Heisenberg recognized the fact hat to the extent the phase w is exactly measurable, to that extent the determination of energy is uncertain, corresponding to the relation

$$Jw - wJ = \frac{h}{2\pi i} \tag{16}$$

where J and w are the angular momentum and phase respectively.

Heisenberg finally drew the following consequences from his analysis: "We have not assumed that the quantum theory, in contrast to classical theory, is a statistical theory in the sense that from precisely given data only statistical conclusions can be drawn. Against such assumptions, the well-known experiments of Geiger and Bothe may be cited. Moreover, in all cases, in which in classical theory relations exist between magnitudes that are exactly measurable, corresponding relations also exist in quantum theory (e.g., the momentum-energy conservation). However, in the exact formulation of the law of causality that, "If we know the present precisely, we can determine the future," it is not the consequence but the initial assumption which is wrong. In principle, we cannot determine all the initial conditions of the present.... The real essence can be best characterized as follows: All the experiments are subject to the laws of quantum mechanics and, thereby, to Eq. [(14)] (Ref. 13, p. 191)."

On returning from his vacation in Norway, Bohr was not immediately satisfied with Heisenberg's formulation of the "intuitive content of the quantum-theoretical kinematics and mechanics."[(13)] As Heisenberg recalled: "When Bohr came back I showed him [my] paper... and I showed him Pauli's reaction. I did realize that Bohr was a bit upset about it because he still felt that it was not quite clear what I had written—not in every way clear. At the same time he saw Pauli's reaction, and he knew Pauli was very critical, so he felt it should, in some way, be right." (W. Heisenberg, AHQP Interview, February 25, 1963, p. 17). Bohr felt that Heisenberg had not treated the thought-experiments with the γ-ray microscope and the investigations on the Compton effect and resonance fluorescence light quite properly; his fundamental objection was something quite different. As Heisenberg recalled: "The main point was that Bohr wanted to take this dualism between waves and corpuscles as the central point of the problem ··· I [on the other hand] would say: 'We have a consistent mathematical scheme and [it] tells us everything which can be observed. [There is] nothing in nature which cannot be described by this mathematical scheme.' It was a different way of looking at the problem because Bohr would not like to say that nature imitates a mathematical scheme, that nature does only things which fit into a mathematical scheme." (W. Heisenberg, AHQP Interview, February 25, 1963, p. 18). Bohr suggested making changes in Heisenberg's paper, but Heisenberg was not at all willing. However, in a postscript added in proof, he incorporated Bohr's suggestions (Ref. 13, pp. 157-198, postscript added in proof).

The discussions between Bohr (assisted by Oskar Klein) and Heisenberg about the thought-experiments and their interpretation continued during the following weeks, often leading to misunderstandings. Heisenberg reported about these discussions in letters to Pauli, Kronig, and Dirac (letters from Heisenberg to Pauli, April 4, 1927, to Kronig, April 8, 1927, to Pauli, April 16, 1927, and to Dirac, April 27, 1927, in Ref. 71). By the end of May the points of view of Bohr and Heisenberg had come closer together, as Heisenberg reported to Pauli (letter from Heisenberg to Pauli, May 31, 1927, in Ref. 71), and after Pauli's visit to Copenhagen in June 1927, harmony between Bohr and Heisenberg had been restored (letter from Heisenberg to Bohr, June 18, 1927, *Bohr Archives*; Heisenberg, AHQP Interview, February 25, 1963, pp. 17, 18). On April 13, 1927, Bohr had already sent the proof-sheets of Heisenberg's paper on the uncertainty principle to Einstein (letter from Bohr to Einstein, April 13, 1927, quoted in Ref. 21, pp. 125, 126). Heisenberg's interpretation, embodied in the uncertainty principle, Eq. (14), could be generalized to any pair of conjugate dynamical variables, and was soon accepted as "the real core of the new theory."(76)

9. THE PRINCIPLE OF COMPLEMENTARITY

Following Niels Bohr's discussions with Schrödinger in Copenhagen in early October 1926, during which the question of "quantum jumps" often came up, and his discussions with Heisenberg in the subsequent winter months, especially after Heisenberg's formulation of the uncertainty principle in February–March 1927, Bohr reflected deeply upon the meaning of the fundamental equations of quantum theory. It has been pointed out by several authors (Ref. 72, p. 345; Stolzenberg, Ref. 14; Ref. 20, Vol. 6, Book I, Part 2, 1988; Ref. 77), that—for Bohr—Heisenberg's uncertainty relations were "a confirmation of the conceptions he had been groping for long before Heisenberg derived his principle from the Dirac–Jordan transformation theory. True, Heisenberg's work prompted Bohr to give his thoughts on complementarity a consistent and final formulation, but these thoughts ··· can be traced back at least to July, 1925." (Ref. 72, p. 345). In the four-page letter, which he sent to Einstein on April 13, 1927 together with Heisenberg's paper (letter from Bohr to Einstein, April 13, 1927, quoted in Ref. 21, pp. 125, 126), Bohr gave his own views concerning it and the general questions of quantum theory, and—in a way—gave expression to his developing ideas on the principle of complementarity (Ref. 21, p. 125).

In the public domain, Bohr discussed for the first time his ideas on complementarity in a lecture on September 16, 1927 at the International Congress of Physics that was convened to commemorate the hundredth anniversary of Alessandro Volta's death.[16,72] (Among the participants at the Volta Conference mentioned in Ref. 72, p. 351, footnote 9, "Bose" was not S. N. Bose of Bose–Einstein statistics, but D. N. Bose—who, while at Calcutha, had received the invitation that had been right fully intended for S. N. Bose to attend the Votta Conference. See S. N. Bose to J. Mehra, in conversations on August 30, 1970, Ref. 78.) Bohr gave a lecture on "The Quantum Postulate and the Recent Development of Atomic Theory."[16] Bohr gave essentially the same lecture at the fifth Solvay Conference in Brussels in October 1927,[17] and at the Danish Academy of Sciences (*Videnskabernes Selskab*).[72] Bohr began his lecture at Como by saying: "I shall try by making use only of the simple considerations and without going into any details of technical mathematical character to describe to you a certain general point of view which I believe is suited to give an impression of the general trend of the development of the theory from its very beginning and which I hope will be helpful in order to harmonize the apparently conflicting views taken by different scientists." (Ref. 16 and Ref. 72, footnote 78, p. 349). Bohr emphasized the distinction between the classical description of a natural phenomenon—based on the assumption that it may be observed without significant disturbance—and the quantum description of atomic processes to which a quantum discontinuity, or "individuality," has to be attributed. In his manuscript (dated September 13, 1927) he had considered the causal space-time description according to the "quantum postulate" as follows: "Characteristic of the quantum theory is the acknowledgment of a fundamental limitation in our classical physical ideas when applied to atomic phenomena. Just on account of this situation, however, we meet intricate difficulties when attempting to formulate the contents of the quantum theory in terms of concepts borrowed from classical theories. Still it would appear that the essence of the theory may be expressed through the postulate that any atomic process open to direct observation involves an essential element of discontinuity or rather individuality completely foreign to the classical ideas and symbolized by Planck's quantum of action. This postulate at once implies a resignation as regards the causal space-time coordination of atomic phenomena." (N. Bohr, Manuscript on "Fundamental problems of the quantum Theory," September 13, 1927, *Bohr Archives*; cited in K. Stolzenburg,

Ref. 14, pp. 156, 157). At Como, Bohr declared: "The very nature of the quantum theory thus forces us to regard the space-time coordination and the claim of causality, the union of which characterizes the classical theories, as complementary but exclusive features of the desription, symbolizing the idealization of observation and definition respectively." (Ref. 16 and Ref. 72, footnote 78, p. 349). Thus Bohr came to the conclusion that the situation in atomic physics could only be described in terms of dual, *complementary* pictures which, in classical physics, exclude each other.[(79)] The uncertainty relations ensure that no contradiction will arise in the exercise of the *principle of complementarity* in nature; they exclude the possibility of situations occurring that exhibit both the wave and particle aspects of a phenomenon simultaneously.[(80)] Bohr's statement about complementarity at Como became the essence of what later on came to be called "the Copenhagen interpretation" of quantum theory.

10. THE FIFTH AND SIXTH SOLVAY CONFERENCES AND BOHR'S DISCUSSIONS WITH EINSTEIN

At the fifth Solvay Conference in Brussels, from October 24 to 29, 1927, quantum mechanics, together with the "Copenhagen interpretation," was publicly presented as a consistent and complete theory of atomic phenomena by its numerous protagonists whose leader was Niels Bohr (see Mehra in Ref. 17). The theme of the Conference was "*Électrons et Photons,*" and among those present were N. Bohr, M. Born, L. Brillouin, L. de Broglie, A. H. Compton, P. Debye, P. A. M. Dirac, P. Ehrenfest, A. Einstein, R. H. Fowler, W. Heisenberg, H. Kramers, W. Pauli, E. Schrödinger, and M. Planck. H. A. Lorentz presided over the Conference. Schrödinger, who was the last speaker, gave a report on the development of wave mechanics. At the end of Schrödinger's talk, Lorentz requested Bohr to give a report on the epistemological problems confronting quantum physics. Bohr's report contained essentially the substance of his contribution at the Como Conference on September 16, 1927.[(16)] Einstein had not attended the Como Conference but was present at the fifth Solvay Conference in Brussels, and he heard a comprehensive account of Bohr's ideas on the interpretation of quantum mechanics. Bohr discussed the question of an appropriate terminology, and stressed the viewpoint of complementarity. Bohr's main argument was that the unambiguous communication of physical evidence required that the experimental arrangement as well as the recording of observations had to be expressed in a common language, "suitably refined by the vocabulary of classical physics." (See Mehra in Ref. 17, p. 151). In all actual experimental work this requirement is fulfilled by using measuring instruments such as diaphragms, lenses, and photographic plates, which are so large and heavy that, notwithstanding the decisive role of the quantum of action for the stability and properties of such bodies, all quantum effects could be disregarded in taking account of their positions and motions.

In classical physics one deals with an idealization, according to which all phenomena can be arbitrarily subdivided. The interaction between the measuring instruments and the object under observation can be neglected, or at any rate compensated, in classical physics. Bohr stressed that such interaction represents an integral part of the phenomena in quantum physics, for which no separate account could be given if the instruments would serve the purpose of defining the conditions under which the observations are obtained.

The recording of observations ultimately rests on the production of permanent marks on the measuring instruments, such as the spot produced on a photographic plate by the impact of a photon or electron. The fact that such recording involves essentially irreversible physical and chemical processes does not introduce further complications, but only emphasizes the element of irreversibility involved in the very concept of observation. In Bohr's view, the characteristic new feature in quantum physics is merely the restricted divisibility of phenomena, which requires a specification of all significant parts of the experimental setup for unambiguous description.

Bohr pointed out that since several different individual effects would in general be observed in one and the same experimental arrangement, the recourse to statistics in quantum physics was therefore unavoidable. "Moreover, evidence obtained under different conditions and rejecting comprehension in a single picture must, notwithstanding any apparent contrast, be regarded as complementary in the sense that together they exhaust all well-defined information about the atomic object. From this point of view, the whole purpose of the formalism of quantum theory is to derive expectations for observations obtained under given experimental conditions." (N. Bohr, quoted in Mehra, Ref. 17, p. 152). Bohr emphasized that the elimination of all contradictions is secured by the mathematical consistency of the formalism; the exhaustive character of the description within its scope is indicated by its adaptability to any imaginable experimental arrangement.

Following Bohr's report on the epistemological questions of quantum theory, there took place an extensive general discussion on the renunciation of pictorial deterministic description and the introduction of probability in the new theory. Almost all the physicists who had been invited to attend the fifth Solvay Conference participated in this discussion. A note has survived (from one of the sessions at this Conference), which

Paul Ehrenfest had passed on to Einstein during one of the lectures saying, "Don't laugh! There is a special section in purgatory for professors of quantum theory, where they will be obliged to listen to lectures on classical physics ten hours every day." To which Einstein replied, "I laugh only at their naïveté. Who knows who would have the laugh in a few years?" (Ref. 35, p. 47; Mehra in Ref. 17, p. 152). Einstein had disliked from the beginning the Göttingen–Copenhagen quantum-theoretical formulation of atomic problems with its emphasis on discontinuity and acausality. Thus, he had written to Ehrenfest: "I look upon quantum mechanics with admiration and suspicion" (letter from Einstein to Ehrenfest, August 28, 1926, Einstein Archives). And, to Born, he had written: "Quantum mechanics is certainly imposing. But an inner voice tells me that it is not the real thing [*der wahre Jakob*]. The theory says a lot, but does not really bring us closer to the secret of the 'old one' [*Geheimnis des Alten*]. I, at any rate, am convinced that *He* is not playing at dice." (letter from Einstein to Born, December 4, 1926; Ref. 23, pp. 90, 91). The exchanges of views that started at the sessions at the fifth Solvay Conference were eagerly continued within smaller groups during the evenings. Bohr had extensive opportunity of having long discussions with Einstein and Ehrenfest. Einstein was particularly reluctant to renounce deterministic description in principle. It was against his scientific belief to accept statistical quantum mechanics as a consistent and complete description of physical reality. He challenged Bohr and the other proponents of the new quantum mechanics with arguments suggesting the possibility of taking the interaction between atomic objects and the measuring instruments more explicitly into account. He proposed cleverly designed thought-experiments to overcome Heisenberg's uncertainty principle, but Bohr was able to recognize the fallacy in Einstein's reasoning and to refute his arguments (Ref. 21, pp. 109-121).

The Bohr–Einstein discussions were resumed at the sixth Solvay Conference, which took place in Brussels from October 20 to 25, 1930. H. A. Lorentz had died in 1928, and Paul Langevin presided over the Conference, which was devoted to the study of the magnetic properties of matter (Ref. 19; Ref. 17, pp. 183-205; Ref. 21, pp. 132-136). Einstein had not been convinced by the answers of Bohr and his colleagues at the previous conference, and the problem of the foundations and interpretation of quantum mechanics became a major subject of discussion outside the official sessions of the sixth Solvay Conference. Einstein again devised clever thought-experiments to demonstrate the violation of the uncertainty principle. Bohr used the conclusion of Einstein's own general relativity theory and the equation $E = mc^2$ to refute Einstein's arguments with his photon-box thought-experiment, showing that he was led to $\Delta E \, \Delta t > h$ in accordance with the uncertainty principle.

In fall 1933 Einstein left for the United States and settled in Princeton, New Jersey, where he continued his scientific activity. By that time he had accepted the "consistency" of nonrelativistic quantum mechanics. In 1935, with Boris Podolsky and Nathan Rosen, Einstein wrote an article in which he raised the question whether quantum mechanical description was "complete",[81] to which Bohr replied immediately.[82] The argument was soon joined in by Schrödinger,[83] and the Bohr–Einstein dialogue concerning classical determinism versus statistical causality was continued until Einstein's death in April 1955. In 1949, in a 70th birthday tribute to Einstein, Niels Bohr gave a beautiful account of his discussions with Albert Einstein over the years (Ref. 27; a full account of the Bohr–Einstein discussions is intended for Ref. 20, Vol. 6, Book I, Part 2, to appear in 1988). Since 1927, when the Bohr–Einstein discussions began in earnest at the fifth Solvay Conference, the questions concerning the physical and philosophical aspects of the interpretation and foundation of quantum mechanics have become the central theme of the investigations of a growing number of physicists and philosophers of science. Initially, however, the discussions between, and the formulations and interpretations of, Bohr, Einstein, Heisenberg, and Schrödinger had sparked this growing debate.

BIBLIOGRAPHY OF PRINCIPAL SOURCES

W. Heisenberg, *Physics and Beyond*, Ref. 41.
M. Jammer, *The Philosophy of Quantum Mechanics*, Ref. 21.
M. Jammer, *The Conceptual Development of Quantum Mechanics*, Ref. 72.
J. Mehra, *The Solvay Conferences on Physics: Aspects of the Development of Physics since 1911*, Ref. 17.
J. Mehra, *The Birth of Quantum Mechanics:* Werner Heisenberg Memorial Lecture, Ref. 35.
J. Mehra and H. Rechenberg, *The Historical Development of Quantum Theory*, Ref. 20.
K. Stolzenburg, "Die Entwicklung des Bohrschen Komplementaritätsgedankens in den Jahren 1924 bis 1929," Ref. 14.

REFERENCES

1. *On the Constitution of Atoms and Molecules*, Papers of 1913 reprinted from *Philosophical Magazine* with an Introduction by L. Rosenfeld (Munksgaard Ltd., Copenhagen; W. A. Benjamin, New York, 1963).
2. W. Heisenberg, "Über quantentheoretische Umdeutung kinematischer und mechanischer Beziehungen," *Z. Phys.* **33**, 879–893 (1925).

3. M. Born and P. Jordan, "Zur Quantenmechanik," *Z. Phys.* **34**, 858–888 (1925); M. Born, W. Heisenberg, and P. Jordan, "Zur Quantenmechanik II," *Z. Phys.* **35**, 557–615 (1926).
4. P. A. M. Dirac, "The fundamental equations of quantum mechanics," *Proc. R. Soc. London A* **109**, 642–653 (1925).
5. E. Schrödinger, *Collected Papers on Wave Mechanics* (Blackie and Son, Ltd, London and Glasgow, 1928).
6. E. Schrödinger, "Über das Verhältnis der Heisenberg–Born–Jordanschen Quantenmechanik zu der meinen," *Ann. Phys.* **79**, 734–756 (1926).
7. W. Pauli, *Briefwechsel/Scientific Correspondence*, Vol. I, (Springer-Verlag, New York, Heidelberg, and Berlin, 1979).
8. C. Eckart, "The solution of the problem of the simple oscillator by a combination of the Schrödinger equation and the Lanczos theories," *Proc. Natl. Acad. Sci. USA* **12**, 473–476 (1926); "Operator calculus and the solution of the equations of quantum dynamics," *Phys. Rev.* **28**, 927–935 (1926).
9. M. Born, "Zur Quantenmechanik der Stoßvorgänge (Vorläufige Mitteilung)," *Z. Phys.* **37**, 863–867 (1926); "Quantenmechanik der Stoßforgänge," *Z. Phys.* **38**, 803–827 (1926).
10. F. London, "Winkelvariable und kanonische Transformationen in der Undulationsmechanik," *Z. Phys.* **40**, 193–210 (1926).
11. P. A. M. Dirac, "The physical interpretation of quantum dynamics," *Proc. R. Soc. London A* **113**, 621–641 (1927).
12. P. Jordan, "Über eine neue Begründung der Quantenmechanik," *Z. Phys.* **40**, 809–838 (1927).
13. W. Heisenberg, "Über den anschaulischen Inhalt der quantentheoretischen Kinematik und Mechanik," *Z. Phys.* **43**, 172–198 (1927).
14. W. Heisenberg, *Physics and Beyond: Encounters and Conversations* (Harper and Row, New York, 1971); J. Mehra and H. Rechenberg, *The Historical Development of Quantum Theory*, Vol. 5, Part 2 (Springer Verlag, New York, Heidelberg, and Berlin, 1977), Chapter IV, Section 5; K. Stolzenburg, "Die Entwicklung des Bohrschen Komplementaritätsgedankens in den Jahren 1924 bis 1929," Doctoral Dissertation, University of Stuttgart, 1975.
15. J. Mehra and H. Rechenberg, *The Historical Development of Quantum Theory*, Vol. 6: *The Completion of Quantum Mechanics*, Book I, Part 2 (Springer-Verlag, New York, to appear in 1988); J. Mehra and H. Rechenberg, *The Historical Development of Quantum Theory*, Vol. 2 (Springer-Verlag, New York, 1982); J. Mehra, "The birth of quantum mechanics," Heisenberg Memorial Lecture (CERN, Geneva, March 30, 1976).
16. N. Bohr, "The quantum postulate and the recent development of atomic theory," in *Atti del Congresso Internazionale dei Fisici, 11–20 Settembre 1927* (Nicola Zanichelli, Bologna, 1928), Vol. 2, pp. 565–568.
17. *Électrons et Photons—Rapports et Discussions du Cinquième Conseil de Physique Tenu à Bruxelles du 24 au 29 Octobre 1927 sous les Auspices de L'Institut International de Physique Solvay* (Gauthier-Villars, Paris, 1928); J. Mehra, *The Solvay Conferences on Physics: Aspects of the Development of Physics since 1911* (D. Reidel, Dordrecht, Holland/Boston, 1975).
18. J. Mehra and H. Rechenberg, *The Historical Development of Quantum Theory*, Vol. 1, Part 1, Section 11.5, and Part 2, Section VI.4 (Springer-Verlag, New York, 1982).
19. *Le Magnétisme—Rapports et Discussions du Sixième Conseil de Physique sous les Auspices de L'Institut International de Physique Solvay* (Gauthier-Villars, Paris, 1932).
20. J. Mehra and H. Rechenberg, *The Historical Development of Quantum Theory*, Vols. 1–6 (Springer-Verlag, New York, Heidelberg and Berlin, 1982–1988).
21. M. Jammer, *The Philosophy of Quantum Mechanics* (Wiley, New York, 1974), Chapter 5.

22. A. Einstein, "Strahlungs-Emission und -Absorption nach der Quantentheorie," *Verh. Dtsch. Phys. Ges.* **18**, 318–323 (1916); "Quantentheorie der Strahlung," *Mitt. Phys. Ges. Zürich* **16**, 47–62 (1916); "Zur Quantentheorie der Strahlung," *Phys. Z.* **18**, 121–128 (1917).
23. *The Born–Einstein Letters* (Macmillan, London and Basingstoke, 1971).
24. A. Einstein, "Grundlage der allgemeinen Relativitätstheorie," *Ann. Phys.* **49**, 769–822 (1916); J. Mehra, *Einstein, Hilbert and the Theory of Gravitation* (D. Reidel, Dordrecht, Holland and Boston, Massachusetts, 1974).
25. A. Einstein, "Bietet die Feldtheorie Möglichkeiten für die Lösung des Quantenproblems?" *Berliner Ber.*, 359–364 (1923).
26. P. Speziali (ed.), *Albert Einstein, Michele Besso, Correspondance 1903–1955* (Hermann, Paris, 1972).
27. N. Bohr, "Discussion with Einstein on epistemological problems in atomic physics," in *Albert Einstein: Philosopher–Scientist*, P. A. Schilpp, ed. (The Library of Living Philosophers, Inc., Tudor Publishing Company, 1949, 1951; Harper and Row, New York, 1959).
28. A. Einstein, "Über die Entwicklung unserer Anschauungen über das Wesen und die Konstitution der Strahlung," *Phys.* **10**, 817–825 (1909).
29. N. Bohr, "Über die Serienspektren der Elemente," *Z. Phys.* **2**, 423–469 1920). English translation: "On the series spectra of elements," in *Collected Works* **3**, 241–260 (1976).
30. *Albert Einstein–Arnold Sommerfeld Briefwechsel*, A. Hermann, ed. (Schwabe and Co., Basel, Stuttgart, 1968), p. 75 (quoted in M. Jammer, Ref. 21, p. 123).
31. A. H. Compton, "Secondary radiation produced by X-rays," *Bull. Natl. Res. Council* **4** (Part 2), No. 20 (October 1922).
32. A. H. Compton, "A quantum theory of the scattering of X-rays by light elements," *Phys. Rev.* **21**, 483–502 (1923).
33. P. Debye, "Zerstreuung von Röntgenstrahlen und Quantentheorie," *Phys. Z.* **24**, 161–166 (1923).
34. A. Einstein, "Über einen die Erzeugung und Verwandlung des Lichtes betreffenden heuristischen Gesichtpunkt," *Ann. Phys.* **17**, 138–148 (1905).
35. J. Mehra, "The birth of quantum mechanics," Werner Heisenberg Memorial Lecture, CERN, Geneva, March 30, 1976 (CERN 76-10, May 14, 1976).
36. J. C. Slater, "Radiation and Atoms," *Nature (London)* **133**, 307–308 (1924).
37. N. Bohr, H. A. Kramers, and J. C. Slater, "The quantum theory of radiation," *Philos. Mag.* **47**, 785–802 (1924). (In German: "Über die Quantentheorie der Strahlung," *Z. Phys.* **24**, 69–87 (1924).)
38. H. A. Kramers, "The Law of Dispersion and Bohr's Theory of Spectra," *Nature (London)* **113**, 673–674 (1924); "The quantum theory of dispersion," *Nature (London)* **114** 310 (1924).
39. W. Bothe and H. Geiger, "Experimentelles zur Theorie von Bohr, Kramers und Slater," *Naturwissenschaften* **13**, 440–441; "Über das Wesen des Compton-effekts, ein experimenteller Beitrag zur Theorie der Strahlung," *Z. Phys.* **32**, 639–663 (1925).
40. G. E. Uhlenbeck and S. Goudsmit, "Ersetzung der Hypothese vom unmechanischen Zwang durch eine Forderung bezüglich des inneren Verhaltens jedes einzelnen Elektrons," *Naturwissenschaften* **13**, 953–954 (1925); "Spinning electrons and the structure of spectra," *Nature (London)* **117**, 264–265 (1926).
41. W. Heisenberg, *Physics and Beyond*, (Harper and Row, New York, Evanston, and London, 1971), p. 38.
42. H. A. Kramers, "Über den Einfluss eines elektrischen Feldes auf die Feinstruktur der Wasserstofflinien," *Z. Phys.* **3**, 199–223 (1920).

43. N. Bohr, *Collected Works, Volume 4: The Periodic System (1920–1923)*, J. Reed Nelson, de. (North-Holland, Amsterdam, New York, and Oxford, 1977), p. 371.
44. W. Heisenberg, "Quantum theory and its interpretation," in *Niels Bohr: His Life and Work as Seen by His Friends and Colleagues*, S. Rozental, ed. (North-Holland, Amsterdam, 1967), pp. 94–95.
45. W. Heisenberg, "Über eine Anwendung des Korrespondenzprinzips auf die Frage nach der Polarization des Fluoreszenzlichtes," *Z. Phys.* **31**, 617–626 (1925).
46. H. A. Kramers and W. Heisenberg, "Über die Streuung von Strahlung durch Atome," *Z. Phys.* **31**, 681–708 (1925).
47. W. Heisenberg, "Erinnerungen an die Zeit der Entwicklung der Quantenmechanik," in *Theoretical Physics in the Twentieth Century—A Memorial Volume to Wolfgang Pauli*, M. Fierz and V. F. Weisskopf, eds. (Interscience, New York, 1960), pp. 40–47.
48. E. Mach, *Die Mechanik in ihrer Entwicklung*, 7th edn. (F. A. Brockhaus, Leipzig, 1912); English translation: *The Science of Mechanics* (Open Court Publishing Company: Chicago, 1893).
49. A. Einstein, "Zur Elektrodynamik bewegter Körper," *Ann. Phys.* **17**, 891–921 (1905).
50. H. Minkowski, "Raum und Zeit," *Phys. Z.* **10**, 104–111 (1909).
51. M. Born and P. Jordan, "Zur Quantenmechanik," *Z. Phys.* **34**, 858–888 (1925).
52. M. Born, W. Heisenberg, and P. Jordan, "Zur Quantenmechanik II," *Z. Phys.* **35**, 557–615 (1926).
53. W. Lenz, "Über den Bewegungsablauf und die Quantenzustände der gestörten Keplerbewegung," *Z. Phys.* **24**, 197–207 (1924).
54. W. Pauli, "Über das Wasserstoffspektrum von Standpunkt der neuen Quantenmechanik," *Z. Phys.* **36**, 236–363 (1926).
55. C. Lanczos, "Über eine feldmäßige Darstellung der neuen Quantenmechanik," *Z. Phys.* **35**, 812–830 (1926).
56. M. Born, *Problems of Atomic Dynamics*, reprint (M.I.T. Press, Cambrige, Massachusetts, 1970).
57. M. Born and N. Wiener, "A new formulation of the laws of quantization of periodic and aperiodic phenomena," *J. Math. Phys.* M.I.T. **5**, 84–98 (1926).
58. M. Born and N. Wiener, "Eine neue Formulierung der Quantengesetze für periodische und aperiodische Vorgänge," *Z. Phys.* **36**, 174–187 (1926).
59. J. Mehra, "The golden age of theoretical physics: P.A.M. Dirac's scientific work from 1924–1933," in *Aspects of Quantum Theory*, A. Salam and E. P. Wigner, eds. (Cambridge University Press, Cambridge, 1972) pp. 17–59.
60. P. A. M. Dirac, "Quantum mechanics and a preliminary investigation of the hydrogen atom," *Proc. R. Soc. London A* **110** 561–573 (1926).
61. P. A. M. Dirac, "On Quantum algebra," *Proc. Cambridge Philos. Soc.* **23**, 412–418 (1926).
62. P. A. M. Dirac, *Quantum Mechanics*, Dissertation, Cambridge University, May 1926.
63. E. Schrödinger, "Quantisierung als Eigenwertproblem. I," *Ann. Phys.* **79**, 361–376 (1926).
64. H. Weyl, "Singuläre Integralgleichungen mit besonderer Berücksichtigung der Fourierschen Integraltheorems," Göttingen dissertation, February 1908.
65. L. de Broglie, *Recherches sur la Théorie des Quanta*, Thèse (Masson & Cie, Paris, 1924); *Ann. Phys.* **3**, 22–128 (1925); Paris dissertation, November 1924.
66. M. Born, "Erwin Schrödinger†," *Phys. Blätter* **17**, 85–87 (1961).
67. W. Heisenberg, "Mehrkörperproblem und Resonanz in der Quantenmechanik," *Z. Phys.* **38**, 411–426 (1926).
68. W. Heisenberg, "Über die Spektra von Atomsystemen mit zwei Elektronen," *Z. Phys.* **39**, 499–518 (1926).

69. E. Schrödinger, "Quantisierung als Eigenwertproblem (Vierte Mitteilung)," *Ann. Phys.* **81**, 109–139 (1926).
70. *Briefe zur Wellenmechanik*, K. Przibram (ed.) (Springer-Verlag, Viennas 1963); English translation (by M. J. Klein) as *Letters on Wave Mechanics* (Philosophical Library, Inc., New York, 1967).
71. W. Pauli, *Scientific Correspondence*, Vol. I (Springer-Verlag, Berlin and New York 1979).
72. M. Jammer, *The Conceptual Development of Quantum Mechanics* (McGraw-Hill, New York, 1966), pp. 285–289; reference to A. Einstein: M. Born, "Albert Einstein und das Lichtquantum," *Naturwissenschaften* **11**, 425–431 (1955).
73. A. Pais, *Inward Bound* (Oxford University Press, New York, 1986), p. 259.
74. G. Wentzel, "Zwei Bemerkungen über die Zerstreuung korpuscularer Strahlen als Beugungserscheinung," *Z. Phys.* **40**, 590–593 (1926).
75. E. Schrödinger, "Der stetige Übergang von der Mikro- zur Makromechenik," *Naturwissenschaften* **14**, 664–666 (1926).
76. E. H. Kennard, "Zur Quantenmechanik einfacher Bewegungstypen," *Z. Phys.* **44**, 326–352 (1927).
77. G. Holton, "The roots of complementarity," in *Thematic Origins of Scientific Thought* (Harvard University Press, Cambridge, 1973), pp. 115–161.
78. J. Mehra, "Satyendra Nath Bose, 1894–1974," *Biog. Mem. Fellows Roy. Soc.* **21** (1975).
79. W. Pauli, "Die allgemeinen Prinzipien der Wellenmechanik," *Handbuch der Physik*, Vol. 24, H. Geiger and K. Scheel, eds. (Springer, Berlin, 1933), and edn., p. 89.
80. C. F. von Weizsäcker, "Komplementarität und Logik," *Naturwissenschaften* **42**, 521–529, and 545–555 (1955).
81. A. Einstein, B. Podolsky, and N. Rosen, "Can quantum-mechanical description of physical reality be considered complete?", *Phys. Rev.* **47** 77–780 (1935).
82. N. Bohr, "Can quantum-mechanical description of physical reality be considered complete?", *Phys. Rev.* **48**, 696–702 (1935).
83. E. Schrödinger, "Die gegenwärtige Situation in der Quantenmechanik," *Naturwissenschaften* **23**, 807–812, 824–828, 844–849 (1935).

Part II: INVITED LECTURES

THE ROLE OF MODERN MATHEMATICS IN THE FORMULATION OF BASIC LAWS IN PHYSICS

K. Bleuler
Institut für Theoretische Kernphysik der Universität Bonn
Nussallee 14-16, D-5300 Bonn, West-Germany

ABSTRACT

In relation to recent most interesting and far reaching developments (e.g. non-abelian gauge theories and heterotic strings) in theoretical physics the importance and the decisive role of an exchange with modern mathematics is emphasized and - at the same time - analyzed from the view-point of a basic understanding of physical laws.

1. INTRODUCTION

In recent years theoreticians have engaged - with the firm hope of reaching a deeper and more consistent understanding of basic physical laws - in a most abstract and far-reaching mathematical research: in particular with algebraic as well as geometric structures[1)] (e.g. supergroups and topological view-points). In contrast to a widespread criticism of this rather formal - i.e. not empirically motivated - development the scope of this contribution is in a first part to stress the basic and in many cases decisive importance of the formal aspect of physical laws as well as the general mathematical view-points all through the recent history of theoretical physics.

In a second part, general arguments for the need, or better the crucial importance of an enlargement or generalization of conventional quantum field theory will be given. In particular, it will be stressed that all theoretical attempts suggested by new empirical results should be seen also in the light of mathematical, i.e. geometrical structural properties of physical laws in general. This view-point which played a decisive role in the past is shown to be a major argument of the recent attempt of finding a deeper and enlarged - i.e. englobing - basic theory.

2. HISTORICAL EXAMPLES

A. The fundamental and generally accepted laws have an intuitive and natural - under general invariance requirements even unique - mathematical, i.e. geometrical structure:

(i) A typical example: The Maxwellian system exhibits two differential operators which, considered separately, have most characteristic intrinsic (e.g. geometric) properties (leading to a unique coordinate-free definition; see sect. III). It is, therefore, amazing to realize that this most impressive mathematical structure covers such an unbelievably large empricial domain, ranging from the Lamb-shift over conventional machinery to modern astrophysics.

(ii) A similar situation holds for Einstein's equations for the gravitational field: The corresponding (i.e. defining) Lagrangian (just given by the curvature scalar and the density factor $\sqrt{-g}$) has - if I might say so - the most natural and simplest thinkable form (see sect. III); again there is, of course, an enormous domain of validity.

These two examples (chosen from a large number of similar cases) illustrate a.o. the deep meaning of basic laws as mediators between the enormously extended real world of physical facts and the abstract realm of natural and intuitive geometric structures. This relation might, however, also be turned around: The 'pre-existence' of certain (logically natural) mathematical structures is, in fact, needed for the very formulation of basic laws in a logically consistent and understandable form (e.g. Riemannian spaces, fibre bundles, Hilbert spaces, Dirac (i.e. Clifford) algebras (etc.). It should, however, been stressed that in some interesting cases abstract structures which later on became an important part of mathematics were, in turn, suggested by physical problems (e.g. superalgebras, Virasoro algebras, a.s.o.).

B. Also the more recent history of theoretical physics exhibits clearly the importance of abstract mathematical view-points: The stepwise extension of the domain of validity of an underlying theory, as well as the elimination of characteristic inconsistencies or even contradictions between so far needed theoretical assumptions are in most cases related to the introduction of additional basic mathematical concepts. In some cases the assumed validity of an abstract mathematical view-point (e.g. characteristic invariance principles) enforces a change, i.e. an adaption, of older physical laws. As an illustration we give here an extremely abbreviated survey of the decisive steps in the development of modern physics:

(i) A typical inconcistency between the Maxwellian equations and the Newtonian law of motion is - within the framework of special relativity - eliminated by enforcing the universal Lorentz invariance in both cases thus readapting or replacing - in perfect agreement with empirical facts - the second one by the well-known Einstein-Lorentz form (see sect. III).

(ii) The still remaining much more serious contradiction between special relativity and the Newtonian law of gravitation practically enforces the most impressive and far-reaching extension of conventional Euclidean geometry to a Riemannian space and calls for the corresponding adaption (or extension) of all basic laws: This formal (i.e. mathematical) view-point thus leads to a new and much deeper insight into the nature of basic laws and the structure of the universe in general.

(iii) A quite similar relation holds, in fact, for the transition from classical to quantum theory: Classical mechanics and Bohr's 'old' quantum condition (with its great merits) are - in general - definitely uncompatible: Again a new concept, i.e. a Hilbert space formed out of 'state vectors' describing a physical system in a perfectly new form, leads to (consistent) non-relativistic quantum theory and thus for a second time to an absolutely new and basic outlook on nature (whereby the former laws represent again a special approximation scheme).

(iv) The same pattern is - in principle - followed in order to obtain consistency between special relativity and the quantum theoretical principles; here the far-reaching - to some extent, however, still mathematically problematic - concept of the quantized field had to be 'invented', i.e. a most impressive but rather formal systematization of basic laws (including automatically the Bose--Einstein or the Fermi-Dirac statistics). A most remarkable improvement (unique group theoretical characterization of all basic interactions, renormalizability etc.) of this field theory was, however, achieved through the basic assumption of an (enlarged) 'gauge invariance' (with 'fibre bundles' as a geometrical basis)[6] inherent already in (classical) general relativity and the Maxwellian field.

(v) A - so far - final step of this - just sketched - survey represents the transition to string theory[2] as shortly discussed above. This most notable and far-reaching step (still in full development[11]) is a.o. motivated by the long standing problem of a consistency between (renormalizable) quantum (gauge) field theory and (<u>un</u>renormalizable) general relativity; it thus represents, in fact, a further (and logically similar) step as the former ones, but, astonishingly enough, aiming at the same time, to the great goal of unification of <u>all</u> basic laws.

3. MATHEMATICAL STRUCTURE

In this section we just give a few explicit examples of overcoming characterisitc - at first sight unsurmountable - inconsistencies through geometric extensions leading - at the same time - to new basic physical concepts.

A. The Maxwellian system and the conventional Newtonian equations of motion exhibit a clearcut contradiction if the transformation to moving frames is considered. For a charged classical point particle under the action of an eletromagnetic field ($\vec{E}$, $\vec{B}$) the latter reads (in Cartesian coord. for the position $\vec{x}(t)$ as function of the time t, where the dots (·) represent the time-derivatives):

$$m\ddot{\vec{x}} = \frac{e}{c} (\vec{E} + \frac{1}{c} [\vec{v}\vec{B}]) \tag{1}$$

and exhibits the Galileian invariance very much in contrast to the Lorentz invariance of the Maxwellian field equations:

$$\partial_\nu F^{\mu\nu} = J^\mu \quad ; \quad \partial_\lambda F_{\mu\nu} \underset{\text{antisymm.}}{} = 0 \tag{2a,b}$$

It should be remarked in this context that this system may be formally obtained through a 'Kaluza-type' dimensional extension from 3-dim. Euclidean systems (i.e. the system of magnetostatics, which has exactly the form (2a,b) with $F^{\mu\nu} \to B^{ik}$, i.e. the appropriate tensor form of the axial vector $\vec{B}$) to the 4-dim. Lorentzian form by adjoining the time t, the electric field $\vec{E}$ and enlarging the metric to g^0 at the appropriate places. Very much the analoguous procedure may be used for the equation of motion thus obtaining instead of (1) the relativistic, i.e. Lorentz-invariant new (and in this case different) law

$$m\overset{\circ\circ}{x}{}^\mu(\tau) = \frac{e}{c} F^\mu{}_\nu \overset{\circ}{x}{}^\nu \tag{3}$$

with the derivations (^{0}) with respect to the proper time τ defined as a 'normed' area on the 4-dim. orbit:

$$\overset{\circ}{x}{}^\mu g^0_{\mu\nu} \overset{\circ}{x}{}^\nu = -c^2 \tag{4}$$

Again the law of motion in a static magnetic field has exactly the same form, but in 3-dim. Euclidean space. Equation (3) differs, however, from (1) because of the (relativistically enforced) use of the proper time τ instead of t. This led, as well-known historically to the first verification of special relativity and exhibits at the same time geometrically a most natural structure (the Lorentz force in the right hand side of (3) appears uniquely determined by invariance and linear velocity dependence which, in turn, and together with the antisymmetry of $F^{\mu\nu}$ is enforced by equation (4)). This problem thus represents an impressive first example how geometry (change of dimension) and invariance (i.e. symmetry) lead directly to the 'simplest possible' form for a physical law which, in turn, leads to most important (but unexpected) effects like the formula for the mass defect through the relativistic form of energy-momentum vector P

$$P^{\mu} = m\overset{o}{x}^{\mu} \tag{5}$$

which in turn led de Broglie to the first step in quantum mechanics by relating P relativistically to the wave-vector k, i.e.

$$hk^{\mu} = P^{\mu} \tag{6}$$

B. Another inconsistency which again was cured geometrically led, in fact, to General Relativity: The Newtonian law of gravitation (historically the basis of an enormous scientifical revolution, but satisfying, of course, the Galilei invariance) is in perfect disagreemet with the absolutely basic Lorentz invariance, i.e. with special relativity. The corresponding geometric enlargement - and definitely the simplest and most natural possibility for overcoming the contradiction is, in fact, given by the transition from Minkowski to Riemannian space containing the space-time dependent metric $g_{\mu\nu}(\zeta)$ as new physical field and enlarging the Minkowskian g^0. First of all the form of the Maxwellian systems just calls for this enlargement: with a slight and geometrically enforced change it obtaines - a point to be stressed - its most natural geometric form: In fact, the two differential operators in the two (i.e. the inhomogeneous and the homogeneous) differential equations (2a,b) considered at a first stage (according to Weyl[3])) independently and applied to a tensor density $\mathfrak{F}^{\mu\nu}$ (replacing $F^{\mu\nu}$) in (2a) and $F_{\mu\nu}$ (unchanged) in (2^b) respectively are the <u>only</u> existing 'natural' differential operators, i.e. the equations (2a,b) are invariant with respect to the general (non-linear) coordinate transformations and thus practically determined through this enlarged (with respect to special relativity) invariance properties. The invariant relation between these two fields containing (for the first time) the metric field $g_{\mu\nu}(\zeta)$ is then, again, uniquely determined by the condition of not containing any derivatives of $g_{\mu\nu}$ (needed in order to obtain the special relativistic case as a decent 'local' approximation) and reads

$$\mathbb{F}^{\mu\nu} = \sqrt{-g}\, g^{\mu\bar{\mu}} g^{\nu\bar{\nu}} F_{\bar{\mu}\bar{\nu}} \tag{5}$$

This (absolutely) invariant system exhibits also a most suggestive analogy to the (ordinary) Maxwellian field in permeable (inhomogeneous and anisotropic) matter: The metric tensor $g_{\mu\nu}$ takes the role of a (general) permeability yielding the bending of light rays through the gravitational field.

The adaption of the equation of motion (3) appears also geometrially determined by the (enforced) replacement of the (proper) time derivative of a 4-vector v^μ along a 'given' orbit $\zeta^\mu(\tau)$ by the <u>covariant</u> derivation

$$D_\tau v^\mu = \dot{v}^\mu + \Gamma^\mu_{\alpha\beta} \dot{\zeta}^\alpha v^\beta \tag{6}$$

which takes into account the 'apparent' change of the components of the vector v^μ through the 'parallel transport' whereby

$$\Gamma^\mu_{\alpha\beta} = \frac{1}{2} g^{\mu\bar{\mu}} \left[-\partial_{\bar{\mu}} g_{\alpha\beta} + \{ {}^{\bar{\mu}\leftrightarrow\alpha}_{\nu\leftrightarrow\beta} \right] \tag{6'}$$

Equation (3) thus reads (replacing $\dot{x}^\nu$ by $\dot{\zeta}^\nu$)

$$m(\ddot{\zeta}^\mu + \Gamma^\mu_{\alpha\beta} \dot{\zeta}^\mu \dot{\zeta}^\nu) = \frac{e}{c} F^\mu_{\ \nu} \dot{\zeta}^\nu. \tag{7}$$

Here the 'geometrically' enforced Γ-term plays - and that represents a geometric generation of a physical interaction, i.e. the leading principle also of gauge theory - the role of the gravitational force (whereby $g_{\alpha\beta}$ represents in view of (6') the enlarged gravitational potential determined, in turn, by the Lagrangian $L = R\sqrt{-g}$ formed geometrically out of the curvature scalar R). This interaction appears thus to exhibit a quadratic velocity dependence (characteristic for obtaining the Newtonian approximation) which cannot be introduced within special relativity because of the inconsistency with eq. (3) which in the new form is automatically enlarged through the introduction of $g_{\mu\nu}(\zeta)$. Within Riemannian geometry (which is suggested, if not enforced, through the physical requirements of measuring in large domains) the basic laws are thus - apparent two additional conditions mentioned above - uniquely determined through the invariance property with respect to the general coordinate transformations

$$\zeta'^{\mu'} = f^{\mu'}(\zeta^0 \ldots \zeta^{\mu} \ldots \zeta^4) \tag{8}$$

containing 4 arbitrary (but regular) functions f^{μ} of 4 variables, i.e. the group of diffeomorphisms determines the classical laws. In particular it follows that only two interaction types (i.e. linear and quadric in the 4 velocities and corresponding to electromagnetism and gravitation, respectively) can exist within this clasical framework.

It thus remains to be seen how the introduction of new, so-called internal variables of the particles - e.g. flavour and colour - to be treated, however, only within the framework of quantum theory will yield (again through geometrical invariance principles) new types of interactions which, in turn, lead to an understanding of the inner structure of heavy particles (so-called hadrons) and their interactions (e.g. electroweak and strong).

4. THE GAUGE THEORIES

The principle which led to this most successful way of introducing additional invariant interaction is due to Yang and Mills[4)] who generalized in a most natural way H. Weyl's (2nd) gauge principle[5] of 1929. The mathematical basis of (enlarged) gauge theory is given by the geometrical concept of 'fibre bundles' which on one side represent the characteristic space structure of modern topology[6], on the other, however, the space naturally generated by the wave function $\Psi^k(x)$ of a particle endowed with an inner degree of freedom k = 1...n, e.g. an n-dim. complex vector space attached to each point x, e.g. isospin (k=2) or colour (k=3). In usual mathematical language the general particle-state thus appears to be a section of a n-dim. 'vector bundle' over the Minkowski (later: Riemannian) space. For a free Dirac particle in Minkowski space (x) one has at first sight n uncoupled field equations (with the Dirace matrices γ_μ and $\kappa = \frac{mc}{h}$):

$$\gamma^\mu \partial_\mu \Psi'^k + \kappa \Psi'^k = 0 \qquad k = 1...n \tag{9}$$

The decisive condition for so-called <u>local</u> SU(n)-gauge invariance now states the invariance of (9) with respect to the space time, i.e. x-dependent n-dim. (special) unitary transformations S(x) (forming in turn a 'principle fibre bundle')

$$\Psi' = S\Psi \tag{10}$$

(i.e. explicitely $\quad \Psi'^{k'}(x) = S^{k'}_k(x)\Psi^k(x)$)

In other words, independence with respect to an arbitrary choice of coordinates within the various 'fibres' attached to each space time point x. This strong but natural condition is again met through the 'usual' replacement of the partial derivation ∂_μ by the corresponding covariant derivation D_μ (in this case called 'connection'-form)

$$\partial_\mu \rightarrow D_\mu \qquad D = \partial_\mu + \Gamma_\mu \tag{11}$$

where Γ_μ is a n-dim. matrix representing the general infinitesimal (i.e. Lie-) element of the group SU(n) corresponding to the infinitesimal displacement δx^μ

$$\delta S = \delta x^{\mu}\Gamma_{\mu}(x) \tag{12}$$

with

$$\Gamma_{\mu}(x) = iA^{a}_{\mu}(x)T_{a} \qquad a = 1...N$$

i.e. the x- and μ-dependent linear combination of the N basic Lie--elements of SU(n) satisfying the Lie-algebra relations

$$[T_{a}T_{b}]_{-} = g_{ab}{}^{c}T_{c} \qquad a = 1...N \tag{13}$$

containing the 'structure constants' g. This x-dependent matrix of coefficients is generally called a 'gauge field' which now appears automatically as an interacting field (i.e. an 'enlarged vector potential) within the Dirac equation q):

$$\gamma^{\mu}(\partial_{\mu} + iA^{a}_{\mu}(x)T_{a})\Psi + \kappa\Psi = 0 \tag{14}$$

Under the gauge transformation (10) the 'gauge potential' aquires, however, the gauge dependent term (using as usual the abbreviation $\mathbb{A}_{\mu} = A^{a}_{\mu}T_{a}$)

$$\mathbb{A}'_{\mu} = \mathbb{A}_{\mu} + S^{-1}\partial_{\mu}S \tag{10'}$$

In order to obtain a gauge <u>invariant</u> field equation one follows the well-known geometrical procedure (already used in G.R.) by forming the curvature (here called field strength) through

$$\mathbb{F}_{\mu\nu} = i[D_{\mu}D_{\nu}]_{-} \tag{15}$$

(leading to the non-linear of characteristic expression)

$$F^{a}_{\mu\nu} = \partial_{\mu}A^{a}_{\nu} - \partial_{\nu}A^{a}_{\mu} + g^{a}_{bc}A^{b}_{\mu}A^{c}_{\nu} \tag{16}$$

and the totally invariant (quadratic) Yang-Mills Lagrangian (which differs from the linear G.R. case*)

$$L = \text{trace}\,(\mathbb{F}_{\mu\nu}\mathbb{F}^{\mu\nu}) \qquad (17)$$

which, in turn, leads to the basic (non-linear) Yang-Mills equation for the gauge field A, i.e. a natural generalization of the Maxwellian system corresponding to the special choice S = U(1) for the gauge group. On the other hand, the 2 cases SU(2) in connection with U(1) and SU(3) form the basis of the 'electroweak' interaction (according to Salam-Weinberg[7)]) and the 'strong' interaction (according to Gell-Mann, Leutwyler and Fritzsch[8)]) covering thus (in combination with G.R. which may be visualized as an enlarged and more refined gauge theory) <u>all</u> basic interactions by which the enormous empirical material of present day physics (i.e. cosmology, solid state, atoms, nuclei, particles) <u>can</u> be interpreted.

5. A SURVEY OF THE PRESENT DAY SITUATION

This short sketch of 'classical' gauge theory was given in order to show its extreme geometrical naturality and (nearly unbelievable) simplicity, as well as its close connection to G.R. In fact, the local gauge invariance (10) appears practically enforced if the Minkowski ground space (used conventionally) is enlarged (this should have been done from the outset!) to the Riemannian manifold and realizing that the Riemannian curvature tensor is also obtained by using the Lorentz group (within the concept of the 'moving frame') as the basic gauge group. (The metric tensor g generating the Christoffel-connection Γ_μ and used a.o. for building the linear Einsteinian Lagrangian $R\sqrt{-g}$ has, however, no counter part here). Considering this analogy it appears, in fact, most impressive or even mysterious that such natural and intimately related geometric structures appear as a basis for the understanding of this enormous body of empirical data ranging from cosmic dimensions to distances related to the inner particle structure.

These insights should be considered, however, just an encourgement for further research, i.e. the starting point of a much deeper and far reaching mathematical investigation: First of all, the definitely needed quantization of the gauge fields has, in spite of some important results (i.e. renormalizability with the help of the BRS formalism or alternatively with 'constructive' field theory) to be reconsidered: There are, a.o. indications that only gauge fields (i.e. the only ones occurring in basic laws) can in fact be taken over into quantum theory: This so far not fully proved fact yields an interpretation of the non-existence of a general 'quantization' rule for arbitrary classical systems and hints - at the same time - for a deep (but far from being understood) relation (or unicity) between classical and quantum theory.

The most important remaining, in fact, long standing problem represents the 'quantization' of the gravitational field definitely needed for reasons of consistency with Heisenberg's principle: Its enlarged geometrical structure mentioned above may be the origine for the failures experienced so far. The formal analogy to (ordinary and 'renormalizable') gauge theories might be another indication for the definite need for a much deeper mathematical insight in the results obtained so far: In fact, BRS renormalization and constructive methods are - in spite of partly spectacular success - far from being satisfactory. In addition, the most beautiful and really impressive geometrization and - to a certain extent - unification of physics just enforces a new and far reaching effort (a.o. foreseen by W. Pauli, W. Heisenberg, P. Dirac) for finding a.o. the mysterious underlying structures selecting just this very special choice of relevant gauge groups discussed here which points to the existence of a deeper more general construction principle behind the empirical facts. In this sense the enormous and enthusiastic effort made within the framework of 'heterotic string theory' should, in fact, be understood and appreciated. From a mathematical view-point this attempt is characterized by a far reaching extension of the stepwise geometrical enlargement seen here (i.e. Euclidian, Riemannian spaces and, eventually, to fibre bundles) leading eventually to the dimensions 10 and 26 and incorporating at the same time in fulfillment of a long standing quest an elementary (smallest) length, i.e. the Planck length. In perfect analogy of the 'classical' development discussed in this short survey the former (older) theories appear to be special (here in the limit of relatively small energies with respect to the Planck mass) or approximations of the full law. In suggestive agreement to a 'prophecy' of W. Heisenberg (in connection of his one attempt for an englobing theory) the new concepts, in particular the corresponding fields, are far away from a direct physical interpretation (describing particles); they are just determined by general symmetry properties expressed by the choice of basic groups. The stepwise discovery of such general symmetries represents, perhaps, the most interesting and astonishing experience of physical research and yields a perfectly new outlook on nature.

6. CONCLUSIONS

This short survey should demonstrate the fact that basic (mainly geometrical) mathematical structures which were - at a first instance - necessary in order to overcome characteristic inconsistencies (even contradictions) appearing in the course of theoretical developments are - at the same time - directly related to new and deeper insights (and connected with perfectly new interpretations) into the basic properties of nature: Apart from the most natural and satisfactory understanding of space-time in general relativity, the introduction of the quantum mechanical state vectors pratically <u>enforces</u> this far reaching revolution which a.o. led to the Heisenberg indeterminacy and the general probabilistic (Copenhagen) interpretation. In this connection it should be stressed that the existence of these 'ideal' probabilities is empirically checked again and again in all kinds of scattering experiments (i.e. a major part of enormously extended empirical investigations) and various correlation experiments of great practical importance (e.g. the correlations observed in successive nuclear γ-γ-transitions leading a.o. to characteristic spin assignments).

A major point emphasized in this extremely short survey is the relative simplicity and naturality of most basic laws if duely formulated within the framework of the corresponding, i.e. extended mathematical basis: e.g. the Lagrangian of general relativity as mentioned above; the most natural geometrical formulation in terms of fibre bundles of gauge theories basic for <u>all</u> interaction schemes of elementary particles; the transition to quantum theory partly in the form of geometric quantization[10)] etc. We are thus tempted to state: Analyzing the apparently structureless and most extended body of experimental data, we find far beyond the surface of directly visible physical facts these beautiful and far reaching abstract structures which - at the same time - point to the existence of more general connections and deeper insights. Several great scientists expressed this fact in (only apparently) different terms: 'Unbelievable applicability of mathematics' (E. Wigner); 'A mysterious relation between empirical facts and mathematical structures' (R. Nevanlinna); 'Physics as the heartblood for mathematical research' (M. Atiyah); 'An analogy between basic research and religious endeavour' (W. Pauli) and, eventually, the famous expression 'Subtle is the Lord' (A. Einstein). Even well-known writers expressed - in very different form - the same idea: 'Are you sure that you will never end up with a perfect contradiction ?' (question asked to a great astrophysicist by F. Dürrenmatt); 'Physics as the basis of huge pillars bearing an enormous dome so far veiled' (C. Zuckmayer after a discussion with W. Heisenberg).

In conclusion a message and a personal belief: We are <u>not</u> 'explaining' nature (according to school books) but we are discovering step by step its inner structure and beauty thus becoming modest but comforted.

7. REFERENCES

1) S.T. Yau, A Survey on the Interaction between Mathematical Physics and Geometry, Int. J. Mod. Phys. A 1, 881 (1986)
2) M.B. Green, J.H. Schwarz, E. Witten, Superstring Theory, two volumes (Cambridge University Press, 1987)
J.H. Schwarz, Review of Recent Developments in Superstring Theory, Int. J. Mod. Phys. A 2, 593 (1987)
3) H. Weyl, Space, Time, Matter, translated by H.L. Brose, Dover Publications, 1950
4) C.N. Yang, R. Mills, Phys. Rev. 96, 191 (1954)
5) H. Weyl, Z. Physik 56, 330 (1929)
6) For an introduction into fibre bundles see e.g.
C. Nash, S. Sen: Topology and Geometry for Physicists, 1983, Academic Press
M.E. Mayer in: Fibre Bundle Techniques in Gauge Theories, Springer Lecture Notes in Physics 67 (1977)
A. Trautman, Rep. Math. Phys. 1, 29 (1970)
For an account of fibre bundles in gauge theory see e.g.:
N.P. Knopleva, V.N. Popov, Gauge Fields, 1981 Harwood Academic Publishers (translated from the Second Russian Edition)
7) Nobel lectures of S. Weinberg, A. Salam and S.L. Glashow, Rev. Mod. Phys. 52 515; 525; 539 (1980)
8) H. Fritzsch, M. Gell-Mann, 16th Int. Conf. on High Energy Physics, Batavia, Vol. 2 (1972), p. 135
H. Fritzsch M. Gell-Mann, H. Leutwyler 47B, 365 (1973)
9) That nuclei can be treated with QCD is shown in:
H.R. Petry et al., 'An Application of QCD in Nuclear Structure', Phys. Lett. 159B, 365 (1985)
H.R. Petry in Springer Lecture Notes in Physics 197, 236 (1984)
K. Bleuler, 'QCD and Nuclear Structure', in: Symmetries in Science II, ed. B. Gruber, R. Lenczewski, Plenum 1986, p. 61
K. Bleuler, in: Fundamental Aspects of Quantum Theory, ed. V. Gorini, A. Frigerio, Plenum 1986
10) See, for example:
Proc. of the Symposium 'Differential Geometrical Method in Mathematical Physics', Bonn 1975, ed. K. Bleuler, A. Reetz, Springer Lecture Notes in Physics, 570, p. 1-108
Proc. DGM in Math. Phys. II, Bonn, 1977, ed. K. Bleuler, H.R. Petry, A. Reetz, Springer LNP 676, p. 295-457
11) Proceedings of the XVI. Differential Geometrical Methods in Theoretical Physics, Como, Italy, to appear at Reidel, Dordrecht

AN ONTOLOGICAL FOUNDATION FOR THE QUANTUM THEORY

DAVID BOHM

Department of Physics,
Birkbeck College, University of London,
Malet Street, London WC1E 7HX
UK

ABSTRACT

In this paper, a systematic and over-all ontology is developed that is consistent with the quantum theory. The paper begins with a review of the causal interpretation of the quantum theory, emphasising several new concepts that have been introduced in this interpretation. These may be summed up in terms of the notion that the wave function determines a quantum potential representing active information, which operates to give form to the motion of particles that however move under their own energy.

The main further new point discussed in this paper is then that the quantum theory itself implies the independent existence of a large-scale manifest reality (i.e., one that is tangible and publicly accessible) in which the new quantum features can be neglected. This is contrasted with the subtle (i.e., intangible) quantum world of the wave function. This subtle quantum world is then shown to be capable of revealing itself in the manifest world, especially in experiments that are sensitive to a quantum level of accuracy. In this way, the problem of measurement is removed, as the human observer is now seen to be related to this public manifest world in essentially the same way as in classical physics. Therefore, no special discussion of the act of observation is needed. Moreover, the ontology itself is seen to imply that the experimental apparatus and the object of the experiment constitute an undivided whole, and so, an ontological explanation is given for Bohr's notion of wholeness of the quantum experiment.

1. Introduction

The quantum theory in its usually accepted interpretations is assumed to give nothing but statistical predictions about the results of measurements. Indeed, without the statistics of measurements, the quantum theory would, in this view, be just a piece of pure mathematics, with no physical significance. This means that a fundamental role is being given to epistemology, in the sense that it is not even possible to express the theory without an essential reference to our knowledge of the world and to our means of obtaining it (e.g., through experimental observations and measurements).

It is the main purpose of this paper to present a consistent ontological interpretation to the quantum theory[1]. Such an interpretation is based on assumptions with regard to the general nature of that which is, considered as a reality that exists essentially independently of our knowledge about it. In a theory of this kind, the actual individual systems, assumed to belong to such an independent reality, are taken as primary. This reality is then shown to be the source of the observed phenomenon, from which we obtain our knowledge of it. Epistemological questions (along with the statistical implications of the theory) are thus given only a secondary kind of significance. As will be brought out throughout this paper, such an ontological basis for the theory has certain important advantages, in that it helps to make the meaning of the theory more clear and easier to grasp, and to suggest interesting and significant new concepts with regard to the nature of the independent reality that is implied by the quantum theory.

It is proposed in this paper that since there are at present no crucial experiments that could decide between the various available interpretations, the best strategy for science would be to allow all of them to be considered seriously together, along with their various advantages and disadvantages. Out of a dialogue between them perhaps a different approach could ultimately emerge, that will move in some new directions going beyond any of the particular interpretations, including the one presented in this paper. In other words, the ontological approach presented here is not to be regarded as a rival claimant for final truth, but rather, as a source of insight that, along with the others, may help lead to further progress of a new kind.

2. The Causal Interpretation Of The Quantum Theory

When I first started to look at the question of the meaning of the quantum theory in 1950, there was a widespread impression that an ontological interpretation of the kind indicated earlier is not actually possible. For various reasons, however, I began at that time to ask if this were really so. In doing this, I began to think of the W.K.B. approximation, which is used in conventional quantum mechanics to deal with the classical limit. To obtain this, we write the wave function, ψ, in polar form; i.e., $\psi = Re^{iS/h}$. We put this into Schrödinger's equation

$$i\hbar\frac{\partial\psi}{\partial t} = -\frac{\hbar^2}{2m}\nabla^2\psi + V\psi \tag{1}$$

and we obtain equations for R and S

$$\frac{\partial S}{\partial t} + \frac{(\nabla S)^2}{2m} + V - \frac{\hbar^2}{2m}\nabla^2 R/R = 0 \tag{2}$$

$$\frac{\partial R^2}{\partial t} + \mathrm{div}\, R^2\frac{\nabla S}{m} = 0 \tag{3}$$

In the classical limit $-\frac{\hbar^2}{2m}\frac{\nabla^2 R}{R}$ is small compared to the terms appearing in eqn. (1), notably $\frac{(\nabla S)^2}{2m} + V$, because there are many oscillations of the wave function in a range in which R changes appreciably. In the W.K.B. approximation, the term involving $\nabla^2 R/R$ is neglected, to yield

$$\frac{\partial S}{\partial t} + \frac{(\nabla S)^2}{2m} + V = 0 \tag{4}$$

The above is however just the classical Hamilton-Jacobi equation. If we put $p = \nabla S$, then Newton's equations of motion follow from (4). The surfaces, S = constant, determine the wave fronts, and the particle always moves normal to the wave front at the point where it is located. If we further assume a statistical ensemble of particles with $P = R^2$, equation (4) then represents the conservation of probability in this ensemble.

All of this had, of course, been well known for a long time. But what I noted in the light of my new question was that those who used the conventionally accepted interpretation of the quantum theory had already surreptitiously slipped into an ontological interpretation, without noticing it. For the W.K.B. approximation was interpreted in terms of an actual individual particle moving in a trajectory normal to an actual set of wave fronts and of an ordinary statistical ensemble of such particles, of the same general kind as that used in classical statistical mechanics, with a conserved probability density, $P = R^2$.

This led me to a second question: Is it possible to use this ontological interpretation more generally, and not merely in the W.K.B. approximation? Inspection of eqn. (2) shows that this is indeed possible For the neglected term, $-\frac{\hbar^2}{2m}\nabla^2 R/R$ is just playing the role of an additional potential, to be added to the classical potential, V. I was therefore led to call this the quantum potential, Q, which is defined as

$$Q = -\frac{\hbar^2}{2m}\frac{\nabla^2 R}{R}. \tag{5}$$

What could such a potential mean? The answer seemed simple. Even the classical potential was originally only something assumed to explain why a particle does not move uniformly in a straight line. Later it was shown that typical potentials (e.g., the electromagnetic) are determined by field equations (e.g., Maxwell's equations). So, the reality underlying a classical potential was generally assumed to be some kind of field of this sort (that obeyed a law of its own). My assumption was then that the known classical potentials are not enough to account for the full behaviour of matter. One also needs a quantum potential, which for the present we simply assume, just as we did originally with classical potentials. We then find by inspection of the equations, that we can consistently assume also that there is a field underlying the quantum potential. This is the wave function, ψ. The "quantum field" satisfies, not Maxwell's equations, but Schrödinger's equation. Just as Maxwell's equations determine the time development of the electromagnetic fields, so Schrödinger's equation determines that of the Schrödinger field, ψ.

Our picture is then that an entity such as an electron, for example, <u>is</u> a particle, always accompanied by its fields, which are essential to understand what it is. In addition to the classical fields, we need the Schrödinger field, ψ. The system of particle plus its set of fields is causally determined; hence the name, causal interpretation. (Although the theory is causal, the whole question of determinism v.s. indeterminism is, as we shall see later, much more subtle than it is in the usual presentation of classical theories.)

We emphasize once again that the primary and fundamental meaning of ψ is then that it is a field, which determines the quantum potential, through the relationship, $Q = -\frac{\hbar^2}{2m}\frac{\nabla^2|\psi|}{|\psi|}$. The interpretation of $|\psi|^2$ as probability density is only of secondary significance. Of course, the conservation equation (3) guarantees that such a probability interpretation of $|\psi|^2$ is consistent; i.e., if it holds at one time, it will hold for all time. But we do not begin with the assumption that $P = |\psi|^2$. Rather, as we shall see, it is in principle possible to begin with an arbitrary probability density, P and to show that because of the chaotic motions implied by the quantum potential, such a probability function will approach $|\psi|^2$. Of course, because of the conservation equation, (3), it will thereafter remain with this distribution.

Using this interpretation (extended to the many-body system, in the way shown in the next section) I found that it was possible to understand all the results of the usual interpretation in a consistent way[(1)(2)]. At first sight, it seemed surprising that a theory could be obtained from such an elementary device as adding a new quantum potential. But closer inspection showed that this is not actually a simple return to classical ideas. For the quantum potential has several strikingly novel features, which do not cohere with what is the generally accepted structure of classical physics. Indeed, as we shall see in more detail later, it implies that the electron is not a structureless particle, but is a highly complex entity that is deeply affected by its quantum field in an extremely subtle and dynamic way. The new quantum properties of matter can then be shown to follow from the qualitatively new features of this interaction, as expressed through the quantum potential.

The first of these new properties can be seen by noting that the quantum potential is not changed when we multiply the field intensity ψ

by an arbitrary constant. (This is because ψ appears both in the numerator and the denominator of Q.) This means that the effect of the quantum potential is independent of the strength (i.e., the intensity) of the quantum field but depends only on its <u>form</u>. By contrast, classical waves, which act mechanically (i.e., to transfer energy and momentum, for example, to push a floating object) always produce effects that are more or less proportional to the strength of the wave.

To give an analogy, we may consider a ship on automatic pilot being guided by radio waves. Here too, the effect of the radio waves is independent of their intensity and depends only on their form. The essential point is that the ship is moving with its own energy, and that the <u>information</u> in the radio waves is taken up to direct the much greater energy of the ship. We may therefore propose that an electron too moves under its own energy, and that the information in the <u>form</u> of the quantum wave directs the energy of the electron.

This introduces several new features into the movement. First of all, it means that particles moving in empty space under the action of no classical forces still need not travel uniformly in straight lines. This is a radical departure from classical Newtonian theory. Moreover, since the effect of the wave does not necessarily fall off with the intensity, even distant features of the environment can profoundly affect the movement. As an example, let us consider the interference experiment(This involves a system of two slits. A particle is incident on this system, along with its quantum wave. While the particle can only go through one slit or the other, the wave goes through both. On the outgoing side of the slit system, the waves interfere to produce a complex quantum potential which does not in general fall off with the distance from the slits but rather, contains deep "valleys" and broad "plateaux". In the regions where the quantum potential changes rapidly there is a strong force on the particle. The particle is thus deflected, even though no ordinary type of force is acting. The movement of the particle is therefore modified in a way that depends on the whole system.

In this explanation of the quantum properties of the electron, the notion of information plays a key role. Indeed it is helpful to extend this notation of information and introduce what could be called <u>active</u> information. The basic idea of active information is that a <u>form</u> having very little energy enters into and directs a much greater energy. The activity of the latter is in this way given a form similar to that of the smaller energy. It is therefore clear that the original energy-form will "inform" (i.e. put form into) the activity of the larger energy.

As an example, consider a radio wave whose form carries a signal. The energy of the sound that we hear from our radio comes, however, not from this wave but from its power plug or batteries. This is essentially an "unformed" energy, that takes up the form or information carried by the radio wave. The information in the radio wave is, in fact, <u>potentially</u> active everywhere, but it is <u>actually</u> active, only where and when its form enters into the electrical energy within the radio. A more developed example of active information is obtained by considering the computer. The information content in a silicon chip can similarly determine a whole range of potential activities which may be actualized through having the form of this information enter the electrical energy coming from a power source. Which of these possibilities will in any given case be actualized depends on a wider context that includes the software programmes and the responses of the computer operator at any given moment.

It is clear that the notion of active information already has a widespread application in many fields. Our proposal is then to extend the notion of the possibility of the objective applicability of the notion of active information to the quantum level. That is to say the information in the quantum potential is <u>potentially</u> active everywhere, but <u>actually</u> active only where this information enters into the activity of the particle.

This implies, however, that as we have already suggested, an electron, or any other elementary particle, has a complex and subtle inner structure (e.g., at least comparable to that of a radio). This notion goes against the whole tradition of modern physics which assumes that as we analyse matter into smaller and smaller parts its behaviour grows more and more elementary. But our interpretation of the quantum theory indicates that nature is far more subtle and strange than previously thought. However, this inner complexity is perhaps not as implausible as may appear at first sight. For example, a large crowd of people can be treated by simple statistical laws, whereas individually their behaviour is immensely more subtle and complex. Similarly, large masses of matter reduce approximately to a simple Newtonian behaviour, whereas the molecules and atoms out of which matter is built have a more complex inner structure.

To make this suggestion yet more plausible, we note that between the shortest distances now measurable in physics (of the order of 10^{-16} cm) and the shortest distances in which current notion of space time probably has meaning, of the order of 10^{-33} cm, there is a vast range of scale in which an immense amount of yet undiscovered structure could be contained. Indeed, this range of scale is comparable to that which exists between our own size and that of the elementary particle.

So far, the double slit experiment has been discussed in terms of a single particle. But a typical experiment, in fact, involves a statistical ensemble of particles all having the same quantum field, ψ. The experiment can be so arranged that each particle, along with its quantum field, goes through the slit system separately and independently. Each of these electrons starts from a different initial position, and so goes through the slit system in a different way, following one of the ensemble of trajectories. As a result, it will arrive at a different point on the screen. After many such particles have gone through, one then obtains a statistical ensemble of spots on the detection screen, which add up to the well-known interference pattern.

Once the quantum field, ψ, is specified, the results of each experiment are in principle determinate and depend only on the initial conditions of the particle. The latter will however fluctuate from one case to the next, because the particles emerge from a source (e.g., a hot filament), in which they are subject to irregular thermal perturbations. Within this source, the actual trajectory of a particle is extremely sensitive to the complex details of the quantum potential, which is in turn sensitive to the small thermal disturbances from other particles in the thermodynamic system. The motion is therefore unstable and chaotic in a sense, that has been defined, for example, by Kadanoff et al.[(4)] This means that complete predictability and controllability of the initial conditions is, in practice, essentially impossible. And so, it will also not be possible to predict just where each electron will arrive on the detection screen. In this fashion, the statistical nature of the predictions of the quantum theory can be explained while causality is still basic to the theory itself.

As to why the probability density in a thermal source should approach $|\psi|^2$, there are several possible explanations. First of all, one can give reasons why it is plausible that this will be the equilibrium distribution resulting from the effects of chaotic thermal disturbances on the quantum potential of any particle that emerges[(5)]. And secondly, it is possible to assume a further stochastic process of random fluctuations of the momentum around an average, $\overline{\overline{p}} = \nabla S$, representing, for example, the action of a subquantum level that would lead to $P = R^2$ in the long run[(6)(7)].

The fact that each particle responds to information from the entire environment gives a simple and tangible account of Bohr's notion of the undivided wholeness of the experimental conditions and the experimental results[(8)(9)]. Thus, each particle goes through only one of the slits, and yet its motion is fundamentally affected by information coming from both slits. One has therefore to consider the whole of the relevant experimental solution, to understand what happens in each case. However, this sort of wholeness implied by the causal interpretation differs in an important way from that of Bohr. For in the causal interpretation, while the entire experiment has to be treated as an undivided whole, this whole is analysable in thought (e.g., through the movement of the particle acted on by the quantum potential). However, in Bohr's approach, the entire experimental situation is an unanalysable whole, about which nothing more can be said at all.

3. Extension To The Many-Body System

We shall now go on to consider the many-body system in which we shall see, in several striking ways, a further development of this difference between classical and quantum ontologies.

We shall begin by considering the two-body system. The wave function $\bar{\psi}(r_1,r_2,t)$ satisfies the Schrödinger equation

$$i\hbar\frac{\partial\psi}{\partial t} = \left[-\frac{\hbar^2}{2m}(\nabla_1^2 + \nabla_2^2) + V\right]\psi \tag{6}$$

where ∇_1 and ∇_2 refer to particles 1 and 2 respectively. Writing $\psi = Re^{iS/h}$ and defining $P = R^2 = \psi^*\psi$, we obtain

$$\frac{\partial S}{\partial t} + \frac{(\nabla_1 S)^2}{2m} + \frac{(\nabla_2 S)^2}{2m} + V + Q = 0 \tag{7}$$

where

$$Q = -\frac{\hbar^2}{2m}\frac{(\nabla_1^2 + \nabla_2^2)R}{R} \tag{8}$$

and

$$\partial P/\partial t + \text{div}_1(P\nabla_1 S/m) + \text{div}_2(P\nabla_2 S/m) = 0. \tag{9}$$

As in the case of the one-body system, eq. (7) can be interpreted as the Hamilton-Jacobi equation with the momenta of the two particles being respectively

$$p_1 = \nabla_1 S \quad \text{and} \quad p_2 = \nabla_2 S.$$

But this time it is the Hamilton-Jacobi equation for a system of two particles responding not only to the classical potential, V, but also the quantum potential, Q. This latter now depends on the position of both particles in a way that does not necessarily fall off with the distance. We thus obtain the possibility of a <u>non-local interaction</u> between the two particles.

Going on to consider the N-body system we will have

$$Q = Q(r_1,r_2,\ldots.r_N,t)$$

so that the behaviour of each particle may depend non-locally on all the others, no matter how far away they may be.

If we take $P = R^2$ as the probability density in the configuration space of the two particles, then clearly eq. (9) is the equation for the conservation of this probability. With this definition of probability, it follows, as has been shown, by means of a detailed and extensive treatment[1][2], that all the statistical results of the quantum theory are (as in the one-body case) the same in the causal interpretation as they are in the usual interpretation.

While non-locality as described above is an important new feature of the quantum theory, there is yet another new feature that implies an even more radical departure from the classical ontology, to which little attention has been paid thus far. This is that the quantum potential, Q, depends on the "quantum state" of the whole system in a way that cannot be defined simply as a pre-assigned interaction between all the particles.

It is clear from this that the wholeness of the entire system has a meaning going beyond anything that can be specified solely in terms of the actual spatial relationships between particles. This is indeed the feature which makes quantum theory go beyond mechanism of any kind. For the essence of a mechanical behaviour is that the parts interact in some pre-assigned way to make up the whole. Even if the interaction is non-local, the system is still mechanical as long as the interaction potential is a fixed and pre-assigned function of the particle variables. But in the causal interpretation of the quantum theory, this "interaction" depends upon the wave function of the entire system, which is not only contingent on the state of the whole but also evolves with time according to Schrödinger's equation. Something with this sort of independent dynamical significance that refers to the whole system and that is not reducible to a property of the parts and their inter-relationships is thus playing a key role in the theory. As we have stated above <u>this is the most fundamentally new ontological feature</u> implied by the quantum theory.

The above-described feature should, in principle, apply to the entire universe. However, as has been shown elsewhere[13], when the wave function of a system factorises into two parts, the corresponding sub-systems will behave independently. In this case, each sub-system can be treated on its own. It is this feature which provides the ground for the possibility of understanding how, in spite of quantum wholeness, the world still behaves in a suitable classical limit as a set of relatively independent parts that interact mechanically.

4. A Brief Resumé Of The Theory Of Measurements In The Causal Interpretation

We shall now give a brief resumé of our treatment of the quantum measurement process[1][2], as this treatment introduces a number of key new concepts that will be carried over into our discussion of the quantum potential approach.

Our first step is, for the sake of convenience, to divide the overall process of measurement into two stages. In the first stage, the "observed system" interacts with the "measuring apparatus" in such a way that the wave function of the combined system breaks into a sum of non-overlapping packets, each corresponding to a possible distinct result of the measurement. Let the initial wave function of the observed system be

$$\Psi_0(x) = \sum_n C_n\psi_n(x). \tag{11}$$

where the $\psi_n(x)$ are the possible eigenfunctions of the operators that are being "measured". The initial wave function of the combined system is then

$$\psi_0 = \phi_0(y) \sum_n C_n\psi_n(x) \tag{12}$$

where $\phi_0(y)$ is the initial wave function of the measuring apparatus. After the two systems have interacted Ψ_0 goes over into

$$\Psi_f = \sum_n C_n\psi_n(x)\phi_n(y) \tag{13}$$

where the $\phi_n(y)$ are the different wave packets of the relevant parameters of the apparatus that correspond to the possible results of the measurement. For a proper measurement to be made, the packets $\phi_n(y)$, must, of course be distinct and non-overlapping.

As shown in Bohm and Hiley[10], during the period of interaction, the wave functions, and therefore the quantum potential, become very complex and rapidly fluctuating functions of the time. When the packets have separated, the "apparatus particles" must have entered one of them (say m) and will have zero probability of leaving (because there is no probability of entering the spaces in between packets). From then on, the quantum potential acting on the particles will be determined only by the packet $\psi_m(x)\phi_m(y)$, because all the other packets (which do not overlap this one) will not contribute to it. So, at least as far as the particles are concerned, we may ignore all the other packets, and regard them as, in the sense discussed in section 2, constituting inactive or physically ineffective information.

One may describe what happens in another way by saying that each packet, $\phi_n(y)$ forms a kind of "channel". During the period of interaction, the quantum potential develops a structure of bifurcation points, such that trajectories of the apparatus particles on one side of one of these points enter a particular channel (say $\phi_m(y)$), while the others do not. Eventually, each particle enters one of the channels to the exclusion of all the others and thereafter stays in this channel. When this has happened, the "observed" particle will behave from then on, as if its wave function were just $\psi_n(x)$, even if $\psi_m(x)$ and the rest of the $\psi_n(x)$ would still overlap. The fact that the "apparatus particles" must enter one of their possible channels and stay there is thus what is behind the possibility of a set of clearly distinct results of a quantum measurement.

At this point, however, one may ask what is the role of the "inactive" packets, not containing the particles. Can we be sure that they must necessarily remain permanently inactive? The answer is that in principle, it is in fact still possible to bring about activity of such packets. For example, one may apply an interaction Hamiltonian to one of these inactive packets, say $\psi_r(x)$, such that it comes to coincide once again with $\psi_m(x)$, while leaving $\phi_m(y)$ unchanged. The two packets together will then give us $\phi_m(y)(\psi_m(x) + \psi_r(x))$. If $\psi_m(x)$ and $\psi_r(x)$ overlap, there will be interference between them, and this will give rise to a new quantum state, in which the previously inactive packet, $\psi_r(x)$, will now affect the quantum potential, so that it will once again be active.

But here, it is necessary to note that thus far, the "measuring apparatus" and the "observed system" have been treated on an essentially symmetrical footing. We have not, as yet, brought into the theoretical description anything that would assign a special role to the state of the measuring apparatus as something that was actually capable of being known by a human being. It was here that we introduced our second stage of the measurement process, which contained a detection or registration device capable of amplifying the distinctions in the states of the "apparatus particles" to a large scale level that is easily observable by ordinary means. Such a registration device will contain a very large (macroscopic) number, N, of particles. When this device interacts with the "apparatus particles", y, its wave function $\lambda(Z_1, \ldots Z_N)$ will have to be brought into the discussion. To each distinct state, n, of the "apparatus particles", there will be a corresponding state, $\lambda_n(Z_1, \ldots Z_N)$ of the registration devices. The wave function of the relevant system will then be

$$\Psi = \sum_n C_n \phi_n(y) \psi_n(x) \lambda_n(Z_1, \ldots Z_N). \tag{14}$$

Each of the λ_n will also not overlap with the others, so that even if the $\phi_n(y)$ should later come to overlap, this would still not affect the quantum potential, as the particles of the registration device will now be in distinct channels.

Could the channels of the registration device in turn be made to overlap again? In Bohm and Hiley[10], it was emphasized that this would have essentially zero probability, because in registration, there has occurred a thermodynamically irreversible process. (So that, for example, to have overlap here would be as improbable as for a kettle of water placed on ice to boil.)

It follows then that once registration has occurred, the packets (or channels) not containing all the relevant particles (including those constituting the registration device) will indeed be permanently inactive or physically ineffective. From this, we can see that there is no need to introduce a "collapse" of the wave function in a measurement. For, because of the very behaviour of all the relevant particles and of the wave functions, the "observed system" will from now on act entirely as if it were in the quantum state, $\psi_n(x)$. This state will correspond to the channels determined by the wave function of all the constituents of the registration device, that are actually occupied by the particles constituting this device. So it follows that everything will from this point on take place as if the wave function had "collapsed" to the actual result, without the need for any such collapse ever to occur.

5. The Classical Limit - Manifest And The Subtle

The above discussion shows the general lines along which one can understand quantum processes as taking place in an independent reality. To understand the essential nature and structure of this reality, however, it is necessary to go more carefully into the approach of the theory to the classical limit.

As we have already seen, whenever the quantum potential, Q, can be neglected, the equations (2) and (7) for S will give rise to the classical Hamilton-Jacobi equation

$$\frac{\partial S}{\partial t} + \sum_i \frac{(\nabla_i S)^2}{2m} + V(X_0, \ldots X_i, \ldots) = 0$$

In reference (1), it is shown by a more detailed analysis that the quantum potential actually will be unimportant under typical conditions that generally prevail in the ordinary macroscopic domain of experience.

The crucial point here is then that in the large-scale limit, the wave function generally no longer affects the motion of the particles in a significant way. The system of particles therefore follows a (Newtonian) law of motion, which refers only to the particle properties themselves. All the new properties implied by the quantum theory, including active information, non-locality, and dependence of the forces between the particles on the over-all quantum state, can now be left out of account. The large-scale limit of the "quantum world" is thus essentially a kind of classical "world on its own". To use Bohr's own mode of description, this is just the ordinary public world of common sense, "refined where necessary, to an expression in terms of the laws of classical mechanics".

To draw attention to this key feature more sharply, I shall call the above domain the <u>manifest world</u>. According to its Latin root, "manifest" means "what can be held in the hand". Literally, it is what is _tangible_. It is what we can _touch_ with our hands and _see_ with our eyes (with so many quanta that quantum properties are unimportant in connecting our senses with the objects in this "manifest world").

In a very accurate description, the world is of course basically an inseparable whole. But at the ordinary large scale level, which concerns the "manifest world" that we sense and experience, everything behaves essentially as if it were constituted of separately existent objects interacting according to classical laws.

To this manifest world, I contrast what I call the _subtle_ world of the wave function. The word "subtle" means _intangible_. Clearly the wave function, which is of the nature of information, cannot be directly "touched" by the senses. Rather it is inferred from the _phenomena_ that can be observed in the manifest world, which arise in our act of touching and seeing, (as extended with our scientific instruments).

The subtle world is not however permanently confined to inferences drawn about a domain of a purely abstract information. Rather, because it operates as _active_ information, it is able to reveal itself in the manifest world. A quantum mechanical measurement is a particular case of a process in which this takes place. Consider, for example, the measurement of angular momentum of an atom by a Stern-Gerlach experiment. The first stage, in which wave packets are separated may be considered as a process of making the distances between them large enough to be measured by methods that are ordinarily applicable in the manifest world. The second stage of registration on detectors then amplifies or magnifies the energy release to a point at which the results in the manifest world can readily be perceived by ordinary means.

We have not as yet gone into detail as to how such a detector may function. Generally, it requires a macroscopic source of energy, in a situation of instability, such that micro-processes can result in a release of energy great enough to show up at the ordinary level of experience. For example, in the Stern-Gerlach experiment, the atom could enter an ionization chamber, containing a gas subject to a strong electrostatic potential. If an atom enters this chamber, it will release ions and electrons. These latter will accelerate in the potential until they too can ionize more atoms, and thus a cascade is initiated. This cascade can grow to such proportions that the variables describing the mean electric current can be treated classically (i.e., through the W.K.B. approximation). At this point, the detection process will have produced an observable effect in the manifest world. And, of course, the differences in this effect for different spins are also great enough to be observed directly in this manfest world.

It is important to emphasize that we have in this way treated the measurement process as a single whole, without any break or "cut" between classical and quantum mechanical levels. In principle, it is thus possible in detail to follow the process by which the subtle quantum level is gradually able to manifest itself. This is in contrast to what happens in conventionally accepted interpretations (e.g., those of Bohr and von Neumann) which begin by assuming a basic distinction between the classical level and what may be called a "quantum reality" that might underlie it, with no possibility of following the connection between them in detail. In the causal interpretation however, there is only one reality. This contains a new kind of distinction between a "subtle world" of the wave function and the "manifest world" of the classical limit. It is important to emphasize that this distinction is not arbitrarily imposed from outside, (e.g., by assumption) but that it has been demonstrated to follow from the over-all theory itself.

Of course, we have so far still been treating the process of manifestation in terms of measurements and observations. However, in reference (1), it is shown how processes spontaneously taking place, such as transitions and ionization, can be treated, without any reference whatsoever to a measuring apparatus. Measurement is indeed just a special case of the general quantum process, which is so arranged that inferences can readily be made about the subtle world of the wave functions on the basis of observations made in the manifest world. Indeed, as Bell[11] has suggested, at a quantum level of accuracy, the very word "measurement" gives a misleading notion of what is happening. In principle a measurement ought to lead to a result enabling us to attribute some property to the measured system that it has independently of the process of being measured. But quantum mechanically, what is called a measurement is actually a thoroughgoing transformation, in which the final result is actually produced, not only at the manifest level, but also at the subtle level (in the sense, for example, that the final wave function of the observed system is transformed into an eigenfunction of the operator that is being "measured"). Bell suggests that at the quantum level, it would be better to talk of an experiment, rather than of a measurement or an observation. And I would add that in this experiment, the object under investigation realizes certain potentialities, according to the experimental arrangement with which it interacts.

One may usefully call attention here to the activity of human beings, which provide what is, in many ways, a good analogy to what happens in quantum processes. For a human being, the body is what is manifest or tangible. But human beings do not generally interact by mechanical forces, i.e., by pushing and pulling on each other. In general, each person is guided by information (rather as the electron is, but of course in a much more complex way). If there are many people together, each may be guided by his own pool of information, or they may be operating from a common pool (analogous to quantum particles with factorizable or non-factorizable wave functions, respectively). The same human being, acting under different conditions, will be guided by different information, and thus realise very different potentialities for action.

In all these cases, the energy comes from the body, but the form given to this energy comes from information existing at a more subtle level (in this case, in the brain and nervous system). Evidently, we could never understand the behaviour of human beings without reference to this more subtle level.

The causal interpretation of the quantum theory can now be seen in principle to remove the whole problem of the observer and the observed from the quantum theory. For the subtle quantum level manifests in the publicly accessible "classical world", which is, as we have seen, itself an implication of the theory as a whole. Human bodies and their sense organs will of course function as parts of this world, because they too follow the laws of classical physics, to an adequate degree of presentation, as they interact within the rest of the manifest world. But then there is a deeper subtle level of experience, perception and consciousness. This deeper level, which may be called "mind", is mediated with the manifest world by the body and by various material processes in the more general context of the manifest world as a whole, in which the quantum of action and all that goes with it can be neglected Therefore, no "cut" is needed between the human being and the rest of the world, any more than is required in classical mechanics.

Although all aspects of the manifest world can be regarded as separately existent, there is nevertheless an effectively continuous flow of energy of interaction between them. The observer interacts with the whole manifest world through this flow, and so, once the experimental arrangement has been set up, there is no need to consider his essential participation in what happens. In agreement with Bohr, however, we may call attention to the undivided wholeness of the object of the experiment and the total experimental arrangement. But in our approach, it is because of the non-factorizable nature of the wave function during the period of interaction that the process is characterized by indivisible wholeness. Nevertheless, in agreement with Bohr[12] the observer himself is clearly seen to be essentially detached from the process of the experiment.

6. Summary and Outlook

Within this article, what has been emphasized is that the causal interpretation implies a distinction between subtle and manifest levels. This distinction is a consequence of the theory itself, and is not the result of an arbitrary imposition. The human observer is part of the whole (although his mental qualities may imply further laws going beyond those of the current quantum theory).

Within this basically ontological approach, knowledge is treated, not as a separate domain, but rather, as a part of the over-all reality. So epistemology plays a secondary part in relation to ontology. The role of epistemology is illustrated with the aid of the diagram in Fig. 1

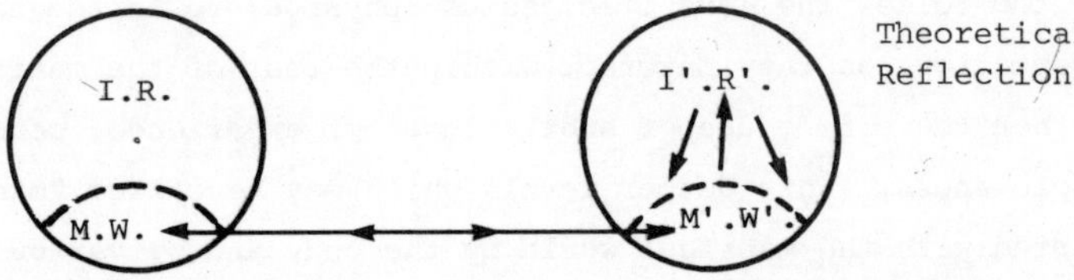

Fig. 1 Relationship Between Independent Reality and its Theoretical Reflection.

The totality of independent reality (I.R.) is basically unknown, but it contains inseparably within it a manifest world (M.W.) that all can observe. We then form a proposed theoretical reflection (I'.R'.), of this reality, which in turn contains a proposed theoretical reflection (M'.W'.) of the manifest world. The theoretical reflection is always at best an abstraction and a simplification of what is basically an unknown reality as a whole.

In an experiment, we observe the phenomena appearing in the manifest world. Our next step is to try to find a conceptual counterpart of these in our theoretical reflection of the manifest world. If we can do this, we then use the theory as a whole to infer a theoretical reflection of the over-all independent reality. This in turn implies further consequences for the theoretical reflection of the manifest world. The test of the theory is to see whether these are found in the actual manifest world.

This approach should be contrasted with most of the current interpretations, especially that of Bohr, who takes epistemology as basic to the meaning of the quantum theory, and who prefers to say nothing at all about ontology, especially about an independent quantum reality, that would be behind the observed phenomena.

REFERENCES

1. This paper is based on a longer article, in which a more extensive treatment of these questions is given. See Bohm, D., and Hiley, B.J., Physics Reports 144, 323 (1987).

2. Bohm, D., Phys. Rev. 85, 166, 180 (1952).

3. Philippidis, C., Dewdney, C. and Hiley, B.J., Nuovo Cimento B52, 15 (1975).

4. Kadanoff, L.P., Götze, W., Hambler, D., Palciaustos, V.V., Réyl, M., Swist, J., Aspres, D. and Kane, J., Rev. Mod. Phys. 30, 395 (1967).

5. Bohm, D., Phys. Rev. 89, 458 (1953).

6. Bohm, D. and Vigier, J.P., Phys. Rev. 96, 208 (1954).

7. Nelson, E., Phys. Rev. 150, 1079 (1966).

8. Bohr, N., Essays on Atomic Physics and Human Knowledge, Wiley, New York (1963).

9. Neils Bohr, A Centenary Volume, eds. French, A.P. and Kennedy, P.T., Harvard University Press, Cambridge, Mass. (1985) p. 153.

10. Bohm, D. and Hiley, B.J., Found. Phys. 14, 255 (1984).

11. Bell, J.S., Epistemological Lett. 9, 11 (1976).

12. Bohr, N., Atomic Physics and Human Knowledge, Science Editions, New York (1961), p. 67.

13. Bohm, D. and Hiley, B.J., Found. Phys. 5, 93 (1975).

UNSHARP REALITY AND THE QUESTION OF QUANTUM SYSTEMS

PAUL BUSCH
Institut für Theoretische Physik
Universität zu Köln
D-5000 Köln, FRG

ABSTRACT

The conception of unsharp quantum reality is reviewed and rederived by explicating conditions for the constitution of physical systems. The question of ascribing properties and states to individual systems is investigated within the convexity approach to quantum physics. In quantum mechanics this approach leads to a generalized notion of observables, represented as positive-operator-valued ("POV") measures. Various quantum physical situations are studied in order to demonstrate the need for this generalization. It is shown that certain POV measures represent *unsharp* observables in the sense of the concept of unsharp reality.

1. INTRODUCTION

This lecture is concerned with an attempt to exploit the possibilities of a *realistic* interpretation of quantum theory. Despite the widespread claim that quantum mechanics deals only with measurement statistics or *ensembles* of systems, probably no physicist can in practice avoid thinking in terms of *individual* systems when discussing quantum experiments. There are various approaches aiming at a reconstruction of a realistically interpreted quantum theory. In particular, the abstract quantum language approach ("quantum logic"), as presented in this conference by P. Mittelstaedt[1)], incorporates in a natural way the famous Einstein-Podolsky-Rosen reality criterion as a necessary and sufficient condition for assuring the presence of an individual system possessing objective properties.

In the Hilbert space realization of this abstract quantum language, observables are represented as projection-operator-valued ("PV") measures. Here we shall have to study the generalized observables described as positive-operator-valued ("POV"), or *effect* measures. This generalization arises in a natural way in the *convexity approach* to quantum theory which emphasizes the *statistical* aspects of physical theories. Yet this framework allows us to formalize the conditions to be imposed on measurement processes so that the constitution of *individual* physical systems becomes possible (Section 3). Beforehand (Section 2) we shall point out the need for POV observables in various fields of quantum physics. The concept of unsharp reality developed in Section 3 will be applied to some typical quantum physical situations which can only be described in terms of proper (non-PV) POV observables. In conclusion (Section 5), the statistical convexity approach can be (partly) endowed with an unsharp reality interpretation which explains the applicability of the quantum logical (*sharp*) reality concept as an idealization.

2. THE NEED FOR POV OBSERVABLES

There are two types of arguments showing that POV observables are indispensable tools in quantum physics. First, new theoretical and experimental possibilities have been discovered which cannot be appropriately described in terms of PV measures; secondly, there are a number of consistency problems arising in the standard (PV) quantum language which can be eliminated by introducing POV observables.

2.1 Some Consistency Problems

We first mention the fact that no measurement can be done with absolute accuracy. Accordingly, the theoretical description of observables will generally be an idealized one. Such idealizations can be (at least partly) estimated by means of a theory of the measuring process. Such an analysis has been carried out in some depth for the Stern-Gerlach experiment[2], leading to an adequate description of measuring inaccuracies in terms of *effect* operators rather than projections.

There are further instances where precise measurements of PV observables are impossible for principal reasons. One kind of restriction follows from the existence of conservation laws.[3] In particular, according to the standard quantum theory of measurement, momentum conservation forbids sharp *predictable* localization measurements whereas "almost" predictable measurements of *unsharp* position observables are well possible.[4]

Yet another kind of problem arises in the case of continuous observables which have been shown to admit no (weakly) *repeatable* measurements.[5] This implies that particles cannot, strictly speaking, be constituted by means of their perhaps most basic property: (sharp) position.[6] Therefore the best one can achieve is *unsharp* reality.

2.2 New Theoretical and Experimental Possibilities

POV observables were discovered independently in various branches of quantum physics. Their first appearance seems to be in an attempt to solve the photon localization problem by Jauch and Piron in 1967:[7] while there exist no PV systems of imprimitivity for the photon (m=0,s=1),[8] a notion of *unsharp* photon localization is possible in the sense of a POV system of covariance.[9]

Another important achievement is the theory of irreducible representations of relativistic and nonrelativistic quantum mechanics on Hilbert spaces of square integrable phase space functions.[10] The corresponding systems of (Poincaré or Galilei) covariance give phase space POV observables representing joint position and momentum measurements.[11,12] This structure arises also in quantum optics, and the phase space POV measures can in fact be constructed from generalized *coherent states*. The measurement theory for those observables was developed in order to justify the the original Heisenberg interpretation of the uncertainty relation.[13,14] Within such a theory it became possible for the first time to outline a precise *quantum mechanical* explanation of *unsharp* path determination as it occurs in cloud chambers, for instance (cf. Section 4). Moreover, it appears that the theory of (the *continuous*) phase space measurements applies also to certain quantum optical *quantum non-demolition measurements* ("QND"):[15] although their definition and measurement are usually only described in terms of *discrete* quantities, the QND observables actually studied are *continuous* quantities closely connected with certain phase space POVs.

Finally, besides position and momentum, also *complementary* quantities like spin $\frac{1}{2}$ components, or path and interference observables, are simultaneously measurable if considered as *unsharp* (POV) observables. One such experiment was actually performed in Cologne[16,17] and is reported on this conference by A. Prieur.[18] Another example will be sketched in Section 4.

3. SYSTEM CONSTITUTION AND UNSHARP REALITY

The problem of system constitution will be investigated in two steps. First we sketch the convexity framework and try to specify that part of its structure that already is implied by preconditions of the possibility of observing individual systems. Next we postulate those properties which measurements must possess so that it becomes possible to interpret experimental data as referring to individual systems as their (unsharp) properties and states. Finally we indicate that the resulting conception of unsharp reality can be adequately applied to the extended Hilbert space quantum theory, including certain POV measures as *unsharp* observables.

3.1 Convexity Framework for Physical Systems

It is impossible within this lecture to give an adequate account of the different versions and motivations of the convexity approach. I mention the monographs of Davies[19], Gudder[20], Kraus[21] and Ludwig[22] where one finds references to the original work of Davies and Lewis, Ludwig, and Mielnik. I shall follow most closely the expositions of Busch and Lahti[23], Lahti and Bugajski, [24] and Ludwig. [22]

The convexity approach aims at a general description of physical experiments as sequences of preparation and registration procedures. Accordingly one introduces as primitive elements of the description sets $\mathcal{P}$ and $\mathcal{R}$ of *preparation* and *registration* procedures. In order to gain physical experience one must be able to prepare the "same" *situation* $p \in \mathcal{P}$ in a reproducible way and make repeated registrations of the *event* $r \in \mathcal{R}$. (We assume the sets $\mathcal{P}$ and $\mathcal{R}$ to be such that any registration procedure can be applied to any preparation.[22]) If some pair (p,r) is to refer to one system, there must exist a law-like connection between p and r, thus a probability $\mathbf{p}(p,r)$ which can be tested statistically by counting relative frequencies.[25] Next, for a realistic description one has to abstract from the sets $\mathcal{P}$ and $\mathcal{R}$ those structures that are independent of the particularities of the measuring devices. This corresponds to the formation of equivalence classes of preparations $[p]\equiv W$ and registrations $[r]\equiv E$; the sets of those classes are the set $\mathcal{S}$ of *states* and the set $\mathcal{E}$ of *effects:*

$$
\begin{aligned}
W &= [p] = \{ p' : \mathbf{p}(p',r) = \mathbf{p}(p,r) \text{ for all } r \in \mathcal{R} \}, \\
E &= [r] = \{ r' : \mathbf{p}(p,r') = \mathbf{p}(p,r) \text{ for all } p \in \mathcal{P} \}.
\end{aligned}
\tag{1}
$$

Then the probabilistic structure is transferred from $(\mathcal{P},\mathcal{R},\mathbf{p})$ to $(\mathcal{S},\mathcal{E},p)$ via

$$\mathbf{p}(p,r) \equiv p(\mathrm{W},\mathrm{E}) \equiv p_{\mathrm{W}}(\mathrm{E}) \equiv \mathrm{E}(\mathrm{W}). \tag{2}$$

If " mixing" of preparation procedures can be regarded as another preparation, the set $\mathcal{S}$ of states possesses a convex structure. But then, since different states can be distinguished by means of the effects, it follows that $\mathcal{S}$ can be identified in a canonical way as a convex subset of a linear space.[20)]In the following we take $\mathcal{V}$ as the vector space generated by the convex set $\mathcal{S}$ as its *base*. The *positive cone*

$$\mathcal{V}^{+} \equiv \{\lambda \mathrm{W} : \lambda \in \mathbb{R}^{+},\ \mathrm{W} \in \mathcal{S}\} \tag{3}$$

is interpreted as the set of (non-normalized) states. It induces a partial ordering on $\mathcal{V}$ via

$$\mathrm{W} \le \mathrm{W}' \ \text{ iff } \mathrm{W}' - \mathrm{W} \in \mathcal{V}^{+}. \tag{4}$$

Furthermore, $\mathcal{S}$ determines a unique strictly positive functional $\mathbf{e}$ on $\mathcal{V}$ such that $\mathcal{S} = \{\mathrm{W} \in \mathcal{V}^{+} : \mathbf{e}(\mathrm{W}) = 1\}$. The functional $\mathbf{e}$ is related to a seminorm on $\mathcal{V}$,

$$\|\mathrm{W}\|_{\mathbf{e}} \equiv \inf\{\lambda > 0 : \mathrm{W} \in \lambda \operatorname{conv}(\mathcal{S} \cup -\mathcal{S})\}, \tag{5}$$

which is characterized by its additivity on $\mathcal{V}^{+}$ and the condition $\|\mathrm{W}\|_{\mathbf{e}} = \mathbf{e}(\mathrm{W})$ for $\mathrm{W} \in \mathcal{V}^{+}$. This seminorm becomes a norm if $\mathcal{V}$ is completed by including *countable* mixtures in the set of states. One can give a physical interpretation to this norm in terms of the following quantities introduced by Gudder[26)]:

$$\begin{aligned} \sigma(\mathrm{W},\mathrm{W}') &= \inf\{0 \le \lambda \le 1 : (1-\lambda)\mathrm{W} + \lambda \mathrm{W}_1 = (1-\lambda)\mathrm{W}' + \lambda \mathrm{W}_2,\ \mathrm{W}_1, \mathrm{W}_2 \in \mathcal{S}\}, \\ \rho(\mathrm{W},\mathrm{W}') &= \sigma(\mathrm{W},\mathrm{W}')\,[1 - \sigma(\mathrm{W},\mathrm{W}')]^{-1}, \end{aligned} \tag{6}$$

which are semimetrics on any convex structure and are metrics on the convex set $\mathcal{S}$ if and only if $\|.\|_{\mathbf{e}}$ is a norm.[27)] More specifically, the *intrinsic* metric ρ induces the *intrinsic* norm

$$\|\mathrm{W}\| = \inf\{\lambda > 0 : \mathrm{W} = \lambda c \mathrm{W}_1 - \lambda d \mathrm{W}_2,\ 0 \le c,d \le 1,\ \mathrm{W}_1, \mathrm{W}_2 \in \mathcal{S}\},\ \rho(\mathrm{W},\mathrm{W}') = \|\mathrm{W} - \mathrm{W}'\| \tag{7}$$

which turns out to be equivalent to the base norm:

$$\|\mathrm{W}\| \le \|\mathrm{W}\|_{\mathbf{e}} \le 2\,\|\mathrm{W}\|. \tag{8}$$

Finally, assuming that $\mathcal{V}^+$ is a *closed* subset of $\mathcal{V}$, one has established the basic structure (C) of the convexity approach:

(C) *Classes of statistically equivalent preparation procedures - the states - are represented as elements of the norm closed generating cone $\mathcal{V}^+$ of the base normed real Banach space $(\mathcal{V},\mathcal{S})$ where $\mathcal{S}$ is the set of normalized states.*

We note that the convexity of $\mathcal{S}$ allows the distinction between *pure* states (extreme elements) and *mixed* states, which will be relevant later on. We shall assume that any mixed state can be prepared by mixing pure states. Then, if a mixed state W is produced in this way, it should represent a collection of mutually independent systems prepared in one of the pure states contributing to W. This implies the convex linearity of the probability functionals $W \longrightarrow E(W)$, $E \in \mathcal{E}$ and therefore also their linearity on $\mathcal{V}$. So the set $\mathcal{E} = \mathcal{E}(\mathcal{V})$ can be identified as a subset of the dual space $\mathcal{V}^*$ of $\mathcal{V}$. More specifically, $(\mathcal{E},\leq)$ is a *partially ordered* set with

$$\tilde{E} \leq E \quad \text{iff} \quad \tilde{E}(W) - E(W) \leq 0 \ \text{ for all } W \in \mathcal{V}^+, \tag{9}$$

so $\mathcal{E}$ is the *order unit interval* $[O,\mathbf{e}] \subset \mathcal{V}^*$.

The occurrence of $\mathbf{e}$ in $\mathcal{E}$ reflects the fact that an experiment generally establishes a *complete* alternative, i.e., a collection of potential registration outcomes (events) such that the probability for obtaining *any* event is equal to unity. Usually the measurement outcomes are described as elements of some Boolean σ-algebra Σ (of subsets of some measurable space (X,Σ)). Then to any $B \in \Sigma$ there corresponds an effect E(B), and the statistical nature of an experiment forces the map $B \longrightarrow E(B)$ to be a normalized *effect valued measure* which is called an *observable.* The so-called *simple* observables are those with ranges $\{E,E'\}$ where $E \longrightarrow E' \equiv \mathbf{e} - E$ denotes the *complementation* on $\mathcal{E}$ satisfying

$$(E')' = E, \quad \text{if } E \leq F \text{ then } F' \leq E'. \tag{10}$$

An important role in the context of system constitution is played by the linear, positive, contracting mappings of $\mathcal{V}$ into $\mathcal{V}$. This set of *operations* $\mathcal{O} = \mathcal{O}(\mathcal{V})$ serves to describe the *state changes* caused through a registration. A *system* is supposed to exist over a period of time during which it may cause an *event sequence.* Accordingly, if physical experiments deal with *systems,* then it must be possible to perform observations without destroying the observed system. In other words, there must exist non-destructive registration procedures -represented as *operations*- which may serve as preparations for further experiments. The linearity of the operations follows from the convexity of the set $\mathcal{S}$ of states and the assumption that the operation acts independently of the state. There is a one-to-one correspondence between the effects and classes of operations:

$$E \longrightarrow \Phi(E) = \{\Phi \in \mathcal{O} : E = e \circ \Phi\}. \tag{11}$$

The set $\Phi(E)$ is not empty since it contains at least the operation Φ_o: $W \longrightarrow \Phi_o(W) = E(W)\, W_o$ for some W_o in $\mathcal{S}$.

The collection of all operations corresponding to the outcomes $B \in \Sigma$ of a measurement constitutes an *instrument* $\mathcal{J} : \Sigma \longrightarrow \mathcal{O}(\mathcal{V})$, $B \longrightarrow \mathcal{J}(B)$ as an operation-valued *measure*, the associated observable being $\mathcal{A} = e \circ \mathcal{J}$.

To summarize, we have tried to show that the convexity framework essentially follows from the fact that certain event sequences refer to single isolated "systems" and thus are connected by probability laws.

Later on we shall concentrate on the Hilbert space realization of the convexity scheme where one identifies $\mathcal{V} = \mathcal{T}_s(\mathcal{H})$, the Banach space of selfadjoint trace class operators on a Hilbert space $\mathcal{H}$, and $\mathcal{S} = \mathcal{T}_s(\mathcal{H})_1^+$, the set of positive normalized trace class operators. The functional e here is the trace functional. The operations are the elements of $\mathcal{O} \equiv \mathcal{O}(\mathcal{H}) = L(\mathcal{T}_s(\mathcal{H}))_{\le 1}^+$, the semigroup of linear positive contractions on $\mathcal{V}$. The set of effects is $\mathcal{E} \equiv \mathcal{E}(\mathcal{H}) = [O,I] \subset L_s(\mathcal{H})$. Therefore the effect observables are also called positive-operator-valued measures; they arise as the natural (and complete) generalization of the standard observable concept which is restricted to the spectral measures. In Section 2 we have shown that this generalization is in fact needed in quantum physics.

It is an unsolved problem to give a step-by-step operational justification of the Hilbert space realization of the convexity scheme. One approach tries to identify a certain sublattice of the lattice of extreme effects with the lattice of Hilbert space projections $\mathcal{P}(\mathcal{H})$, starting by characterizing pure, ideal, first kind operations and requiring the validity of the projection postulate.[24)]

3.2 System Constitution and the Reality Criterion

We shall attempt to formalize the following preconception of a *physical system.* A system is a part of the "whole" of physical reality which, in spite of its connections with the rest of the world (its environment) may be recognized as a separate entity: it must be *observable* due to its influence on its environment, i.e., by producing sequences of *events.* As was argued above, these events must be connected at least by a probability law. Furthermore, in order to be able to "learn" about stable entities in the physical world, it must be possible to count relative frequencies of "similar" events in "similar" situations. This requires the existence of symmetries and finally is reflected in the description of elementary particles as irreducible representations of some *symmetry group,* thus introducing the distinction between *essential* (like mass, spin, charge) and *accidental* properties. In this way a base norm space $(\mathcal{V},\mathcal{S})$ will host a representation of the Galilei or Poincaré group by automorphisms of states and effects and thereby denote a certain system *type.* Strictly speaking, the above argument leads to space-time translation invariance only; the full kinematical group comes in through the requirement of *objectivity* in the sense of *observer-independence.*[1,28]

To observe an *individual* system, there must be event sequences that can be interpreted as *actual* (unsharp) properties of "the" system that produced the sequence. Properties will be represented by certain *effects.* According to the above description of a system, an *actual* property should be observable without changing the system. This is the content of the Einstein-Podolsky-Rosen reality criterion (cf. Ref. 1). Insofar as the conception of physical systems entails their characterization as carriers of potential or actual properties (*elements of (unsharp) reality*), this reality criterion turns out to be not only a *sufficient* but also a *necessary* condition for system constitution.

In order to formalize these ideas, a number of definitions is needed which are listed in Ref. 28. For convenience we recall the following. Let $\varepsilon \in [0,1]$ be a "small" number, $\Phi \in \mathcal{O}(\mathcal{V})$, $\Phi \neq O$, $E \in \mathcal{E}(\mathcal{V})$, $E \neq O$. Then we call Φ ε-*predictable* (ε-pred) if there exists a state $W \in \mathcal{S}$ such that $e \circ \Phi(W) \geq 1-\varepsilon$; the corresponding effect $e \circ \Phi$ is called ε-*actualizable* and is ε-*actual* in W. Φ is ε-*reproducible* with respect to E if E is ε-actualizable and $E(W) \geq 1-\varepsilon$ implies that $e(\Phi W) \neq 0$ and $E(\Phi W) \geq (1-\varepsilon) e(\Phi W)$ for all $W \in \mathcal{S}$. Such effects E are called ε-*reproducible* (ε-rep). Finally, Φ is ε-preparatory (for $\varepsilon = 0$ sometimes called *weakly repeatable*, cf. Refs. 5 and 19) with respect to some ε-actualizable E if $E(\Phi W) \geq (1-\varepsilon)\mathbf{e}(\Phi W)$ for all $W \in \mathcal{S}$. Such effects are called ε-preparable (ε-prep). For the following definition we note that the function on $\mathcal{S} \times \mathcal{S}$

$$(W,W') \longrightarrow \mathbf{d}(W,W') \equiv \sup\{ |F(W)-F(W')| : F \in \mathcal{E} \} \leq \|W-W'\|_{\mathbf{e}} \tag{12}$$

is a *distance* (since the effects completely *separate* different states) and as such makes the set $\mathcal{S}$ of states a *metric space*. The metric $\mathbf{d}$ was first introduced by Jauch et al[29] within the context of Hilbert space scattering theory. While Gudder's intrinsic metric is a concept which is most appropriately adapted to the convex structure, the above *statistical distance* (also called *natural metric*[26]) is directly related to the operational possibilities of *statistically* determining differences between pairs of preparations. For this reason we use the metric $\mathbf{d}$ in the following:

Definition 1: An effect E is an accidental (unsharp) property of some system of type $(\mathcal{V},\mathcal{S})$ if it satisfies the following conditions for some $\varepsilon \in [0,\frac{1}{2}]$:

(P_ε1) Both E and the complement E' are ε-actualizable.

(P_ε2) Both E and E' admit "almost" *non-disturbing* tests, that is:

(ND) $E(W) \geq 1-\varepsilon$ implies $E(\Phi W) \geq (1-\varepsilon)\mathbf{e}(\Phi W)$ and $\mathbf{d}(\Phi W/\mathbf{e}(\Phi W), W) \leq \mu(\varepsilon)$ for some "small" $\mu(\varepsilon) \geq 0$,

and similarly for E'.

We finally obtain the reformulation of the EPR condition as a

Criterion of unsharp reality:

There exists an individual system $\mathbf{S} = (W,E)$ of type $(\mathcal{V},\mathcal{S})$, possessing in state $W \in \mathcal{S}$ the (ε-)*real (unsharp) property* $E = E(\mathbf{S})$ if $E \in \mathcal{E}$ is an unsharp property and is ε-actual in W. Such ε-actual properties are also called ε-*real*, or *elements of unsharp reality*.

Setting $\varepsilon = 0 = \mu(\varepsilon)$ recovers part of the characterizations of *sharp* properties as they are formulated in terms of ideal, first kind measurements in the convexity approach to quantum logic.[24]

The above reality criterion does not touch the question whether the prepared *statistical* state W pertains to the *individual* system $\mathbf{S}$ as *its state* $W(\mathbf{S})$. This can only be ascertained if the system represents the "isolated" entity alluded to above: in an experimental situation W the presence of an isolated individual can be inferred from the statistics *alone* only if the prepared state is a *pure* state. Otherwise, if W is mixed, there will in general be many different decompositions of W into pure states (unless the set $\mathcal{S}$ is a simplex which it is only in the *classical* case). In such a case there is no way to decide from the statistics alone, without knowing the previous *history* (preceding preparations and registrations) of the system, whether the individual is actually in one of those ("admissible") pure states occurring as *components* in one of the decompositions of *W*, or whether W is the only adequate description of the individual system state. (We call a state P a *component* of W if $P \in \mathcal{C}(W) \equiv \{W' \in \mathcal{S} : \lambda W' \leq W \text{ for some } \lambda \in (0,1]\}$.) From quantum physics we have learned that, for instance in EPR-correlated systems, such an *ignorance interpretation of mixtures* as *Gemenge* is inadmissible. So *if* such a kind of preparation is known to be carried out on the system under investigation, *then* the mixed state W refers to the individual. Otherwise, only *pure*, or *almost pure states* *W* can be ascribed to an individual system.

Definition 2: A state $W \in \mathcal{S}$ is called (ε-) *almost pure* if it is "close" to a pure state $P \in Ex(\mathcal{S})$, i.e., $\mathbf{d}(W,P) \leq \varepsilon$ for some "small" positive $\varepsilon \in [0,\frac{1}{2})$.

If individual system states are prepared by means of measuring *properties*, there must exist (almost) non-disturbing, (ε-) preparatory operations Φ such that $\hat{W} \equiv \Phi W/\mathbf{e}(\Phi W)$ is "almost" independent of $W \in \mathcal{S}$ as long as $\mathbf{e}(\Phi W) \neq 0$:

$$\mathbf{d}(W,\hat{W}) \leq \varepsilon \quad \text{for some "small" } \varepsilon,\ 0 \leq \varepsilon \ll 1. \tag{13}$$

Due to the preparatory nature of Φ we have for $E \equiv \mathbf{e} \circ \Phi$: $E(\hat{W}) \geq 1-\varepsilon'$ for some small ε'. It is easy to show that this implies the existence of some *pure* state $P_o \in \mathcal{C}(W)$ such that $E(P_o) \geq 1-\varepsilon'$. Since Φ is almost non-disturbing, condition (ND) of Definition 1 yields $\mathbf{d}(P_o,\hat{P}_o) \leq \mu(\varepsilon')$. Taken together, these inequalities give:

$$\mathbf{d}(\hat{W},P_o) \leq \mathbf{d}(\hat{W},\hat{P}_o) + \mathbf{d}(\hat{P}_o,P_o) \leq \varepsilon + \mu(\varepsilon'). \tag{14}$$

Assuming that the sum of the small numbers ε, $\mu(\varepsilon')$ is small again, we conclude that for every $W \in \mathcal{S}$, $\hat{W}$ is *almost pure;* so we obtain the following result:

> ***The only states which can be prepared as states of isolated individual systems by means of measuring properties are almost pure states.***

It is clear that in the *sharp* limit the *pure* states are singled out. Analogous result have been obtained within the quantum logic approach[1,25] as well as in the algebraic quantum mechanics.[30]

3.3 Constitution Conditions in Hilbert Space Quantum Mechanics

It has been shown elsewhere[28] that any Hilbert space effect E with spectrum extending into both $[0,\varepsilon]$ and $[1-\varepsilon,1]$ is an ε-unsharp property in the sense of Definition 1. In particular, the generalized Lüders operation

$$W \longrightarrow \Phi_L W = E^{1/2} W E^{1/2} \tag{15}$$

is ε-reproducible with respect to E and non-disturbing with $\mu(\varepsilon)$ of order $\sqrt{\varepsilon}$. However, this operation is *not* ε-preparatory; but it is easy to construct ε-prep *tests* for E: for instance, the generalized Lüders operation corresponding to

$$F = \int_{1-\varepsilon}^{1} \lambda \, dE_\lambda \qquad (\text{where } E = \int_0^1 \lambda \, dE_\lambda) \tag{16}$$

has this property.

In the following we give a characterization of the *sharp* reality limit which turns out to single out the *extreme* effects. The following notions are found in the literature.[5,19] An operation Φ is *weakly repeatable* if $\mathbf{e}(\Phi\Phi W) = \mathbf{e}(\Phi W)$ for all $W \in \mathcal{S}$; Φ is *strongly repeatable* if $\Phi\Phi W = \Phi W$ for all $W \in \mathcal{S}$. Φ is *ideal* if $\mathbf{e}(\Phi W) = 1$ implies $\Phi W = W$ for all $W \in \mathcal{S}$. Note that weak repeatability and ideality imply strong repeatability. We have the following

Proposition 1: Let Φ be an operation on $\mathcal{U} = \mathcal{T}_s(\mathcal{H})$.

(1) Φ is weakly repeatable iff it is ε-prep for $\varepsilon = 0$.

(2) Φ is ideal iff it is non-disturbing for those W in which $E = \mathbf{e} \circ \Phi$ is ε-actual for $\varepsilon = 0$.

Proof: (1) is obvious; (2) follows from the fact that in the Hilbert space realization of the convexity scheme the statistical distance $\mathbf{d}$ induces a norm which is equivalent to the base (i.e., the trace) norm (see below).

In the following we shall understand operations as *completely positive* linear contracting mappings on $\mathcal{T}_s(\mathcal{H})$.[21] This assumption is physically well-justified since complete positivity is necessary and sufficient for the existence of a quantum mechanical unitary interaction model for the operation (cf. the "second representation theorem" of Kraus, Ref. 21.). It allows to prove a theorem which improves a result formulated in Ref. 19. There a technical non-degeneracy condition is needed that can be dropped here; further, we do not need to refer to the richer structure of instruments as is the case there.

Theorem 1: If an operation Φ (completely positive!) is weakly repeatable and ideal then $E = \mathbf{e} \circ \Phi$ is an extreme effect (a projection) and Φ is the (pure) Lüders operation, $\Phi W = \Phi_L W = E W E$.

Proof (sketch): The proof makes use of the first representation theorem of Kraus [21] which states that every completely positive operation can be expressed as a countable sum of pure operations:

$$\Phi W = \sum_{\lambda \in \Lambda} A_\lambda W A_\lambda^+ . \tag{17}$$

One first shows that weak repeatability implies the A_λ to be of the form

$$A_\lambda = \sum_k |\varphi_k\rangle\langle\psi_k^\lambda| , \tag{18}$$

where the φ_k are an orthonormal basis of the 1-eigenspace of $E = \mathbf{e} \circ \Phi$ and the ψ_k^λ are a collection of elements of $\mathcal{H}$ of norm less than unity such that

$$E = \sum_\lambda A_\lambda^+ A_\lambda = \sum_{\lambda,k} |\psi_k^\lambda\rangle\langle\psi_k^\lambda| . \tag{19}$$

Then applying the ideality easily leads to the statement that (17) is indeed the Lüders operation with E from (19) being a projection operator.

The remainder of this section is devoted to a characterization of the statistical distance $\mathbf{d}$ and of almost pure states. First we note that the statistical distance $\mathbf{d}$ in general induces a seminorm on $\mathcal{V}$, $\|W\|_{\mathcal{E}} = \sup\{|F(W)| : F \in \mathcal{E}\}$. We may define a family of seminorms $\|W\|_{\mathcal{A}}$ where $\mathcal{A} \subset \mathcal{E}$ is substituted for $\mathcal{E}$.

Proposition 2: Let $\mathcal{V} = \mathcal{T}_s(\mathcal{H})$, $\mathcal{S} = \mathcal{T}_s(\mathcal{H})_1^+$, $\mathcal{E} = [0,1]$, $\mathcal{A} \subset \mathcal{E}$.

(1) For $\mathcal{A} = R(A)$, the range of an observable (Σ, A), $\|.\|_{\mathcal{A}}$ is a norm iff A is informationally complete (i.e., for $W, W' \in \mathcal{S}$, $W = W'$ whenever $tr[W A(X)] = tr[W' A(X)]$ for all $X \in \Sigma$).

(2) $\|.\|_{\mathcal{E}}$ is a norm which coincides with the intrinsic norm $\|.\|$. It is therefore equivalent with the trace norm, $\frac{1}{2}\|W\|_{\mathbf{e}} \le \|W\|_{\mathcal{E}} \le \|W\|_{\mathbf{e}}$.

(3) $\|W\|_{\mathcal{E}} = \max\{\|W_+\|_{\mathbf{e}}, \|W_-\|_{\mathbf{e}}\} = \|W\|_{\mathcal{P}(\mathcal{H})}$, $\mathcal{P}(\mathcal{H}) = Ex(\mathcal{E})$ being the lattice of Hilbert space projections. Here W_+, W_- are the unique positive and negative parts of W determined by the spectral theorem, i.e., $W = W_+ - W_-$, $W_+, W_- \in \mathcal{S}$, $W_+ \cdot W_- = 0$.

(4) $\|W\|_{\mathcal{E}} = \|W\|_{\mathbf{e}}$ iff $W \ge 0$. $\|W\|_{\mathcal{E}} = \frac{1}{2}\|W\|_{\mathbf{e}}$ iff $tr[W] = 0$.

Proof: (1) is straightforward. Gudder[27)] determined the intrinsic norm to be of the form given in (3). That this is actually the same for $\|.\|_{\mathcal{E}}$ follows immediately from the definition, $\sup\{|\mathrm{tr}[W\cdot F]| : F\in\mathcal{E}\} = \sup\{|\mathrm{tr}[F\cdot(W_+ - W_-)]| : F\in\mathcal{E}\}$, by choosing F as the projection onto the closure of the range of W_+. (4) follows from (3).

Now we may list some properties of the statistical distance.

Proposition 3: Let $\mathcal{S} = \mathcal{T}_s(\mathcal{H})_1^+$, then $(\mathcal{S},\mathbf{d})$ is a complete metric space with **d** satisfying:

(1) $\mathbf{d}(W,W') = \|W-W'\|_{\mathcal{E}} = \frac{1}{2}\|W-W'\|_{\mathbf{e}}$.

(2) For $P, P' \in \mathrm{Ex}(\mathcal{S})$, $\mathbf{d}(P,P') = [1-\mathrm{tr}(P\cdot P')]^{1/2}$.

(3) $\mathbf{d}(W,W') \leq 1$, and $\mathbf{d}(W,W') = 1$ iff $W\cdot W' = 0$.

(4) For fixed W with largest eigenvalue λ_0, $\inf\{\mathbf{d}(W,P) : P\in \mathrm{Ex}(\mathcal{S})\} = \mathbf{d}(W,P_0)$ where $P_0 \in \mathrm{Ex}(\mathcal{S})$ is any pure state occurring in the spectral decomposition of W as an "eigenstate" for λ_0, i.e., $W\cdot P_0 = \lambda_0 P_0$. One may write $W = \lambda_0 P_0 + (1-\lambda_0)W'$ for some $W'\in\mathcal{S}$ such that $P_0 \cdot W' = 0$.

Proof: (1) results from Proposition 2 (4). (2) is obtained by calculating the spectrum of P-P'. (3) is proved by using the spectral decompositions of W and W', applying the triangle inequality and finally using (2): $\mathbf{d}(W,W') = \|\Sigma_i w_i P_i - \Sigma_k w'_k P'_k\|_{\mathcal{E}} = \|\Sigma_{i,k} w_i w'_k [P_i - P'_k]\|_{\mathcal{E}} \leq \Sigma_{i,k}\, w_i w'_k\, [1-\mathrm{tr}(P_i\cdot P'_k)]^{1/2} \leq 1 = \Sigma_i w_i \cdot \Sigma_k w'_k$. Equalities are obtained iff for all i,k, $P_i \cdot P'_k = 0$, thus $W\cdot W' = 0$. To prove (4), we first note that $\lambda_0 = 1$ trivially leads to $W = P_0$ and thus $\mathbf{d}(W,P_0) = 0$. Assuming $\lambda_0 \neq 1$, one verifies that $W - \lambda_0 P_0 \geq 0$; therefore $W' \equiv (1-\lambda_0)^{-1}\cdot [W-\lambda_0 P_0] \in \mathcal{S}$. This state satisfies $W'\cdot P_0 = 0$ since P_0 occurs in the spectral decomposition of W. Then $\mathbf{d}(W,P_0) = \|P_0 - W\|_{\mathcal{E}} = (1-\lambda_0)\|P_0 - W'\|_{\mathcal{E}} = 1-\lambda_0$, where the last equality follows from (3). Next, for any $P\in\mathrm{Ex}(\mathcal{S})$, $\mathbf{d}(W,P) = \sup\{|\mathrm{tr}[F\cdot W] - \mathrm{tr}[F\cdot P]| : F\in\mathcal{E}\} \geq 1-\mathrm{tr}[P\cdot W] \geq 1-\lambda_0 = \mathbf{d}(W,P_0)$, where we have chosen F = P. (Note: $\mathcal{S}$ is *complete* because $\mathcal{V}$ is a Banach space.)

We conclude this section with a characterization of almost pure states.

Theorem 2: The following are equivalent for $W \in \mathcal{S} = \mathcal{T}_s(\mathcal{H})_1^+$, $\varepsilon\in[0,\frac{1}{2}]$:

(i) W is almost pure, i.e., there exists a pure state $P\in\mathrm{Ex}(\mathcal{S})$ such that $\mathbf{d}(W,P) \leq \varepsilon$.

(ii) There exist $P\in\mathrm{Ex}(\mathcal{S})$, $W'\in\mathcal{S}$, $1-\varepsilon \leq \lambda \leq 1$ such that $W = \lambda P + (1-\lambda)W'$.

(iii) The largest eigenvalue λ_0 of W satisfies $\lambda_0 \geq 1-\varepsilon$.

Proof: The equivalence of (i) and (iii) is a direct consequence of Proposition 3, statement (4). The calculations in the proof of that statement also show that (iii) implies (ii). Finally, (ii) implies (i): $\mathbf{d}(W,P) = (1-\lambda)\,\mathbf{d}(P,W') \le 1-\lambda \le \varepsilon$, where we have used (3) of Proposition 3.

4. EXAMPLES: POV MEASURES AS UNSHARP OBSERVABLES

So far we have developed a concept of unsharp reality and have demonstrated its applicability within the Hilbert space quantum mechanics. Now we give some physically important examples of POV observables which represent elements of unsharp reality.

4.1 Phase Space Observables

Quantum mechanics forbids sharp localization in phase space. In particular, there are no non-negative phase space distribution functions which are linear in the states W and yield *sharp* position (Q) and momentum (P) as marginals. For a long time one had to be content with the Wigner pseudo distribution as a useful technical tool.[31]

On the other hand it is easy to formulate in a completely satisfactory way a covariant phase space localization concept by means of the following POV (for the sake of simplicity we assume one space dimension, so that $\Gamma = \mathbb{R}^2$; $\hbar = 1$):

$$\mathcal{A}: \mathcal{B}(\Gamma) \longrightarrow \mathcal{E}(\mathcal{H}),\ \ \Delta \longrightarrow \mathcal{A}(\Delta) \equiv (2\pi)^{-1} \int_\Delta W_{qp}\, dq\, dp\ , \tag{20}$$

$$W_{qp} = U_{qp} W_0 U_{qp}^+,\ W_0 = \sum_k w_k |\varphi_k\rangle\langle\varphi_k| \in \mathcal{T}_s(\mathcal{H})_1^+,\ U_{qp} = \exp(i[pQ - qP]).$$

The marginals are covariant position and momentum POV measures:

$$X \longrightarrow \mathcal{A}(X \times \mathbb{R}) = \chi_X * f(Q) \equiv Q_f(X),\ Y \longrightarrow \mathcal{A}(\mathbb{R} \times Y) = \chi_Y * g(P) \equiv P_g(Y),$$
$$f(q) \equiv W_0(q) = \sum_k w_k |\varphi_k(q)|^2\ ,\quad g(p) \equiv W_0(p) = \sum_k w_k |\tilde{\varphi}_k(p)|^2\ . \tag{21}$$

The convolution form of these POVs justifies speaking of *unsharp* observables as it shows that they are obtained by "smearing" the Q and P spectral measures with the "confidence" functions f and g. From the definitions of f, g as position and momentum distributions in the state W_0 it follows that the variances satisfy the uncertainty relation, that is, $\Delta f \cdot \Delta g \ge \hbar/2$ which must be interpreted as an *indeterminacy relation* for *individual* measurement outcomes.[13,14)]

A more rigorous motivation for interpreting $\mathcal{A}$, Q_f, and P_g as *unsharp* observables comes from the fact that for sufficiently large but *bounded* $\Delta \in \mathcal{B}(\Gamma)$, $X, Y \in \mathcal{B}(\mathbb{R})$ bounded and such that $\Delta \subset X \times Y$, the effects $\mathcal{A}(\Delta)$ and $Q_f(X)$ $(\geq \mathcal{A}(\Delta))$, $P_g(Y)$ $(\geq \mathcal{A}(\Delta))$ are *unsharp properties*: $\mathcal{A}(\Delta)$ is trace class with trace $\mu(\Delta)\cdot(2\pi)^{-1}$, and its largest eigenvalue is bounded by $\min\{1, (2\pi)^{-1}\cdot\mu(\Delta)\}$ and approaches unity with increasing $\mu(\Delta)$. Therefore, for some sufficiently "large" Δ satisfying at least $\mu(\Delta) > 2\pi h$, $\mathcal{A}(\Delta)$ is an unsharp property. Note that 1 cannot be an eigenvalue of $\mathcal{A}(\Delta)$ since this would force f and g to have bounded supports, which is excluded by the Paley-Wiener theorem.[32)]

The phase space measurements have been analyzed with respect to their preparatory features in Ref. 13. In that paper a measurement is studied which leads to a phase space *instrument* $\mathcal{I}_0$ that can be used for quasi-classical path determinations in the following sense: given a finite (bounded) $\Delta \in \mathcal{B}(\Gamma)$, there is a class of states $W = |\varphi\rangle\langle\varphi|$, $\varphi \in \mathcal{H}$, for which $\mathcal{I}_0(\Delta)$ acts as an almost non-disturbing, ε-repeatable and ε-preparatory operation. A necessary condition for this is

$$\Delta_\varphi Q \ll \Delta f\,, \quad \Delta_\varphi P \ll \Delta g\,, \quad \text{and therefore} \quad \Delta f \cdot \Delta g \gg h\,. \tag{22}$$

Thus applying this operation to such states W amounts to a phase space measurement which does not disturb the object and yields outcomes that allow an (unsharp) prediction of the trajectory. The measurement "inaccuracies" are "macroscopically" large in the sense of equation (22). This model may serve as a possible explanation of cloud chamber tracks.

There is another *preparatory* phase space instrument[12)] that is suited for *almost pure state* preparations:

$$W \longrightarrow \mathcal{I}(\Delta)W = (2\pi)^{-1}\int_\Delta T_{qp}\,\mathrm{tr}[WT_{qp}]\,dq\,dp. \tag{23}$$

Here T_{qp} are the pure states W_{qp} obtained as in equation (21) from a *pure* state $W_0 = T_0$. As is easily verified, for instance for a Gaussian state T_0, the operation $\mathcal{I}(\Delta_0)$ constitutes an ε-preparatory test for an unsharp phase space localization property $\mathcal{A}(\Delta)$ for any sufficiently large bounded Δ containing Δ_0. For $\Delta_0 \subset X \times Y$ such that $\mu(X) \ll \Delta f$ and $\mu(Y) \ll \Delta g$ so that f and g are practically constant on intervals of the size of X, Y, the states $W(\Delta_0) \equiv \mathcal{I}(\Delta_0)W/\mathrm{tr}[\mathcal{I}(\Delta_0)W]$ are almost pure states in the sense of being close to any $T_{q_0p_0}$

with $(q_0,p_0) \in \Delta_0$. (Use the statistical distance to show that $\mathbf{d}(W(\Delta_0),T_{q_0p_0}) \leq \sup\{\|T_{qp} - T_{q_0p_0}\|_e : (q,p) \in \Delta_0\}$ which can be made arbitrarily small by choosing a suitably small Δ_0.

4.2 Unsharp Spin Measurements

POVs can also be used to describe measurements which yield only *unsharp* outcomes due to some perhaps inevitable imperfections of the devices.[2,33] For instance, in the Stern-Gerlach experiment for a spin $\frac{1}{2}$ system, the passage through an inhomogeneous magnetic field establishes a coupling between the spin and the center of mass of the atom, leading to a state transformation

$$\Psi_0 = \varphi \otimes \xi_0 = (c_+\varphi_+ + c_-\varphi_-) \otimes \xi_0 \longrightarrow \Psi = c_+\varphi_+ \otimes \xi_+ + c_-\varphi_- \otimes \xi_- \tag{24}$$

with ξ_+ and ξ_- representing the "upper" and "lower" beams, respectively. Now the unsharpness comes in due to the fact that the states ξ_+ and ξ_- cannot be strictly separated, e.g. because of the spreading of wave packets. In particular, they are not (strictly) orthogonal. Measuring a localization observable $\mathbb{1} \otimes E^Q(X)$ (the position spectral measure) leaves the spin system in an outgoing (reduced) state $W(X)$ which is determined by the equation

$$\mathrm{tr}[a\cdot W(X)] = \mathrm{tr}[a \otimes E^Q(X)\cdot |\Psi\rangle\langle\Psi|] \qquad \text{for all } a \in L_s(\mathcal{H}), \tag{25}$$

that is, the process is summarized in the *operation*

$$\begin{aligned} W \longrightarrow \quad W(X) &= \int_X A_q\, W A_q^+\, dq\,, \qquad\qquad A_q = \textstyle\sum_k \xi_k(q) T_k \ \text{(a.e.)} \\ &= \textstyle\sum_k p_{kl}(X)\, T_k W T_l \end{aligned} \tag{26}$$

$$\text{where} \quad W = |\varphi\rangle\langle\varphi|, \quad T_k = |\varphi_k\rangle\langle\varphi_k|, \quad p_{kl}(X) = \int_X \xi_k(q)\,\xi_l^*(q)\, dq\,.$$

The measured effect is given by the relation $\mathrm{tr}[W(X)] = \mathrm{tr}[W\cdot E(X)]$:

$$E(X) = p_{++}(X)\,T_+ + p_{--}(X)\,T_-\,. \tag{27}$$

If we take $X \in \{\mathbb{R}_+,\mathbb{R}_-\}$ and assume $p_{kk}(\mathbb{R}_l) \approx 1$, $p_{kl}(\mathbb{R}_j) \ll 1$ $(k \neq l)$ then the $W_k = W(\mathbb{R}_k)/\mathrm{tr}[W(\mathbb{R}_k)]$ are (mixed) states "close" to the pure states T_k. More quantitatively, we may represent the positive operators of trace one on $\mathcal{H} = \mathbb{C}^2$ in the Poincaré sphere via

$$A = A_{\mathbf{a}} = \tfrac{1}{2}[\mathbb{1} + \mathbf{a}\cdot\boldsymbol{\sigma}]\ , \ \boldsymbol{\sigma} = \text{the Pauli operators},\ \mathbf{a} \in \mathbb{R}^3,\ \|\mathbf{a}\| \leq 1. \tag{28}$$

Then the statistical distance $\mathbf{d}$ on $\mathcal{S}$ coincides with the Euclidean metric on

$\mathbb{R}^3$ (up to a factor of 2):

$$\mathbf{d}(W_{\mathbf{n}}, W_{\mathbf{m}}) = \tfrac{1}{2}\| W_{\mathbf{n}} - W_{\mathbf{m}} \|_{\mathbf{e}} = \tfrac{1}{2}\| \mathbf{n} - \mathbf{m} \| . \tag{29}$$

In this way one can readily check that the operation (26) $W \longrightarrow W(\mathbb{R}_k)$ has an ε-preparatory influence at least (and only) on those states W for which the probability $\mathrm{tr}[E(\mathbb{R}_k)\cdot W]$ is not smaller than the "error" probability $p_{jj}(\mathbb{R}_k)$ $(j \neq k)$.

Instead of using a Stern-Gerlach device, one could imagine of producing a correlated state of the form (24) by coupling the spin $\frac{1}{2}$ system to another "two-dimensional" system and then measuring a PV observable A with eigenbasis $\{\phi_1, \phi_2\}$ and spectral projections P_1, P_2. As above the outgoing spin state is obtained from the relation $\mathrm{tr}[a \cdot W_k] = \mathrm{tr}[a \otimes P_k \cdot |\Psi\rangle\langle\Psi|]$:

$$\begin{aligned} W \longrightarrow \mathbf{W}_k &= A_k W A_k^{+} = \textstyle\sum_{jl} p_{j\,l}(k)\, T_j\, W T_l \quad (k \in \{1,2\},\ j,l \in \{+,-\}), \\ \text{where} \quad A_k &= \langle\phi_k|\xi_+\rangle T_+ + \langle\phi_k|\xi_-\rangle T_-, \quad p_{jl}(k) = \langle\xi_j|\phi_k\rangle\langle\phi_k|\xi_l\rangle \end{aligned} \tag{30}$$

The measured effect now is

$$E_k = A_k^{+} A_k = p_{++}(k)\, T_+ + p_{--}(k)\, T_- \tag{31}$$

If the $\langle\phi_k|\xi_j\rangle$ happen to be real, then the A_k are equal to $E_k^{1/2}$, so in that case the operation (30) is the generalized Lüders operation. In the general case we see that A_k commutes with E_k which implies that this operation (30) is ε-reproducible and, like (26), is ε-preparatory for incoming states W with reasonably large probabilities $\mathrm{tr}[E_k \cdot W]$ (e.g., $\geq \frac{1}{2}$).[28,34] For further quantitative discussions of the properties of these unsharp measurement operations, cf. Ref. 2. Here it may suffice to emphasize that such measurements are *not* just *inaccurate* ones; on the contrary, the unsharpness corresponds to an *inherent quantum indeterminacy* of the measurement outcomes.

The generalized Lüders measurements have also been employed in a treatment of the EPR experiment for unsharp spin observables presented on the preceding Joensuu conference two years ago.[34]

4.3 Unsharp Joint Spin Measurements

As we have already seen in the phase space example, non-commuting observables may be *coexistent*, that is, *jointly measurable* when treated as *unsharp* (POV) observables. We also mentioned the Mittelstaedt-Prieur-Schieder experiment[16,18] as an example for the case of path and interference observables; a theoretical description in terms of joint POV observables is given in Ref. 17. Another instructive way of using POVs for inference questions in the context of interferometer experiments was outlined by Schroeck.[35]

In Ref. 17 I have also proposed an experimental setup for simultaneously measuring non-commuting spin or polarization components (Fig. 1). In the sketched setup the following process is assumed to take place. The spin (state $\varphi = c\varphi_+ + e^{i\chi}\sqrt{(1-c^2)}\varphi_- \in \mathcal{H}_S = \mathbb{C}^2$, $c \in [0,1]$) is coupled to the path $\psi_0 \in \mathcal{H}_p = \mathbb{C}^2$ by means of a Nicol prism P so that

$$\Psi_0 = \varphi \otimes \psi_0 \longrightarrow \Psi = c\varphi_+ \otimes |s\rangle + e^{i(\chi+\delta)}\sqrt{(1-c^2)}\,\varphi_- \otimes |\bar{s}\rangle . \tag{32}$$

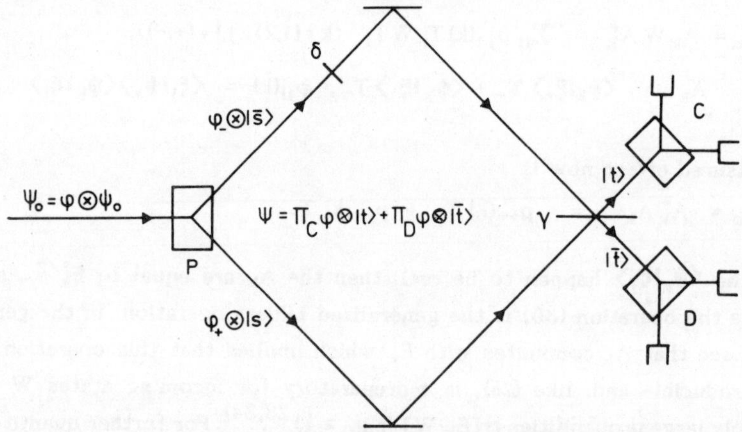

Fig. 1: Schematic setup for a joint measurement of spin components.[17]

Here the parameter δ stands for all phase shifts due to reflections. The split beam is recombined and split again at a mirror with transparency γ. Introducing orthogonal states $|t\rangle$, $|\bar{t}\rangle$ such that $|s\rangle = \sqrt{\gamma}\,|t\rangle + \sqrt{(1-\gamma)}\,|\bar{t}\rangle$, $|\bar{s}\rangle = -\sqrt{(1-\gamma)}\,|t\rangle + \sqrt{\gamma}\,|\bar{t}\rangle$, we may write

$$\Psi = \Pi_C\,\varphi \otimes |t\rangle + \Pi_D\,\varphi \otimes |\bar{t}\rangle ,$$

$$\Pi_C = \sqrt{\gamma}\,|\varphi_+\rangle\langle\varphi_+| - e^{i\delta}\sqrt{(1-\gamma)}\,|\varphi_-\rangle\langle\varphi_-| , \quad \Pi_D = \sqrt{(1-\gamma)}\,|\varphi_+\rangle\langle\varphi_+| + e^{i\delta}\sqrt{\gamma}\,|\varphi_-\rangle\langle\varphi_-| . \tag{33}$$

Now one measures the effects $\tilde{C}_i \otimes |t\rangle\langle t|$ and $\tilde{D}_j \otimes |\bar{t}\rangle\langle\bar{t}|$ $(i, j \in \{+,-\})$ which are pairwise commuting projections. The measured effects are read off by abstracting from the path degree of freedom:

$$\begin{aligned} \langle\varphi|C_i|\varphi\rangle &= \langle\Psi|\tilde{C}_i \otimes |t\rangle\langle t|\,|\Psi\rangle = \langle\varphi|\Pi_C^+\,\tilde{C}_i\,\Pi_C|\varphi\rangle \\ \langle\varphi|D_j|\varphi\rangle &= \langle\Psi|\tilde{D}_j \otimes |\tilde{t}\rangle\langle\tilde{t}|\,|\Psi\rangle = \langle\varphi|\Pi_D^+\,\tilde{D}_j\,\Pi_D|\varphi\rangle \end{aligned} \tag{34}$$

The four effects C_i, D_j constitute the range of a discrete, four-valued observable on the spin Hilbert space. It entails three two-valued POV observables as its marginals which are determined by the three effects

$$E_1 = C_+ + D_+ \,, \quad E_2 = C_+ + D_- \,, \quad E_3 = C_+ + C_- . \tag{35}$$

All of these effects turn out to be of the form (28) and thus represent unsharp spin properties; and the present scheme constitutes a joint measurement of these non-commuting observables.

5. CONCLUSION

Let us summarize the main results. We developed a conception of unsharp reality by explicating conditions for the constitution of individual physical systems. By this procedure we have shown that the *statistical, operationalistic* convexity approach to physical systems is open to a *realistic* interpretation. The notions of (unsharp) properties and (almost) pure individual system states were applied to quantum mechanical POV measures, thus proving them to represent *unsharp* observables. This, finally, explains the stability of the idealized quantum logical *sharp* reality concept against small unsharpness. The concluding examples made clear that the well-known quantum mechanical complementarity verdict can be "relaxed" from a "sharp" statement forbidding *sharp* joint measurements into a statement about the possibilities of *unsharp* joint measurement.

To conclude, I wish to point out that parts of the work reviewed in this lecture were carried out in collaborations with Pekka Lahti and Franklin Schroeck who also influenced other parts in innumerable discussions during the last years. The unsharp reality program was strongly inspired by and elaborated in the spirit of the Cologne quantum logic school established by P. Mittelstaedt and E.W. Stachow.

6. REFERENCES

1. Mittelstaedt, P., "Language and Reality in Quantum Physics", this volume.
2. Busch, P., and Schroeck, F.E., "On the Reality of Spin and Helicity", in preparation.
3. Wigner, E.P., "Die Messung quantenmechanischer Operatoren", Z. Physik **133**, 101-108 (1952).
4. Busch, P., "Momentum Conservation Forbids Sharp Localization", J. Phys.A: Math. Gen. **18**, 3351-3354 (1985).
5. Ozawa, M., "Quantum Measuring Processes of Continuous Observables", J. Math, Phys. **25**, 79-87 (1984).
6. Lahti, P.J., private communication (1986).
7. Jauch, J.M., and Piron, C., "Generalized Localizability", Helv. Phys. Acta **40**, 559-570 (1967).
8. Wightman, A.S., "On the Localizability of Quantum Mechanical Systems", Rev. Mod. Phys. **34**, 845-872 (1962).
9. Ali, S.T., and Emch, G.G., "Fuzzy Observables in Quantum Mechanics", J. Math. Phys. **15**, 176-183 (1974).
10. Prugovecki, E., *Stochastic Quantum Mechanics and Quantum Spacetime*, D. Reidel, Dordrecht-Boston-London (1984).
11. Schroeck, F.E., "A Model of a Quantum Mechanical Treatment of Measurement with a Physical Interpretation", J. Math. Phys. **22**, 2562-2572 (1981).
12. Schroeck, F.E., "The Transitions Among Classical Mechanics, Quantum Mechanics, and Stochastic Quantum Mechanics", Found. Phys. **12**, 825-841 (1982).
13. Busch, P., "Indeterminacy Relations and Simultaneous Measurements in Quantum Theory", Int. J. Theor. Phys. **24**, 63-92 (1985).
14. Lahti, P., "Complementarity and Uncertainty - Some Measurement and Information Theoretical Aspects", this volume.
15. Caves, C.M., "Quantum Nondemolition Measurements", in: *Quantum Optics, Experimental Gravity, and Measurement Theory*, eds. P. Meystre, M.O. Scully, Plenum, New York, 1983, pp. 567-626.
16. Mittelstaedt, P., Prieur, A., and Schieder, R., "Unsharp Particle-Wave Duality in a Photon Split Beam Experiment", Found. Phys. **17** (1987) 891-903.
17. Busch, P., "Some Realizable Joint Measurements of Complementary Observables", Found Phys. **17** (1987) 905-937.

18. Mittelstaedt, P., Prieur, A., and Schieder, R., "Unsharp Joint Measurement of Complementary Observables in a Photon Split Beam Experiment", this volume.
19. Davies, E.B., *Quantum Theory of Open Systems*, Academic, New York, 1976.
20. Gudder, S.P., *Stochastic Methods in Quantum Mechanics*, North Holland, New York, 1978.
21. Kraus, K., *States, Effects, and Operations*, Springer, Berlin, 1983.
22. Ludwig, G., *Foundations of Quantum Mechanics I*, Springer, Berlin, 1983.
23. Busch, P., and Lahti, P.J., " On Various Joint Measurements of Position and Momentum Observables in Quantum Theory", Phys. Rev. **D29**, 1634-1646 (1984).
24. Lahti, P.J., and Bugajski, S., "Fundamental Principles of Quantum Theory", Int. J. Theor. Phys. **24**, 1075-1104 (1985).
25. Stachow, E.W., "Structures of Quantum Language for Individual Systems", in: *Recent Developments in Quantum Logic*, eds. P. Mittelstaedt, E.W. Stachow, Bibliographisches Institut, Mannheim, 1985, pp.129-145.
26. Gudder, S., "State Automorphisms in Axiomatic Quantum Mechanics", Int. J. Theor. Phys. **7**, 205-211 (1973).
27. Gudder, S., "Convex Structures and Operational Quantum Mechanics", Commun. math. Phys. **29**, 249-264 (1973).
28. Busch, P., "Unsharp Reality and Joint Measurements for Spin Observables", Phys. Rev. **D33**, 2253-2261 (1986).
29. Jauch, J.M., Misra, B., and Gibson, A.G., "On the Asymptotic Condition of Scattering Theory", Helv. Phys. Acta **41**, 513-527 (1968).
30. Primas, H., "Contextual Quantum Objects and their Ontic Interpretation", this volume.
31. Hillery, M., O'Connell, R.F., Scully, M.O., and Wigner, E.P., "Distribution Functions in Physics: Fundamentals", Phys. Rep. **106**, 121-167 (1984).
32. Busch, P., and Lahti, P.J., "A Note on Quantum Theory, Complementarity, and Uncertainty", Philos. Sci. **52**, 64-77 (1985).
33. d'Espagnat, B., "Are the Quantum Rules Exact? The Case of the Imperfect Measurements", Found. Phys. **16**, 351-361 (1986).
34. Busch, P., "Elements of Unsharp Reality in the EPR Experiment", in: *Symposium on the Foundations of Modern Physics*, eds. P. Lahti, P. Mittelstaedt, World Sci. Publ. Co., Singapore, 1985, pp. 343-357.
35. Schroeck, F.E., "Experimental and Philosophical Foundations of the Formalism of Stochastic Quantum Mechanics", in: *The Physics of Phase Space*, eds. Y.S. Kim, W.W. Zachary, Springer, New York, 1987, pp. 426-428.

THE SPACE, TIME AND FIELD CONCEPTS IN WOLFGANG PAULI'S WORK

Charles P. Enz
Department of Theoretical Physics
University of Geneva
1211 Geneva 4, Switzerland

ABSTRACT

This paper discusses the philosophical background of some of the ideas which are prominent in Pauli's work. These are the field concept, the atomicity of charge, the conservation laws and the notions of vacuum and of space and time. Finally, the central role plaid by the unconscious is illustrated and argued to be at the root of Pauli's tradition-minded and history-oriented attitude in physics and in philosophy. No attempt is made to place these ideas into a biographical framework.

1. INTRODUCTION

Any attentive reader of Pauli's work [1] must be struck by the persistence with which some of his very early ideas reappear in many of his later publications. In essence they have to do with the field concept, the atomicity of electric charges as expressed by Sommerfeld's finestructure constant $\alpha \sim 1/137$ and with the notion of vacuum. Illustration of this fact which I have stressed before [2] is one of the aims of the present paper.

This remarkable constancy of opinion is yet another aspect of Pauli's thinking which has more often been praised for its critical quality and its rigour. (See, e.g. the Preface in Ref. 1, Vol. 1, p. V, the Foreword by Niels Bohr in Ref. 3, p. 1, the recollections of W. Heisenberg in Ref. 3, p. 40, and in Ref. 4, Vol. CII, p. 437. See also the contributions in Ref. 5.) It reflects a certain conservativism which is also evidenced in the well-known attribute of "living conscience of theoretical physics" mentioned in the Preface to Ref. 1. Indeed, as Pauli himself observed in retrospect, he was more a classicist than a revolutionary (see the Preface in Ref. 6, Vol. 1, p. xxiv). But according to a saying by Heisenberg [7] "Only a true conservative can be a true revolutionary". This sensitivity to tradition had its origin in Pauli's broad cultural background which he had acquired in his highschool years. However, I hope to show at the end of this paper that there was also a deeper psychological root.

Yet, as Heisenberg himself emphasizes in his recollections in Ref. 3, Pauli also plaid an essential part in the revolutionary development of quantum mechanics. The reason why he did not get full credit for his contributions is explained by Heisenberg in his Trieste Evening Lecture of June 1968 "Theory, criticism and a philosophy" reproduced in his collected works (Ref. 4, Vol. CII, p. 437). There Heisenberg says of Pauli :

> *"He tried first of all to be inspired by the experiments and to see in a kind of intuitive way how things are connected, and at the same time he tried to rationalize his intuitions and to find a rigorous mathematical scheme so that he really could prove everything what he said. Now this is I think, simply too much, and therefore Pauli has through his whole life published much less than he could have published if he had abandoned one of these two postulates".*

This sympathetic characterization of Pauli's working method points to the necessity of using other sources, mainly his extensive correspondence [8], to grasp his ideas and to see his real contributions, particularly to quantum mechanics.

Unfortunately, only 3 letters by Pauli to Heisenberg from the twenties have survived (Ref. 8, Vol. I, letters [86], [118] and [143]). But these letters as well as those by Heisenberg to Pauli, of which there are 64 from this period, and the correspondence with Bohr and others paint a vivid picture of Pauli's intense participation. Thus, for example, Pauli's letter [143] to Heisenberg of 19 October 1926 gives a qualitative description of the uncertainty principle (l.c., p. 347, translation):

> *"One may view the world with the p-eye and one may view if with the q-eye but if one opens both eyes simultaneously then one gets crazy."*

And his role as the uncompromising critic of Bohr's formulation of the Copenhagen interpretation - which Bohr first exposed in his famous Como-lecture of 16 September 1927 - is contained in the letters [169], [172], [177], [180] by Bohr to Pauli as well as in Pauli's letters [173], [179] to Bohr.

Pauli's broad cultural background mentioned above naturally lead him into truely philosophical questions which in his later years he formulated in deep and beautiful essays [9]. Some of these questions such as the notion of space, the quality of time and the analogy between an observer in quantum theory and in psychology, the role of the unconscious in creative thinking will be mentioned in this paper.

2. THE FIELD CONCEPT AND POSITIVISM

On 3 November 1919, at age 19, Pauli submitted his third paper for publication from Sommerfeld's institute in Munich and at the same time became a member of the German Physical Society (see Ref. 6, Vol. 1, footnote 643, p. 382). This paper dealt with "the perihelion motion of Mercury and the deflection of light rays in Weyl's theory of gravitation" [10]. The only passage of interest here is the last paragraph whose importance I had stressed before (Ref. 2, p. 767). It is quoted here in the translation by Mehra (Ref. 6, Vol. 1, p. 382) :

> *"In Weyl's theory one operates constantly with the field strength in the interior of the electron. For the physicist, however, this quantity is defined only as a force acting on a test-body; and, since there exist no smaller test-bodies than the electron itself, the concept of an electric field strength in a mathematical point appears to be an empty, meaningless fiction. In physics, one would like to stick to introducing only those quantities which can, in principle, be observed. Might we not altogether be on the wrong track in considering <u>continuum theories</u> for the field in the interior of the electron ?"*

The postulate expressed in the above passage, namely to introduce only observable quantities into physics - which, incidentally, predates quantum mechanics by six years - has been interpreted in Ref. 2 as due to the positivistic influence of Pauli's godfather Ernst Mach (See also Ref. 6, l.c.). While this is very plausible, the fact that, as will be shown in the next section, Pauli stuck to the above criticism to the end of his life, sheds some doubt on a purely positivistic interpretation.

Indeed, as discussed in Section 5 of Ref. 11, Pauli in his later years did "not belong to any philosophical direction which carries a name ending with the syllable ⟨⟨-ism⟩⟩"; his general direction, rather, was "to keep a certain median line between extreme directions" (translation from the essay of 1954, "Phänomen und physikalische Realität", Ref. 9, p. 93; see also p. 80). But already in his twenties he was quite openminded towards positivism which he considered "to be a perfectly consistent and noncontradictory world view. But of course, it is not the only possible one" (translation from a postcard of 1922 to M. Schlick, Ref. 8, Vol. II, p. 692).

In 1925, probably influenced by Pauli (see the Introduction), Heisenberg also made use of the above postulate. The result was the breakthrough of his famous paper "Ueber quantenmechanische Umdeutung kinematischer und mechanischer Beziehungen" [12] which in the abstract proposes an attempt "to gain foundations for a quantum theoretic mechanics based exclusively on relations between principally observable quantities" (translation of Ref. 12). The wording of this abstract is strikingly similar to the postulate of Pauli's third paper; it takes over the formulation in the letter of Heisenberg to Pauli of 24 June 1925 (Ref. 8, Vol. I, letter [93], p. 227) quoted by Heisenberg in Ref. 3, p. 43 and reproduced here in the translation of Ref. 2, p. 772 :

> *"The principle is : In the calculation of any quantities such as energy, frequency etc, only relations among essentially observable quantities should occur".*

This principle, Heisenberg believed, was the basis of Einstein's theory of relativity. However, when in sping 1926 he was invited to give a colloquium talk on the new quantum mechanics in Berlin, Einstein in a private discussion at his appartment told Heisenberg afterwards that this was nonsense, even if he may have used it himself, and that only theory decides what can be observed. When Heisenberg referred to Mach as the originator of this principle, Einstein developped a detailed criticism of Mach's "principle of economy of thought". - This at least is Heisenberg's recollection in Ref. 13, p. 91 ff.

Pais in his Einstein-biography [14] essentially confirms this attitude of Einstein. There on p. 283 Einstein is quoted to have said in 1922 :

> *"Mach's system [consists of] the study of relations which exist between experimental data; according to Mach, science is the totality of these relations. That is a bad point of view; in effect, what Mach made was a catalog and not a system. Mach was as good at mechanics as he was wretched at philosophy. This short-sighted view of science led him to reject the existence of atoms...." [15]*

A few lines later I.B. Cohen is quoted saying that just before his death

> *"Einstein said he had always believed that the invention of scientific concepts and the building of theories upon them was one of the creative properties of the human mind. His own view was thus opposed to Mach, because Mach assumed that the laws of sicence were only an economical way of describing a large collection of facts." [16]*

And in a footnote on the same page Pais writes :

> *"In his autobiographical sketch, Einstein mentioned that the critical reasoning required for his discovery of special relativity was decisively furthered by his reading of Mach's philosophical writings [17]."*

In conclusion it thus seems that both, Heisenberg and Pauli were more influenced by Einstein's theory of relativity than by Mach's principle of economy of thought in adopting the postulate of introducing only observable quantities into physics. As to Pauli's attitude expressed in the passage from his third paper [10] quoted at the beginning of this section, it seems to me that the correct philosophical interpretation is contained in Heisenberg's characterization of Pauli's working method quoted in the Introduction, namely that "he tried first of all to be inspired by the experiments". In other words, his was not so much an attitude of Machian positivism than a demand for strictly operational definitions without, however adopting a generalized operationalism.

3. THE DUALITY BETWEEN ELECTRIC FIELD AND ELEMENTARY CHARGE

The criticism of Weyl's theory expressed in the passage of Pauli's third paper [10] quoted in the last section concerned the contradiction between the continuity of the field used by Weyl and the atomicity of the electric charge used by Pauli in his operational definition of the field. According to Mehra (Ref. 6, Vol. 1, p. 382) Weyl essentially agreed with this criticism; in a letter to Pauli, dated 9 December 1919 he wrote : "Field physics, I believe really plays only the role of ⟨⟨world geometry⟩⟩." The same criticism is repeated again at the very end of Pauli's famous review of relativity theory in the "Encyklopädie der mathematischen Wissenschaften" [18]. There in footnote 394 Pauli also refers to the "Nauheimer Diskussion, Phys. Ztschr. 21 (1920), p. 650", the scientists' meeting at Bad Nauheim of September 1920 where Einstein is reported to have had a more cautious attitude (see Ref. 6, Vol. 1, p. 383).

With respect to the contradiction mentioned above, the beginning of the last section, no. 67, of the "Relativitätstheorie" is of even more fundamental interest. In the translation of Ref. 19 this passage of 1921 reads (p. 205) :

> *"It is the aim of all continuum theories to derive the atomic nature of electricity from the property that the differential equations expressing the physical laws have only a discrete number of solutions which are everywhere regular, static and spherically symmetric.... It is clear that differential equations which have this property must be of a particularly complicated structure. It seems to us that this complexity of the physical laws in itself already speaks against the continuum theories. For it should be required, from a physical point of view, that the existence of atomicity, in itself so simple and basic, should also be interpreted in a simple and elementary manner by theory and should not, so to speak, appear as a trick in analysis."*

As to the last sentence of this passage Pauli indeed had some idea. But he must have considered it too speculative to be printed. Thus, similarly as a decade later in the case of the neutrino-idea (see the famous open letter of Pauli to the "radioactive ladies and gentlemen" and his letter to Oskar Klein, both from December 1930, Ref. 8, Vol. II, nos [259] and [261]), Pauli expressed this idea only in a letter to A.S. Eddington dated 20 September 1923 (Ref. 8, Vol. I, no [45], p. 117); it reads in translation :

> *"Much more satisfactory would it be if measurement of the field strength could be based solely on enumerations [Abzählungen] without the use of mechanical notions. Instead of introducing discontinuous functions, it would perhaps be useful to complete the latter arbitrarily into continuous functions, and, as the laws of relativity theory are invariant against coordinate transformations, so the laws of the future quantum physics will perhaps also be invariant against modifications of the field functions themselves, if only the latter leave invariant certain outcomes of enumeration.... Because of the position of the definition of the electromagnetic field strength within the classical theory relative to mechanics I believe that the same detachment of the field notion from mechanics, with simultaneous introduction of some kind of atomism (countable sets instead of continuum), will bring the elucidation of the quantum facts and of the electrical elementary particles."*

The optimism expressed in the last sentence, unfortunately, did not come true. As I have discussed at the end of Ref. 2 (see also my contribution II, 11 in Ref. 5) the atomicity of the electric charge became for Pauli a test for the success of any theory of elementary particles and in the form of Sommerfeld's finestructure constant $\alpha \cong 1/137$ even took magic-symbolic significance (see also Ref. 11, Sec. 5). In addition to the passages mentioned in Ref. 2, this question also appears in Pauli's essay of 1949 (1955) "Einsteins Beitrag zur Quantentheorie" (Ref. 9, p. 62), in his talk of 1954 "Matter" (Ref. 9, p. 5) and his essay of 1957 "Phänomen und physikalische Realität" (Ref. 9, p. 101).

The connection with the divergence problems of quantum field theory is explicitly mentioned in the famous Pauli-Weisskopf paper on the scalar relativistic wave equation [20] where on p. 713 the following is said about the theory developed in this paper (translation):

> *"Incidentally, it also leads to an infinite selfenergy, not only for electrical particles but also to an infinite material selfenergy of light quanta. A further progress in these questions could therefore probably only be expected through a theoretical understanding of the numerical value of Sommerfeld's finestructure constant."*

The reason for this connection between the atomicity of the electric (and any other) charge with the field-theoretic divergence problem again lies in the continuity of the field concept. Pauli, in what is probably his last statement on this problem, writes in the essay "Albert Einstein in der Entwicklung der Physik", first published in the "Neue Zürcher Zeitung" of 12 January 1958 (Ref. 9, p. 90, quoted in the translation of Ref. 2, p. 791),

> *"The relation of the applicability of the ordinary space-time concept in the small with the properties of the smallest physical objects, the so-called ⟨⟨elementary⟩⟩ particles is not disclosed."*

4. FIELD THEORY AND QUANTIZATION

Quantum mechanics, although it changed fundamentally the perception of the physical world, did not resolve (Ref. 19, Supplementary Note 23, p. 225)

> *"the fundamental character of the duality (or, as one says since 1927, complementarity) between the measured field and the test body used as measuring instrument. This question has since been discussed by N. Bohr at the eight Solvay Conference of*

Physics in 1948".

Indeed, at that conference, Bohr pointed to a new complementarity of fields and particles and said that (Ref. 21, p. 260)

> *"In a future, more comprehensive theory of elementary particles, the relation between the elementary unit of electric charge and the universal quantum of action may play a more fundamental role than in present theories." [22]*

The reason why quantization has not brought the solution of the field-charge duality problem is because it did not carry the discretisation of the field-concept in the direction Pauli had hoped for in his letter to Eddington of 1923 : The field quanta are parametrized in the old continuous space-time ! And discretisation of space or, equivalently introduction of a momentum cut-off, does not help neither. Indeed, in relation with the principally unlimited precision of determining the position of an electron, Pauli observes in his "Wellenmechanik" of 1933 (Ref. 23, footnote on p. 246, translation) :

> *"Incidentally, from this one sees that an universal length certainly cannot exist, as is also evident from reasons of relativistic invariance."*

Pauli's last attempt to get out of this impasse was his unpublished work with Heisenberg on the nonlinear spinor equation with degenerate vacuum and indefinite metric written towards the end of 1957. For a while he hoped that the breakthrough could come there; indeed in a letter to L.D. Landau written in Berkeley 9 month before his death he wrote (quoted from contribution II, 11 in Ref. 5) :

> *"In any case it is clear to me, that the indefinite metric in Hilbert-space is the last attempt to save a theoretical description using field operators, which depend continuously on space and time."*

I am astonished that Pauli apparently never got the idea of the Planck length which is the intersection of the Compton wavelength $\hbar/mc$ with the Schwarzschild radius Gm/c^2 and therefore is the ultimate limit of localization [24]. Hence any spacial structure, such as strings, whose extention is limited to the Planck length ceases to contradict causality and leads to a natural discretisation of space-time.

Another difficulty with relativistic quantum mechanics concerns the well-known asymmetry between space and time. It is due to the fact that stability of a system implies the existence of a minimum in the energy spectrum, i.e., a ground state. But as Pauli observes in his "Wellenmechanik" of 1933 (Ref. 23, footnote 1, p. 140), this property prevents the introduction of time as an operator satisfying a canonical commutation relation with the Hamiltonian, analogous to the one between position and momentum. Since the ground state of the radiation field is the vacuum, that is empty space without quanta, Pauli argued that it should also have zero energy. However, historically since Planck, a zero-point energy of $h\nu/2$ had been associated with each field oscillator of frequency ν in the vacuum state. Otto Stern, with whom Pauli had extensively discussed this question during his Hamburg years (1923-1928), had argued that the zero-point energy of the phonons in a solid was real because it explained certain experiments on the vapour pressure of isotopes [25]). In a letter to me dated 21 January 1960 (reproduced in Ref. 25) Stern recalls these discussions :

> *"I for my part always tried to convert Pauli to the zero-point energy against which he had the gravest hesitations."*

Indeed, for the reasons outlined above Pauli insisted on the distinction between material oscillators and radiation; on p. 250 of Ref. 23 he argues (translation) :

> *"At this point it may be observed right away that it is more consistent not to introduce here a selfenergie of $\frac{1}{2}\hbar\nu_r$ per degree of freedom, in contradistinction to the material oscillator. Indeed, because of the infinite number of degrees of freedom the latter would give rise, on the one hand, to an infinitely large energy which, on the other hand, would be unobservable in principle because it is neither emitted nor absorbed or scattered and therefore cannot be contained within walls, and because, as is evident from experience, it does not produce a gravitational field."*

The zero-point energy is a divergence problem of the free Hamiltonian and may trivially be discarded by Wick-ordering the field operators. More serious are the well-known divergence problems caused by the interactions between quantum fields. In 1938 Pauli had made a model calculation with the Fermi-interaction of β-decay. It was published in Russian (German translation in Ref. 5, Section X, 4) and proposed to show the inappropriateness of higher-order perturbation calculations with the 4-fermion interaction. However, this problem has since become obsolete with the introduction of intermediate bosons.

Both intermediate-boson theories, the Yukawa theory of strong interactions and the electro-weak theory of Salam and Weinberg may be enlarged into gauge field theories. This, together with their "Grand Unification", has been accomplished only in recent years, based on the Yang-Mills theory of non-Abelian gauge fields [26]. What is not generally known is that Pauli simultaneously and independently of Yang had developped the same $SU(2)$-gauge field theory as Yang and Mills. However, Pauli had worked out the whole differential geometry of the theory and had made the connection with the 5-dimensional Kaluza-Klein formalism, features that are both missing in the Yang-Mills paper. Pauli never published this work. In Section 4 of Ref. 11 which contains more details concerning this manuscript, two possible reasons were given for this fact, first the uncertainty about the mass of the gauge field, and second its vector character which contradicted the known properties of the pion.

Pauli's last work on quantum field theory was his collaboration with Heisenberg on the nonlinear spinor equation mentioned above. The new ingredients that finally had seduced Pauli to collaborate in this endeavour of Heisenberg was, on the one hand, the inclusion of an $SU(2)$ symmetry also called "Pauli-group" which, as Gürsey had shown, could be written as isospin-symmetry. On the other hand, it was the idea of the indefinite metric which Pauli had introduced in the Lee model and which now took the form of a doubly degenerate vacuum (see contribution II, 11 in Ref. 5). This last idea is particularly noteworthy because it is the origin of spontaneous symmetry breaking and of Goldstone bosons which are omnipresent in modern physics. In the letter to Landau quoted above which Pauli had written a month before his withdrawal from the collaboration with Heisenberg he observes (quoted from Ref. 11, Sec. 4) :

> *"The problem of the connection between the vacuum-expectation values [of two or more factors of localized field operators] on the one hand and the Lagrangian on the other hand is a very fundamental one. The idea seemed to me seductive, that in an indefinite metric it may be possible, that the former could have a lesser symmetry (group-invariance) than the latter and that in this way one could reach an interpretation of the gradual breakdown of invariances by passing from the strong interactions to the electromagnetic ones and again from these to the weak interactions."*

5. CONSERVATION LAWS AND NUCLEAR STRUCTURE

Doubt about energy (and momentum) conservation have arisen twice during the development of quantum theory, first at the atomic level and later at the nuclear level. Already in 1923 Bohr had pointed to the incompatibility between the continuous nature of electrodynamics and the discreteness of atomic levels, and he had concluded that the only way to reconcile the two was to give up strict energy-momentum conservation in atomic radiation processes and to replace it by conservation in the statistical mean. (Note that this was before the Copenhagen interpretation !) Pauli and also Einstein were strongly opposed to this idea which later was proven wrong by the experiments of Geiger and Bothe and of Compton.

In 1930 the question of energy conservation arose from the continuous β-decay spectrum of RaE, and Bohr and Pauli reacted similarly as they had seven years before (see, e.g., Ref. 27). Of course, this problem had nothing to do directly with the one raised by Bohr in 1923 since it concerned nuclei which, in principle, could obey different physical laws (see the second footnote on p. 312 of Ref. 28). But still, the analogy between the two situations is obvious and it is gratifying that Pauli's view prevailed again. It may therefore be illuminating to give a closer look on Pauli's arguments which utlimately had led him to postulate the neutrino. These ideas are exposed most clearly and in great detail in a talk on "The conservation laws in the theory of relativity and in nuclear physics" that Pauli gave on 27 october 1937 in Moscow and which I.E. Tamm translated into Russian. This account is reproduced in German translation in Sec. X,3 of Ref. 5.

In the first part of this talk Pauli shows in an unconventional derivation that under quite weak assumptions about the dependence of energy E and momentum $\vec{p}$ on the velocity $\vec{u}$ of a point particle and assuming invariance of the respective conservation laws under an arbitrary boost $\vec{u} \to \vec{u}+\vec{v}$, the usual expressions for E and $\vec{p}$ are valid both in classical and in relativistic mechanics. More interesting, however, are the second and third parts.

Part 2 shows that both, charge conservation in electrodynamics and energy-momentum conservation in general relativity may be formulated in a closely analogous manner as the vanishing of the surface integral over the respective currents through an arbitrary closed surface. These integrals, incidentally, may be combined into a 5-dimensional theory (see also Pauli's essay "Raum, Zeit und Kausalität in der modernen Physik" of 1934, Ref. 9, p. 73). The importance of this surface-integral form of the conservation laws is taht no knowledge about the physics valid in the interior of the closed surface is needed. From this result Pauli then concludes in Part 3 that without any knowledge of nuclear physics the known validity of the physics involved at the atomic level guarantees the conservation laws at the nuclear level.

As to Bohr's idea of 1923, Pauli argues that it is hard to see how statistical conservation could hold at all without a conservation law, and he recalls the Compton effect which agrees with strict energy-momentum conservation. Thus, he argues, quantum theory is a complete and closed theory and then adds, probably to the address of the Soviet ideologists, that philosophy should refrain from making prescriptions upon physical theory. Twice in this talk Pauli emphasizes his belief that the analogy between conservation of energy and of electric charge possesses a deep meaning. And since charge conservation has never given rise to difficulties he concludes that this is another good reason to believe in energy conservation. This together with the collected experimental information had led him, he says, to postulate a new particle, the neutrino.

6. THE NOTIONS OF SPACE AND OF TIME IN HISTORICAL VIEW

Pauli's view of empty space was more radical than just a vacuum state with zero energy and zero particle number. Indeed, already in his letter to Eddington of 1923 mentioned earlier he writes (footnote on p. 117 of Ref. 8, Vol. I) what 32 years later he repeats in almost the same words in the closing address of the president to the Berne-congress on relativity (Ref. 29, p. 266, translation):

> "*That it is, for instance, logically possible to write down Maxwell's equations in the vacuum without introducing a charge, this bothers me; for if there were no charges, it were obviously not possible to measure this field.*"

There is a striking resemblance of this idea with Mach's principle, which claims that the inertial mass of a body should be due entirely to the matter contained in the universe; a material body in an empty universe should not show any inertia. The equations of general relativity do not satisfy what Einstein called Mach's principle. In his presidential address to the jubilee congress on relativity in Berne Pauli said (translation of Ref. 9, p. 79):

> "*According to these equations the existence of a non-vanishing G-field in a space-time world free of matter remains logically possible. And in so far as the G-field exists, space and time are not empty.*"

Thus Pauli's view about electrodynamics may be characterized as the hope for a Mach's principle.

The philosophical background of Mach's principle is to consider space as recipient of matter. Historically, this notion of space as an entity distinct and independent of matter is in line with Newton's idea of absolute space which Mach and Einstein later liberated of its abosluteness. But it is in opposition to the view held by Descartes who had identified space, that is extension, as the true nature of bodies [30]. Both, Pauli in his essay of 1955 "Die Wissenschaft und das abendländische Denken" (Ref. 9, p. 107-108) and in his correspondence with Markus Fierz and also Fierz in his paper "Ueber den Ursprung und die Bedeutung der Lehre Isaak Newtons vom absoluten Raum" [31] have described the fascinating history of the notion of space. In these writings (see also Ref. 32, particularly the footnote on p. 133) the notion of independent, absolute and infinite space is traced back to the Italien philosophers, mainly Ficino and Patrizzi, and further to the hermetic and the cabbalistic tranditions.

As to the notion of time, Pauli thought that quantum mechanics had modified it in the sense that the observer - which had been introduced into theoretical physics by Einstein in his discussion of moving clocks and sticks - has become an active player : Because of complementarity the observer has the freedom of choice between mutually exclusive experiments. With his decision he interferes with the quantum world but at the same time he has to sacrifice information. In his essay of 1954 "Wahrscheinlichkeit und Physik" Pauli says (Ref. 9, p. 21, translation) :

> *"Herewith the observation acquires the character of the <u>irrational, unique actuality</u> with a non-predictable result."*

In this attitude of choice and sacrifice (see the Jung birthday-essay quoted below, Ref. 9, p. 115) time acquires the creative quality of "becoming" which is fundamentally different from the automatic evolvement of the time in classical deterministic physics where there is no room for surprise.

In his essay "Die Wissenschaft und das abendlandische Denken" quoted before (Ref. 9, p. 110) and also in his Kepler article [32] Pauli compares this creative role of the observer with the experience of the redeeming transmutation [erlösende Wandlung] of the alchemist through the opus (see also Ref. 11, Sec. 5). In this description of the role of the quantum-physical observer Pauli emphasizes, however, that the psychic state of the observer does not enter into the physical <u>laws</u>. Indeed, in his essay of 1954 dedicated to the 80th birthday of C.G. Jung "Naturwissenschaftliche und erkenntnistheoretische Aspekte der Ideen vom Unbewussten" he writes (Ref. 9, p. 115, translation) :

> *"Once the observer has chosen his experimental setup, he has no influence anymore on the result of the measurement which, objectively registered, is available to everybody. Subjective properties of the observer or his psychic state enter the laws of nature of quantum mechanics no more than those of classical physics."*

Thus the relation of the quantum-physical observer to the surrounding world differs from the one of the alchemist in that (Ref. 9, p. 115, translation) :

> *"With the physical measurement the ⟨⟨gift of the sacrificer [Gabe des Opfernden]⟩⟩ is not a part of himself but a piece of the outside world, whereupon there is also no transmutation [Wandlung] of the observer. After the measurement the observed system remains, again separated from the observer, left to itself."*

In contrast to Bohr who in the mid-fifties was ready to retract from the strict Copenhagen interpretation by referring to a detached observer - a notion introduced by Einstein (see Ref. 9, p. 22) - in his lecture at the bicentenial celebration of Columbia University (Ref. 33, 1958 : p. 75, 1985 : p. 83), Pauli was convinced that, in the contrary, a future theory would connect the observer even more intimately to experiment. The reason for this conviction, most likely, is expressed in the following passage which is the continuation of the quotation at the beginning of this section (Ref. 29, p. 266, translation) :

> *"Therefore I was satisfied when quantum mechanics was created since I still hoped that the pair of opposites, testbody and field, could perhaps one day be formulated in analogy to the other pair of opposites, momentum and position."*

Another aspect of time that had occupied Pauli in the fifties is that of biological evolution. He agreed with "Jung's perception that the archetypes are an hereditary imprint of the ancestral lineage" (from the Jung birthday-essay, Ref. 9, p. 122, translation from Ref. 11, Sec. 5) which, according to Max Delbrück (Ref. 34, p. 118), is also the opinion of Konrad Lorenz who, however, replaces "archetype" by "Kantian a priori". In a more fundamental dispute with Max Delbrück (who in 1932 had declined an offer by Pauli to become his assistant, accepting instead a position in Lise Meitner's institute in Berlin), Pauli criticised the extreme Darwinist view of random mutations and selection of the fitest, writing in a letter to Delbrück of 4 February 1954 reproduced in Ref. 35 (here quoted from Ref. 11, Sec. 5) :

> *"Probably the solution is a <u>complex</u> one, and <u>besides</u> the holy chance there exist processes with a <u>directed goal</u> and <u>also</u> causal influences of the environment."*

7. THE ROLE OF THE UNCONSCIOUS

The finalistic touch of the "processes with a directed goal" mentioned at the end of the last section most probably again expresses a deep conviction of Pauli. At another point of the Jung birthday-essay (Ref. 9, p. 123) he mentions the birds of passage as example for inherited instinctive actions, an argument similar to Jung's interpretation of the archetypes as hereditary imprint mentioned above. This points to the unconscious.

In fact, from his essays, particularly the one for Jung's 80th birthday, one gains the conviction that for Pauli the notion of the unconscious as developed by Jung had a deep meaning. It is only before this background that Pauli's keen interest in dreems can be understood which goes back to 1930 when he first met Jung as a patient but which continued to the end of his life. Again on the basis of the unconscious he examined in his Kepler-article [32] the beginning of the scientific era in the 17th century, drawing a comparison between the pre-modern scientist Johannes Kepler and the medieval alchemist Robert Fludd. The polemic between these two contemporaries might still be of interest today, Pauli notes (Ref. 32, p. 162, translation) :

> *"An indication for this we see in particular also in the fact that the ⟨⟨quaternary⟩⟩ attitude of Fludd corresponds psychologically to a larger <u>completeness of experience</u> [Erleben] as compared to the ⟨⟨trinitarian⟩⟩ attitude of Kepler."*

This means that, according to Pauli, the split of the natural sciences from the other endeavours of the human mind which started in the 17th century, has had an empoverishing effect on modern western civilisation. In fact, the central theme of Pauli's essays is the hope that the new way of thinking introduced by quantum mechanics, and above all by Bohr's idea of complementarity, will lead to a richer world view unifying the sciences, the humanities and the religions.

Thus Pauli's hope for the future symbolically lay in the pythagorean holy number Four which, he observes in the Jung birthday-essay (Ref. 9, footnote 22, p. 128), is the number of nucleic acids forming the genetic code, but which is also the amount of quantum numbers Pauli had needed to formulate his exclusion principle ! As Jung observes in his synchronicity-article [36] which appeared under the same cover as Pauli's Kepler-article [32], the Fourth is the One to be added to the Three. Thus (Ref. 36, p. 99-101) time has to be added to the 3 spatial dimensions, synchronicity to space, time and causality and the virgin Mary to the Christian Trinity. In Pauli's case the spin quantum number had to be added to those of energy, angular momentum and its projection.

Consciousness was for Pauli like the portion of the psychic landscape which is momentarily illuminated by the rotating beam of a lighthouse but recedes into the darkness of the impersonal unconscious as soon as the light beam has passed. In this sense the psychological process of observation brings about a change in the **conscious** which in turn reacts back on the unconscious. Pauli draws a fascinating parallel of this process with an observation in physics, writing in his Jung birthday-essay (Ref. 9, p. 115, translation of Ref. 11, Sec. 5) :

> *"Since the unconscious is not quantitatively measurable and hence, is not mathematically describable, and since every extension of the consciousness (⟨⟨rendering conscious⟩⟩) must retroactively alter the unconscious, a ⟨⟨problem of observation⟩⟩ in relation with the unconscious has to be expected which does indeed show analogies to the one in atomic physics but which contains still considerably larger difficulties."*

In the same sense, finally, must be understood Pauli's motto of his later years (translation from the letter to M. Fierz of 13 October 1951 quoted in Ref. 37) :

> *"The still older is always the new",*

or, expressed differently in "Die Wissenschaft und das abendländische Denken" (Ref. 9, p. 110) :

> *"Was die Alten sungen, das zwitschern die Jungen."*

This motto, I believe, also explains the psychological reason for Pauli's tradition-minded, history-oriented attitude in physics and in philosophy, mentioned in the Introduction.

REFERENCES

[1]. **Wolfgang Pauli, Collected Scientific Papers** ed. R. Kronig and V.F. Weisskopf (Interscience, New York, 1964).

[2]. Enz, C.P., "W. Pauli's Scientific Work", in **The Physicist's Conception of Nature**, ed. J. Mehra (Reidel, Dordrecht-Holland, 1973), pp. 766-799.

[3]. **Theoretical Physics in the Twentieth Century. A Memorial Volume to Wolfgang Pauli**, ed. M. Fierz and V.F. Weisskopf (Interscience, New York, 1960).

[4]. **Werner Heisenberg, Gesammelte Werke**, ed. W. Blum, H.-P. Dürr and H. Rechenberg, Series C, Philosophical and Popular Writings (Piper, Munich, 1984).

[5]. **Wolfgang Pauli 1900-1958. "Das Gewissen der Physik"**, ed. C.P. Enz and K. v.Meyenn (Vieweg, Braunschweig, 1987), to be published.

[6]. Mehra, J. and Rechenberg, H., **The Historical Development of Quantum Theory** (Springer, New York, 1982), Vols. 1-4.

[7]. See C.F. von Weizsäcker, this volume.

[8]. **Wolfgang Pauli, Scientific Correspondence with Bohr, Einstein, Heisenberg a.o.**, Vol. I : 1919-1929, ed. A. Hermann, K. v.Meyenn and V.F. Weisskopf (Springer, New York, 1979), Vol. II : 1930-1939, ed. K. v. Meyenn et al. (Springer, Berlin, 1985).

[9]. Pauli, W., **Physik und Erkenntnistheorie** (Vieweg, Braunschweig, 1984).

[10]. Pauli, W., Verhandl. Deut. Physik. Ges. 21, 742-750 (1919).

[11]. Enz, C.P., "Wolfgang Pauli, Physicist and Philosopher", in **Symposium on the Foundations of Modern Physics**, ed. P. Lahti and P. Mittelstaedt (World Scientific, Singapore, 1985), pp. 241-255.

[12]. Heisenberg, W., Z. Phys. 33, 879-893 (1925).

[13]. Heisenberg, W., **Der Teil und das Ganze** (Piper, München, 1969); reprinted in Ref. 4, Vol. CIII, pp. 3-334; English translation by A.J. Pomerans, **Physics and Beyond** (Allen and Unwin, London, 1971).

[14]. Pais, A., **"Subtle is the Lord ...". The Science and the Life of Albert Einstein** (Oxford University Press, 1982).

[15]. Einstein, A., Bull. Soc. Fran. Phil. 22, 91 (1922); see also Nature 112, 253 (1923).

[16]. Cohen, I.B., Sci. Amer., July 1955, p. 69.

[17]. Einstein, A., in **Albert Einstein : Philosopher-Scientist**, ed. P. Schilpp (Tudor, New York, 1949), p. 21.

[18]. Pauli, W., **Relativitätstheorie**, Encykl. math. Wiss. (Teubner, Stuttgart, 1921), Bd. V, Art. 19. Reproduced with modernized symbols and with addition of the supplementary Notes translated into German from Ref. 19 by Klaus Schilling (Boringhieri, Torino, 1963).

[19]. Pauli, W., **Theory of Relativity**, "Translated from the German by **G. Field**. With Supplementary Notes by the Author" (Pergamon, London, 1958). Also : **Teoria della Relatività**, traduzione di Paolo Gulmanelli (Boringhieri, Torino, 1958).

[20]. Pauli, W. and Weisskopf, V., Helv. Phys. Acta 7, 709-731 (1934).

[21]. Mehra, J., **The Solvay Conferences on Physics. Aspects of the Development of Physics since 1911** (Reidel, Dordrecht-Holland, 1975).

[22]. Bohr, N., in **Les Particules Elémentaires, Rapports et Discussions du Huitième Conseil de Physique tenu à l'Université Libre de Bruxelles du 27 septembre au 2 octobre 1948** (R. Stoops, Brussels, 1950), pp. 376-380.

[23]. Pauli, W., **Die allgemeinen Prinzipien der Wellenmechanik**, Hdb. der Physik, eds H. Geiger and K. Scheel, 2. Aufl. (Springer, Berlin, 1933), Vol. 24, Part 1, pp. 83-272.

[24]. Enz, C.P., Epistemologia 10, 187-206 (1987).

[25]. Enz, C.P., "Is the zero-point energy real ?" in **Physical Reality and Mathematical Description**, eds C.P. Enz and J. Mehra (Reidel, Dordrecht-Holland, 1975), pp. 124-132.

[26]. Yang, C.N. and Mills, R.L., Phys. Rev. 96, 191 (1954).

[27]. Enz, C.P., Helv. Phys. Acta 54, 411 (1981).

[28]. Pais, A., **Inward Bound** (Clarendon, Oxford, 1986).

[29]. Pauli, W., "Schlusswort durch den Präsidenten der Konferenz", in **Fünfzig Jahre Relativitätstheorie**, Helv. Phys. Acta, Suppl. 4, 261-267 (1956).

[30]. Descartes, R., **Les principes de la philosophie** (Paris, 1724), thèse II.4.

[31]. Fierz, M., **Ueber den Ursprung und die Bedeutung der Lehre Isaac**

Newtons vom absoluten Raum, Gesnerus 11, 62-120 (1954).

[32]. Pauli, W., "Der Einfluss archetypischer Vorstellungen auf die Bildung naturwissenschaftlicher Theorien bei Kepler", in C.G. Jung, W. Pauli, **Naturerklärung und Psyche**, (Racher, Zürich, 1952); also Ref. 1, Vol. 1, pp. 1023-1114.

[33]. Bohr, N., **Atomphysik und menschliche Erkenntnis** (Vieweg, Braunschweig 1958 and 1985).

[34]. Delbrück, M., **Mind from Matter ?** Ed. G.S. Stent et al. (Blackwell Scientific, Palo Alto, 1986).

[35]. Enz, C.P., "Bohr, Delbrück, Pauli and Biology", in **Essays in honour of Kalervo Vihtori Laurikainen on the occasion of his seventieth birthday**, ed. P. Lahti, Ann. Acad. Sci. Fennicae, Series A,VI. Physica 431 (Helsinki, 1986), pp. 23-32.

[36]. Jung, C.G., "Synchronizität als ein Prinzip akausaler Zusammenhänge", in C.G. Jung, W. Pauli, **Naturerklärung und Psyche** (Rascher, Zürich, 1952).

[37]. Laurikainen, K.V., **Wolfgang Pauli and Philosophy**, Gesnerus 41, 213 (1984).

VEILED REALITY

B. D'ESPAGNAT
Laboratoire de Physique Théorique et Particules Elémentaires
Université de Paris XI
91405 Orsay
FRANCE

ABSTRACT

Each of us, presumably, has his own way of looking at the problem of reality. I was kindly asked by Professor Laurikainen, when he organized this symposium, to describe to you my approach, leading to the notion of a veiled reality (which, I understand, has some similarity with Pauli's notion of reality). I am grateful for this nice opportunity. I shall do my best to comply with the request.

The concept of reality is basically a philosophical, not a physical, concept. Hence the subject of this talk lies, to some extent at least, outside physics, a fact which will be illustrated by the circumstance that I shall show you no equations. But the methods I shall apply are those of physics rather than those of philosophy. In particular, I shall try to define unambiguously the concepts I use. Also I shall not demand that some general view should be accepted from the start. Rather I want to proceed as we do in science, that is, by formulating some working assumptions and looking at their consequences. With the idea that we shall reject the assumptions if the consequences turn out to be wrong.

Preliminary remark

These assumptions make no claim at originality so that it would be pointless to invent new words to name them. Unfortunately the old words used here (realism etc...) all were given many acceptations in the literature. To avoid ambiguities let it be specified that in the present lecture the words in question are always taken in one and the same sense, which is the one specified by the following set of definitions. I hope that in this way consistency, at least, will be preserved.□

1. THE STANDPOINT OF REALISM

Definition : it consists in considering that "something" exists, the nature of which (a set of material objects, an entity, God, or what not) is left for the time being completely unspecified. This "something" is assumed not to depend on the cognitive powers of mankind. But of course it influences decisively the results of our measurements.□

Henceforth this "something" will be called "independent reality" or, for short "reality".

Remark a

This definition is very weak in that only the existence, not the nature, of reality is postulated. Hence realism as here defined covers most of the great philosophies of the past. On the other hand the independence condition is very strong. I here reject even unavowed dependences on human cognitive powers, such as references to such notions as that of "systems incapable of being isolated", and the rest. For who, if not man, is supposed to "isolate" or to fail being able doing so ?□

Remark b

Many philosophers use the word "reality" to signify a concept partly defined by means of a reference to the abilities of men. This convention is unobjectionable as such. But by definition it is not the one made here. In my convention the concept in question, if used, should therefore be given another name.□

Is reality describable ? Opinions on this may and do differ. We must consider various views. Here I only formulate hypotheses corresponding to the answer yes.

2. THE HYPOTHESIS OF PHYSICAL REALISM

Definition : it consists in considering that reality (in the above sense of the word) can in principle be nonambiguously described by physics (plus other "hard" sciences if necessary).□

As I said, this is but a working hypothesis. At the start I request neither that it should be accepted nor that it should be rejected : we must decide on the basis of a comparison with actual physics.

Now, to describe we need words, that is, concepts. Physical realism therefore implies that some at least of the concepts used in physics do label elements of reality. What are these concepts ? Here there are two possibilities that are *a priori* worth considering. Let them be expressed in the form of two alternative hypotheses.

2.1 The Hypothesis of Near Realism

Definition : it consists in considering that the concepts labelling the elements of reality are all familiar, everyday ones, such as very small solids (grains, tiny particles of substance), forces etc.□

Roughly speaking near realism identifies reality with things. That is, it reduces it to be a collection of localized objects such as electrons, quarks and so on, conceived of as being objects (in the popular sense, thus, localized) interacting by means of forces. We all know however that such a hypothesis cannot be upheld any more. Even though some physicists still deem it proper to use such a language when writing popular books, they all know that they thereby cheat (and, in my view, on an essential point). For example relativity has taught us long ago that it is definitely not by means of the familiar concepts of space and time that we can hope to describe reality. And elementary particle physics goes even much farther, with such phenomena as pair creation and annihilation and so on. Hence, already near the end of the "classical age" of physics, physical realism had taken, in the mind of its physicists supporters, the form of the alternative hypothese 2.2, below.

2.2 The Hypothesis of Mathematical Realism

Definition : independent reality cannot be described exclusively by means of familiar, everyday concepts. But it is nevertheless scientifically describable "as it really is". Its description must be made by means of nonfamiliar, noneveryday concepts, borrowed from mathematics.□

Let me just mention that this hypothesis has a long history associated with such names as Plato, Galileo, Descartes, Spinoza and, nearer to us, the Einstein of the later years. Let me also forcefully stress that according to this hypothesis physics does not merely describe a human point of views on reality. It describes reality as it is.

Being interested with the question of reality we shall have to examine how the consequences of these hypotheses compare with facts. But before that I must of course mention that neither the hypothesis of physical realism (near or mathematical) nor even the standpoint of realism were the only views that were considered, even by scientists. An alternative was proposed in the form of logical or empirical positivism, according to which both science and ordinary knowledge are indissolubly linked with human experience. So that their purpose can only be a description of the phenomena, that is, of things, events and so on, as they are perceived and combined by collective human experience.

Now, as you all know of course, the advent of quantum mechanics coincided with a great increase of the consideration given by physicists to the positivistic trend. This increase looks even more considerable if we appreciate it not on what these physicists said but on what they actually did. And indeed there were good reasons for the increase in question. The point can be made very clear if we compare classical and quantum probabilities. Let us, for example, compare the tossing of a coin and the passing of a spin 1/2 particle polarized along O_x through a Stern-Gerlach apparatus directed along O_z. In both cases, imagine that some time after the event took place we "look at the result" (directly or with the help of some other apparatus). In both cases, the probability that we observe a head (respectively a $S_z = + 1/2$ result) is 1/2. In both cases if we repeat the experiment a large number n of times, the number of favorable results is roughly n/2. But in the case of the coin we may interpret this by saying that just before we looked, each coin already was in the state (head or tail) in which we now find it, whereas in the spin case we may not (since a beam recombination would prove us wrong). In other words, in the quantum case we are deprived of the possibility of a naively realistic interpretation that was straightforward in the classical case. We remain only, or at least so it seems at first sight, with the "positivistic" interpretation that physics only allows us to predict what we shall observe.

Hence there are some good grounds for the positivist-sounding statements of people like Bohr, Heisenberg, Wigner and the like. To make things short let me just show you a statement made by the latter.

> The laws of quantum mechanics only furnish probability connections between results of subsequent observations carried out on a system. It is true, of course, that the laws of classical mechanics can also be formulated in terms of such probability connections. However they can also be formulated in terms of objective reality. The important point is that the laws of quantum mechanics can be expressed only in terms of probability connections [between results of subsequent observations].

E. Wigner, Symmetries and reflections, Indiana University Press. The phrase inside square brackets, not present in the original text, is but a reiteration of the specification formulated in the first sentence, inserted here by me to make my claim even more transparent.

Now there are two points to be made with respect to this quotation. The first one is, in fact, just the one I produced it for. The quotation underlines the positivistic, or phenomenist nature of orthodox quantum physics. The second one is as follows. Nobody would seriously deny the fact that Wigner's main achievements lie in the field of mathematical physics. Hence the fact that he wrote this text shows in a specially clear way a fact that is extremely important. This fact is that mathematical realism is by no means the only conception that allows mathematics to play a basic role in physics. Indeed, contemporary theorists have developed what we might call a "mathematical positivism" whose mathematics are just as highbrow as those of the mathematical realism of, say, an Einstein. But they do not aim at a description of reality as such. Essentially their aim is to connect past and future observations. As usually described in advanced textbooks, algebraic quantum field theory is, I would say, a stronghold of this mathematical positivism.

There is, I think, a clear way of expressing the difference between mathematical realism and mathematical positivism. It is to distinguish between strongly objective and weakly objective statements. I am very much surprised that physicists practically never make this distinction, even when they discuss basic questions. They just say "science is objective". But this is a source of misunderstandings since the word objective has at least two senses. Some scientific statements are objective in the sense that they make no reference to the concept of observation, or measurement, as such. Hence they can be understood as bearing on reality. Such, according to Wigner's foregoing quotation, are the statements of classical physics (think, for example of Newton's gravitation law). These I call strongly objective. Other statements do make reference to the concept of observation or measurement. But they still may be called "objective" since what they assert is supposed to be valid for any observer. They are therefore unambiguous. Most of them take the form of rules : "if we do this we shall observe that". These statements I call weakly objective, or intersubjective. In this language we may say that what Wigner points out here is :

a) that a strongly objective statement can always be converted into a weakly objective one but that the converse is not always true.

b) that some at least of the quantum laws can only be expressed by weakly objective statements, leading to mathematical positivism.

Now the stage is set and we may start a discussion. I want to make it "from the bottom to the top" rather than conversely. This is because I am a physicist as you are. Contrary to professional philosophers, physicists like to proceed that way, namely by starting with rather concrete, down-to-earth ideas, like Galileo with his balls, and proceeding to more abstract ones only when their data force them to do so.

So we start with a man-in-the street idea, namely realism and even naive, or "near" realism. But this part of the discussion was already done, in fact. We know that near realism must be rejected, not on a priori grounds but because it is incompatible with the facts we learn from physics.

The next candidate is mathematical realism. Now I already sketched a criticism of this idea, but this was merely a sketch. I did not claim that the criticism was final. So, let us have a closer look.

When I, for one, have such a closer look I think I can distinguish three or four approaches by means of which mathematical realism is defended. The first one, I think, is simply not correct. It consists in focalizing on the considerable range and power of the mathematical formalism and in forgetting, or overseeing, the fact that, because of the existence of mathematical positivism, this does not at all prove that mathematical realism is true. Some very mathematically minded people believe they justify mathematical realism by pointing to the tremendous role of abstract mathematics in helping us to find out the elementary laws that, in field theory, superstring theory etc... play the same role as the Schrödinger equation in non-relativistic quantum mechanics. But they seem not to be aware of the fact that, already in nonrelativistic quantum theory, the Schrödinger equation is not the end of the story --and that the same is true in the relativistic case-- because the problem is "can we interpret the state-vector (or the density matrix or what not) as an element of reality ?". For quite a number of reasons the answer is _no_. Hence it cannot be said that the present mainstream of theoretical physics, (unification theories, superstring theory and so on) truly discloses what reality is.

The second approach is a variant of the first. It is based on the generalizations of the quantum formalism called "quantum logic" and "algebraic theories". It is quite a fascinating approach and one that I think leads naturally to an "empirical reality" concept to which we shall return in a moment. But I think that its most serious supporters would agree with what I shall now say : because of nonseparability (alias "EPR entanglement") the elements of this empirical reality, and in particular the localized objects, merely describe _human points of view_ on reality as such. Hence such theories definitely fall outside what I called mathematical realism.

Another approach by which, if not exactly _mathematical_ realism, at least a (somewhat weakened form of) physical realism is defended is _macroobjectivism_. The proponents of this approach grant that wave functions and the like do not necessarily correspond to what is actually happening in the real world, but they try to save the idea that, at least, when we say that the pointer of an instrument lies in such and such a graduation interval we assert something that bears on independent reality. Well, time does not allow me to enter into a discussion of the measurement problem, especially since the subject is intricate. Let me just try to briefly make two points.

a) Some theorists assume that the instrument pointer can be attributed a wave function so that before we look at the pointer the wave function of the composite system (particle plus pointer) is (in the general case) a superposition of several macroscopically distinct states. After we have looked it is only composed of one of them, of course. But these theorists simply deny that there should be a problem there for, they say, the wave function collapse "is merely an increase of our information". Such a statement clearly implies that before we looked our information was not maximal. Indeed it implies more. It implies that our information could have been maximal under no conceivable conditions, not even if we then had known the exact wave function of the composite system (a theoretical possibility which is compatible with the initial assumption). The statement therefore conflicts with the assumption that quantum theory is complete. Hence I can accept it only in conjunction with a hidden-variables theory, which then ought to be made explicit and studied for its own sake (see below).

b) Most theorists do not treat instruments that way. But although I read several proposed solutions to the measurement problem that I found quite enlightening, nevertheless none of them made it possible for me to say that the localization of the pointer after it interacted with the particle is an element of reality. My main reason is that in all these theories a reference is made to human abilities, be it in a highly indirect way. Hence I can only grant that some of these theories nicely describe empirical reality, i.e. nicely account for the observed phenomena (see e.g. ref. 1).

Next let me come to the L. de Broglie, Bohm, Bell and others hidden-variables theory. And let me first of all specify that, as regards Bohm's work what I have in mind is not the content of his book, but just that of his 1952 papers[2] and of his recent papers with Hiley. There, the situation is different. I do agree that this theory does meet the requirements of strong objectivity. In other words I do agree that, in Bohm and Hiley's words, it is a possible ontology. I do agree also that this is a big step since, thanks to it, we now see that there is no blunt contradiction between quantum physics (I mean its predictive rules) and physical realism.

If somebody tells me "as a physical realist I believe that such a theory does describe independent reality 'as it really is'"I shall not be able to prove to him that he is being inconsistent. But this does not imply that I, myself, share such a belief. The points are as follow. First there is no experiment showing that this is the right theory. This is significant since this theory has at least one competitor, namely Everett's relative state theory[3], which has its own weak points but which is also compatible with physical realism I would say. Next, there are, I think, difficulties with manifest relativistic covariance. For example, although a recent hidden-variables model by John Bell[4] operationally reproduces all the relativistic predictions, still its author was forced to base it on the idea of independently existing (but nonobservable) Euclidean space and Newtonian time. I do not know whether this particular difficulty is there to stay. But I think that the mere fact it appeared illustrates a feeling of artificiality that I, for one, also get when I am told about the existence of, for example, a swarm of empty wave packets and such things. If such theories are true, then whatever sophisticated instruments we use we shall always have access to but a very tiny part of the iceberg of independent reality and even that part will always be seen by us in a highly distorted way. Under these conditions it seems to me that we are very much over-ambitious when we try to reconstruct the whole of the non-observed part. For we must then water down very much a guiding rule that was most useful, for example, to Einstein, and which is to soberly model the basic features of our interpretation of the theory on the mathematical elements of our theoretical synthese of appearances, and to use Occam's rasor to cut away all other features...

For these reasons, while I willingly say "it is _possible_ that independent reality be structured that way" I am very much reluctant at saying "it _is_ structured that way" and even at saying "we may reasonably hope that in some future we shall have a scientific certainty that it is structured that way". Hence I do not consider that such theories meet or will ever meet the requirement of nonambiguity which is part of those of what I called "physical realism".

Well, on the occasion of explicating my standpoint with respect to the hidden-variables theories I hope I succeeded in conveying to you something of my balanced judgement concerning the hypothesis of physical realism. I do not strictly disprove it, but for a variety of a posteriori reasons, derived from physics and that I find quite convincing, I do not feel able to believe it.

What I still have to do is to try explaining to you my position as regards the standpoint of realism. This question splits in two parts,

a) should we retain the idea ?

b) if we answer _yes_, what can we say concerning our relationship with the reality in question ?

Let us look successively at these two points. As regards the first point, quite definitely my answer is _yes_. For example, when a theorist tells me that the wave function is just a measure of our information, I feel bound to ask "information on what ?", without being able to consider that the question is meaningless (if he answers "on a system", I must ask "what is a system ?"). I must grant of course that there is no way of proving realism in the way one proves a theorem or the truth of an empirical statement. My answer is _yes_ nevertheless, mainly for reasons that are implicit in most people's mind, that have been stated by a number of philosophers and that I find quite convincing. They can be briefly sketched as follows.

a) the opposite assumption is that the verb "to be" does not apply to anything. But this is obviously wrong. I do not know whether I exist as an absolute existant, maybe I am only the dream of some superior Being, or what not, but I simply cannot say without indulging in a contradiction that nothing at all is speaking presently to you. So that, something at least "is". And it seems preposterous to me to think that this "flicker of consciousness" should be the only existant.

b) As scientists, we know that there is a difference between building up a physical theory and deciding what will be the rules of a new game or the basic ideas of a new philosophy. In the cases of game and philosophy our freedom is exclusively limited by the non-contradiction principle. In the case of physics there are additional limitations. A theory can be fully self-consistent and yet entail consequences that are proved wrong by experiment, and in that case we are forced to reject it. In other words something offers resistance to us, so that it is not our own creation. There must therefore be "something" prior to us. I mean some basic element that is not under our control.

c) Finally we observe regularities in our impressions and we all agree on the elementary facts we observe. The usual explanation is because these things are due to some external reality that has a structure, and I think this "trivial" explanation is sound and should be kept. I grant that here I am not using the notion of a causal link in the manner in which it is used in science, since in science the cause and the effect should be of the same nature, that is, they should both be phenomena. But I do not think this is an objection. I simply say that while reality is not a cause-in-the-scientific-sense, it is nevertheless the <u>explanation</u>, the <u><<raison d'être>></u> of the observed phenomenological regularities, and of the intersubjective agreements as well.

Now let me come to the point about our relationship with reality. We already reached the conclusion that it is not fully knowable and we need not come back on this. But shall we say that it is utterly unknowable ? Instead I say that this reality is "veiled". Why, and what do I mean ?

To try to convey this to you let me first refer to Kant's idea concerning the "thing-in-itself". It is often said that Kant considered it to be utterly unknowable. But in fact I am not so sure. Somewhere he compares our relation with the thing-in-itself to that of a judge with respect to somebody who is brought up for trial. And Kant says that we force the thing-in-itself to answer questions <u>in our own</u>

language, pretty much as the judge obliges the defendant to answer the questions he asks. We see everything through the a priori structure of our own sensibility and intelligence so that we cannot know the thing-in-itself as it really is. This is often taken as an argument purporting to show that we know nothing about the latter. But I maintain that the analogy with the court does not go that far. It is true that the judge asks his own questions in his own way. But still he tries to be informed of something that is external to him. And he is quite convinced that this expectation is not a complete delusion. Similarly I maintain that while physics is admittedly but a human view on reality, still it is not an arbitrary human product but does convey, albeit in a distorted form, an undefinable "something" of the true structure of reality.

One point that comforts me in this opinion is the following one. Relativistic spacetime is obviously not an inborn concept. In Kant's time, physics was based on Euclidean space and universal time and it was possible to maintain, as Kant did, that these concepts were just inborn ; that they came not from the thing-in-itself but exclusively from us. Now, however, physics uses spacetime which, as I said is not an inborn concept. So it is more difficult now to maintain Kant's position in this matter. We are therefore more or less forced to consider that it is something external to us that forced upon us the use of relativistic spacetime. I do not say that spacetime as we know it is, as such, a property of reality. On the contrary I think it is not. But I think that indirectly it does reflect an uncontrollable "something" of the structure of reality. Similarly I consider that due to the Bell theorem we can assert with confidence that reality is not "local". More generally I would say that physics plays in the intellectual realm somewhat the same role as parables and some fables play in the moral realm : the role of conveying in a disguised form a truth that cannot actually be expressed otherwise.

So now I think I can make clear what I mean when instead of saying that reality is unknowable I merely say that it is veiled. What I mean is in fact two things.

First I consider that the standpoint of realism is correct and even unavoidable. The verb "to be" is meaningful. And at the same time I know --this time from contemporary physics-- something that the realists of yore could not know, namely that reality does not boil down to what is describable, that is to the phenomena. Object is not Being. Thus, just by putting these two things together, it seems I know that there is something beyond the phenomena. This, I think, is real knowledge and very important (especially since it contradicts the implicit idea of most nonphysicist scientists...).

Second, I consider, as I just said, that the great laws of physics, although they do not describe reality as it is, yet constitute a human point of view on reality and therefore do capture an undefinable something of its structure. For me this is a very important argument for rationally opposing scientific reductionism. For I say : if physics both fails at giving us a sure, exhaustive description of reality, and yet succeeds in capturing something of it, why other approaches, such as music, painting, poetry should not ?

Although "empirical reality" is not my subject, I must here say just one or two words about this concept, just for contrasting it with that of independent reality. As I mentioned, and as Professor Primas will probably tell us again, things are not elements of independent reality. Rather, they are views we take on it. They are phenomena in the etymological sense. Well, what I call empirical reality is just the set of all the phenomena. In a way, it is what we get when we reify the set of recipes that mathematical positivism consists of. And we are forced to proceed to this conceptual reification if we want to speak of objects at all ! Even if we want to speak of molecules. In my view it is to this empirical reality that ideas such as complementarity apply. It constitutes the stuff on which our actions bear and it is the very subject-matter of the rational knowledge called science. But it must not be confused with independent reality, the faithful description of which presumably exceeds the powers of human reason.

Returning then for the last time to independent reality, I would like to conclude by referring again to what I said just a moment ago, namely that, as a result of the analysis we made, it now seems plausible that approaches other than science (I mentioned music, painting and poetry) should also, concurrently with science, be capturing some undefinable something of the structure of independent reality. In our age of totalitarian techno-science I think this is very important for a reenchantment of the world that has become necessary. And it is gratifying that science itself should give us indications along these lines.

REFERENCES

1. d'Espagnat, B., Foundations of Physics 17, 507 (1987).

2. Bohm, D., Phys. Rev. 85, 166-180 (1952).

3. Bell, J.S., Beable in Quantum Theory, Quantum Implications, B.J. Healey ed., Routledge and Kegan Paul, London (1987).

NIELS BOHR'S CONCEPT OF REALITY

Henry J. Folse
Department of Philosophy
Loyola University
New Orleans, LA 70118
U.S.A.

I. Bohr's Realism: Complementarity and the Realism Debate

One approach to Bohr's concept of reality is from the perspective of the philosophers' debate over the issue of realism vs. anti-realism. Although this debate over the correct interpretation of scientific theories has come to assume center stage in recent philosophy of science, its relation to the quantum revolution and Bohr's viewpoint of complementarity remains controversial and ill understood. Thus it is common to find the anti-realist arguing that the success of quantum theory has dealt the death blow to realism, while at least some realists herald the same event as providing the grounds for reconstructing the conceptual framework through which we express our knowledge of physical reality.[1]

No doubt the problems in the interpretation of quantum theory have contributed to this controversial relationship between the quantum revolution and realism. Therefore, since Bohr's framework of complementarity presents an interpretation of quantum theory, it must be concerned with the issue of realism. It is meaningful to ask what is Bohr's concept of microphysical reality only if complementarity demands a *realist* interpretation of scientific knowledge. If Bohr's interpretation of science is judged to support anti-realism, then any concept of physical reality derived from complementarity must be regarded as at best another

----

1. I have defended this view in *The Philosophy of Niels Bohr: The Framework of Complementarity*, Amsterdam: North Holland Physics Publishing, 1985.

speculative metaphysician's "castle in the air" destined to stand in ruins like its predecessors along the route of history.[2] Thus the realism/anti-realism issue presents us with the choice of deciding whether the concept of reality inherent in complementarity is but a "castle in the air" or provides a bit, even perhaps a rather fundamental bit, of knowledge about those microphysical entities which produce the phenomena that serve as the empirical evidence for quantum theory.

In my earlier work I have argued that the common perception of Bohr as an anti-realist is unsupportable for two different sorts of reasons. *First*, it is at odds with the historical record of Bohr's journey towards complementarity, for if one thing at all emerges from that history, it is that complementarity was the product of a tortuous passion to comprehend how it is possible to develop a consistent objective description of the atomic domain. Had Bohr espoused an anti-realist interpretation of theory, there would have been no need to relate the predictively successful devices of the mathematical formalism to the physical basis for this description of nature. Bohr's pressing need to interpret the formalism would then be a complete *non sequitur* and his struggle to formulate the new framework of complementarity would never have occurred. Moreover, Bohr's viewpoint was hardly the product of arm-chair speculation, as the phrase "castle in the air" would indicate. Instead, it was the result of a brilliant dialectic between the demands of logical consistency imposed by theory and the demands of empirical adequacy imposed by experiment.

Second, not only is an anti-realist understanding of complementarity incompatible with the historical record of Bohr's journey to complementarity, but also it would make Bohr's framework for the description of nature an incoherent framework.[3] In my previous work I have supported this claim by following something of a *via negativa*, in other words, I have tried to show only that without presupposing there to be *some* quantum mechanical object apart from the observed phenomena which are the physical evidence for quantum theory, it is quite meaningless to speak of the quantum descriptions of such phenomena as providing *complementary* evidence about the *same* object. Bohr most certainly does want to speak in this way; therefore, complementarity must presuppose that the objects described by quantum mechanics can in some sense be said to have reality apart

---

2. A. Fine, *The Shaky Game: Einstein, Realism, and the Quantum Theory* (Chicago: The University of Chicago Press, 1986), p. 116.

3. Folse (1985)

from their phenomenal appearances. Furthermore it is clear that Bohr wants to claim that in the quantum theoretical description we have knowledge of these objects, and this knowledge is "complete" in the sense that there is no alternative method of describing these objects such that a more precise determination of their phenomenal appearances could be predicted.

Nevertheless, following Bohr's lead, hitherto I have not attempted any positive account of the nature of microphysical reality from the viewpoint of complementarity. Furthermore, I do endorse the consensus conclusion that Bohr certainly does mean to argue against any realist interpretation of the wave or particle pictures which enable us to interpret the phenomenal evidence. Hence it seems appropriate to ask that I take a more affirmative stance and proceed to characterize Bohr's concept of reality in a straightforward way. But here my case for Bohr's realism confronts a rather serious obstacle: Bohr persistently evades any direct engagement with the question of "reality". Indeed, as you fully well know, many have understood him to have banned "reality" from physics!

Thus it seems that Bohr might well be in the position of maintaining simultaneously that we must presuppose that the objects described by quantum mechanics *exist*, but at the same time it is impossible to *say* anything about them. My reply to this objection is to take the usual philosopher's tactic of drawing a distinction. The distinction to which I mean to call attention has become rather common in the recent philosophical literature on realism; it is the distinction between the *ontological* thesis of "entity realism" which asserts the real existence of unobserved entities and the *epistemological* thesis that a particular theory provides knowledge about such entities, the thesis I shall call "epistemic realism." *Entity* realists think that when we accept a theory we have warrant for believing that the objects and events spoken of in non-observational theoretical terms really do exist as the unobserved causes of observed phenomena; whereas anti-realists hold that acceptance of a theory justifies no such belief.

It is useful to begin our discussion with "entity realism," because although we are often blocked by Bohr's refusal to give ontological pronouncements about the *nature* of the reality belonging to the quantum mechanical object, there can be no dispute over the fact that Bohr took the atomic systems described by quantum mechanics to be real entities existing independently of our knowledge of them. Bohr could not have stated his view more clearly than in this passage from 1929:

> Natural phenomena, as experienced through the medium of our senses, often appear to be extremely variable and unstable. To explain this, it has been assumed, since early times, that the phenomena arise from the combined action and interplay of a large number of minute particles, the so-called atoms, which are themselves unchangeable and stable, but which, owing to their smallness, escape immediate perception. Quite apart from the fundamental question of whether we are justified in demanding visualizable pictures in fields which lie outside of the reach of our senses, the atomic theory was originally of necessity of a hypothetical character; and, since it was believed that a direct insight into the world of atoms would, from the very nature of the matter, never be possible, one had to assume that the atomic theory would always retain this character. However, what has happened in so many other fields has happened also here; because of the development of observational technique, the limit of possible observations has continually been shifted. We need only think of the insight into the structure of the universe which we have gained by the aid of the telescope and the spectroscope, or of the knowledge of the finer structure of organisms which we owe to the microscope. Similarly, the extraordinary development in the methods of experimental physics has made known to us a large number of phenomena which in a direct way inform us of the motions of atoms and of their number. We are aware even of phenomena which with certainty may be assumed to arise from the action of a single atom. However, at the same time as *every doubt regarding the reality of atoms has been removed* and as we have gained a detailed knowledge of the inner structure of atoms, we have been reminded in an instructive manner of the natural limitations of our forms of perception.[4]

We must recall that the realist issue which shaped the period prior to Bohr's entrance into atomic physics was the old Machian debate about the reality of atoms. Bohr comes to scientific maturity in a day when it seemed experimental evidence had settled this question once and for all in favor of entity realism with respect to atoms. This belief forms the very cornerstone of his life's work: the exploration of how we can describe atomic systems. It is simply ludicrous to hold that Bohr imagined his lifelong involvement with the question of how to describe atomic systems was concerned merely with inventing a predictive formalism; it was a concern over how to describe *nature* , especially in the light of those new discoveries made about nature that had precipitated the quantum revolution.

----

4. N. Bohr, "The Atomic Theory and the Fundamental Principles Underlying the Description of Nature," in *Atomic Theory and the Description of Nature* (Cambridge: The University Press, 1934, reprinted 1961), pp. 102-103, italics mine.

Nevertheless, it is one thing simply to *presuppose* the existence of an ontologically independent real object, and quite another to defend the claim that we *know* something about it. Thus my defense of Bohr's realism remains incomplete until I justify how it is possible to have some knowledge about quantum mechanical objects. Unless Bohr can be understood to be holding some form of epistemic realism, one might object that any role for the quantum mechanical object as an entity apart from the phenomena, on Bohr's account is limited to a philosophically doomed concept, a Lockean "I know not what" or a Kantian "noumenon". Without some epistemic defense for claims about the entities that populate the microphysical domain, such a bit of metaphysical baggage is likely to go the way of all that is unknowable. Thus a complete realist defense of quantum physics entails that the ontological thesis of entity realism must be buttressed by some form of epistemic realism.

The belief that Bohr denies epistemic realism may seem to be a natural conclusion from Bohr's refusal to speak about the nature of quantum mechanical reality, but this belief rests on two presuppositions which have been so widespread that we tend to forget our implicit acceptance of them. The first is that an epistemic realist interpretation of any scientific theory must defend a *correspondence view of truth*. The second is that to have knowledge about an object one must be able to make *statements* about this object which can be defended as true in a correspondence sense. In particular when these presuppositions are combined with the widespread acceptance of a substance/property ontology, we are led to expect that to have knowledge about an object one must be able to make true descriptive statements that predicate properties of the subject in a manner that corresponds to properties objectively possessed by a substantial entity. It is this nest of presuppositions which justifies the objectivity of the "pictorial description" characteristic of classical physics. And this is obviously a descriptive ideal that Bohr wants to reject.

Within the classical framework it was quite possible to defend *this* form of epistemic realism through a correspondence notion of truth. To do this it was sufficient to construct an internally consistent conceptual model in which the theoretical parameters defining the classical mechanical state of the system stood in a one to one correspondence with the properties of the model. Although Berkeley argued that the mind had no power to frame such an idea, the judgment of history was to ignore him and proceed to build a world-view based on an ontology of *rei extensae* having only those "primary" properties accorded to them in order to define their classical mechanical state. Typically the anti-realist could hold that such a model is merely a construct, but in spite of the empiricist scruples which objected that experience could never justify beliefs about a world causing experience and the Kantian restriction of knowledge to phenomena, the complete realist of classical physics knew reasonably well what physical reality was supposed to be like.

However, any attempt to build a correspondence between the parameters of the quantum theoretical description and the properties imagined to be possessed by independently existing entities seems doomed to inconsistency. Indeed, Bohr insists that the properties of waves and particles tracing their careers through space and time cannot be ascribed to the quantum mechanical object apart from our observational interaction with it. Thus although Bohr is clearly a realist with respect to entities, when we come to the issue of epistemic realism, his defense of the completeness of the quantum theoretical description of these entities implies that he must abandon a correspondence account of truth.

Nevertheless, in order not to leave his entity realism from hanging without an epistemic defense, Bohr must hold some alternative position concerning epistemic realism without a correspondence theory of truth. This is of course quite impossible as long as one holds to the traditional positivist epistemological framework which presupposes that a theory provides knowledge about an object if an only if it justifies making *true descriptive statements* predicating properties of some substantial entity. Indeed, the linguistic consensus which treats theories as linguistic entities makes such a presupposition inevitable. Applying this presupposition to quantum state descriptions we get the familiar paradoxes of the collapse of the wave-packet, such as that involved in the claim that Schrodinger's cat is in a superposition of live and dead states. Such paradoxes suggest that here the quantum realist must simply take leave of the philosophical fashions in epistemology. Indeed, in philosophy itself this consensus has already began to unravel over precisely this point: if correspondence is taken to be the criterion of truth, then the philosopher's historical quest for knowledge of reality becomes a fool's errand.[5]

---

5. The follies of the "linguistic consensus" have been so named and analyzed by Edward Pols, "After the Linguistic Consensus: The Real Foundation Question," *Review of Metaphysics 40* (1986), 17-40.

Our reflections thus far can be summarized as follows: Bohr's realism is not concerned with arguing the ontological thesis of entity realism; this position is one which he simply *assumes* as established experimentally. His problem is this: having assumed that there really are atomic systems, how can we alter our understanding of the use of concepts in describing nature to justify claiming that the quantum theoretical account provides knowledge about such entities. Although he never put it this way, Bohr's "epistemological lesson" was addressed directly to this problem. Inevitably, critics who see Bohr as an anti-realist fail to heed this lesson and presuppose the correspondence view of truth and knowledge that Bohr was concerned to replace with a new thesis. Bohr's outlook requires a turning away from truth in the correspondence sense of "picturing" and requires a revolution in our beliefs about the nature of empirical knowledge. In its most general form, Bohr's new version of epistemic realism could be expressed as follows: acceptance of a scientific theory as providing a complete description of a specified range of phenomena warrants holding that theory provides *knowledge* about the unobserved entities which cause these phenomena that form the empirical evidence for the theory. I propose, then, that in order to understand *Bohr's* realism, we must drop correspondence with its associated emphasis on reference and, instead, quest for knowledge in the quantum domain by asking in what such knowledge could consist? Let truth fall where it will (it is hardly a word which enters science at all) and let us not ask about the reference of theoretical terms until we know in what knowledge of atomic systems could consist.

II. Reality and the Description of Nature: The Bohr-Einstein Debate

At the center of the debate on reality and the quantum description of nature is of course the monumental encounter between Bohr and Einstein, certainly by any standard one of the most significant intellectual confrontations in the history of science. Yet the precise relation of this particular debate to the question of realism is also, perhaps not surprisingly, as obscure and ill-understood as the relationship between quantum theory and realism generally.

To begin, I want to emphasize two points about this debate as a concrete episode in the history of physics. The *first* is that today it has become increasingly well recognized that though the debate may have appeared to its participants as concerned with the correct description of selected *Gedankenexperimenten*, the differences between Bohr and Einstein reflected a deep division in the question of what an acceptable objective description of physical reality requires. Yet in spite of holding opposing views about the nature of reality, *both men were able to make essential contributions to the same almost miraculously successful theory*. What inference might we make from this fact? One obvious possibility would be that the individual physicist's conception of reality (and what constitutes an objective description of it) is a bit of excess metaphysical baggage that has no logical bearing on whether or not one can participate meaningfully in the scientific enterprise. But another possibility, one that we might tend to ignore, is that in sharpening the contrast between our opposing heroes, we have forgotten the extent of their shared commitment to the common goal of understanding the physical basis for the theoretical description of nature in physics. Few discussions of Bohr bother to note that in many of his presentations of complementarity, Bohr devotes considerable attention to arguing that the conceptual revolution required by quantum physics is essentially analogous to the one which Einstein had already championed with respect to relativity. The fact that Bohr clearly believed that his reasoning in many ways paralleled Einstein's made his failure to convert Einstein all the more frustrating to his patient but persistent spirit. But it was all the more frustrating because he believed that Einstein was one with whom he shared many fundamental philosophical values.

This leads to a *second* point about this debate: although the two men certainly held different concepts of reality, it is a misunderstanding to see their opposition in terms of the opposition of realism and anti-realism. Although the nature of Einstein's realism remains controversial, virtually all sides would agree that he was in later years some kind of realist, and that this realism was the source of his disagreement with Bohr. Moreover, as I claim to have shown, Bohr must be at least an entity realist for complementarity itself makes no sense unless we accept the reality of atomic objects. *Thus with respect to the reality of the objects described by quantum physics, both men were realists*. Nevertheless, the misinterpretation that Bohr was championing an anti-realist stand against Einstein's realism has become so entrenched that though it had its origin in part in the misunderstanding of their differences in which the participants in the debate shared, it has now come to color our whole way of looking at what is at issue. Here I want to undo some of that misinterpretation by stressing what they held in common.

To begin with the broadest point of shared commitment: Bohr and Einstein were both virtuosi at the philosophy of nature. It is common to point out that both were in some sense *philosophers*, but I suspect that we fail to recognize that both were doing philosophy in the sense of the philosophy of nature.[6] By this expression I mean the endeavor to frame a conception of physical reality which justifies the claims that natural science provides objective knowledge of physical reality. Both Bohr and Einstein were motivated by nothing less than a *passion* to seek a consistent understanding of the description of nature through the methods of physical science, which impelled them to perform feats of mental pyrotechnics at which we lesser spirits can only stand in awe. Their yearning to describe a dynamic reality as an ordered cosmos led *both* men to recognize the need to restructure the conceptual framework through which the natural philosopher hopes to describe physical reality. This passion is the motive impulse for *realism* in science; however, both men also shared as well in that other passion which motivates the *anti-realist*. Thus they sought to fashion new perspectives for the description of reality by following the drive to reduce fundamental presuppositions to the simplest set by pruning away claims to the truth of excess metaphysical assumptions. Although this desire to reduce "excess" metaphysical assumptions has characterized many an anti-realist, such an ideal need not be hostile to the realist passion for philosophy of nature as well. To eliminate the "excess" assumptions in the sense of removing the more restricting in favor of the more general presupposition is in fact Bohr's ideal of how the conceptual framework employed in the description of nature grows. Thus the Newtonian description presupposing a framework of absolute space and time may be regarded as a special case of the more general relativistic framework of space-time which awards no such privileged status to any one frame of reference. In the special case of velocities much lower than c, the more general relativistic description is said to "reduce" to that of the classical mechanical scheme. Similarly, in Bohr's eyes, the

----

6. I do not mean to refer to the nineteenth century tradition of romantic *Naturphilosophie*; instead I mean to place Bohr and Einstein in that enterprise which so concerned the great Enlightenment framers of the classical world-view, the tradition of Descartes, of Newton and Leibniz, and of Kant. The practitioner of this mode of philosophy *par excellence* is of course Aristotle, but its source goes back through the late Plato at least to the Milesians.

quantum revolution eliminated the special assumption that in the description of the interaction between systems the parameter of action changes continuously. In the obvious parallel to the relativistic case, Bohr's correspondence principle asserted that in the special case of interactions involving large scale exchange of action, relative to Planck's quantum, the quantum description must "reduce" to the classical description.

Thus Bohr and Einstein shared with the anti-realist the desire to eliminate excess assumptions en route to a more general conceptual framework. However, neither did so because he desired to purge the concept of reality from the description of nature. Both regarded theory to be far more than merely a scheme for reasoning from one set of observed phenomena to the prediction of another set of phenomena. They both saw that the description of nature requires the constant revision of the conceptual framework within which such a description is expressed precisely because the continued exploration of experience reveals the limitations of previous presuppositions about that reality. As the description of nature advances, the physicist's concept of reality evolves precisely because that concept of reality is discovered to be inadequate to the further description of nature. The issue between Bohr and Einstein concerns not whether or not physics describes reality; rather it concerns what concept of reality the growth of experience has taught the physicist he must employ in the description of nature.

Here, then, we have come upon a strategy for getting around the obstacle posed by Bohr's reluctance to speak about the nature of reality in the quantum description of nature. Bohr's arguments for complementarity clearly do have a good deal to say about what must be *rejected* of the classical account of this reality. The concepts of causality and space and time are to play significantly different roles, although in their use to describe the phenomena which count as evidence for the theory, they will retain their original meanings, as given in everyday experience, suitably refined by classical physics. While this portion of Bohr's argument does not establish what *can* be said about quantum reality, it does establish what *cannot* be said about it. Knowing that Bohr rejects Einstein's presuppositions about the nature of reality, by isolating these presuppositions, we can see what complementarity implies we *can* know about reality. In this way, we may lay out the boundary conditions of our problem.

In this connection recent scholarship on Einstein's view of the celebrated EPR experiment is particularly revealing. Fine, Jammer, and Howard have shown that the customary interpretation is based on the natural but *erroneous* assumption that Einstein's view was the view of the EPR paper itself. In fact, it is now well established by Einstein's June 19, 1935 letter to Schrodinger that Einstein did not write the paper and was disappointed with the way *Podolsky* presented the argument of the paper.[7] It is customarily assumed that the debate turns on the question of the *completeness* of the quantum description. Einstein tried to justify his belief in the incompleteness of the quantum description by showing that it was possible to design experimental situations in which we could learn more about the *nature of physical reality* than the quantum formalism permits. To do so he advanced a concept of physical reality that held fast to two claims about what the nature of physical reality was like: the assumption of *locality*, imposed by relativity, that effects from spatially separated systems cannot be propagated faster than the speed of light, and the assumption of *separability*, or the belief that "separated systems possess separate real states."[8] Bohr in turn defended completeness by arguing that Einstein's view of the nature of physical reality "contains an essential ambiguity" and that if one recognizes that physics describes phenomena, at the atomic level the wholeness of observational interactions implies that observing position and observing momentum are *two different phenomena* and so these observations describe different phenomenal objects. Thus it does not follow that we have a "free choice" of which of two possible observations to make on the *same* object. In thus challenging Einstein's requirements on the nature of physical reality, Bohr does *not* argue to "throw out reality," instead, in effect he puts forward a *rival conception of reality*.

Therefore today it has become well recognized that the confrontation between Bohr and Einstein was a collision of rival concepts of physical reality, but what both participants ignored and what has been ignored until recently is what *grounded* each natural philosopher's position about how physical reality could be described. Howard points out that

> Einstein did believe that all physical systems at all times possess definite, observer-independent properties which are revealed to us by

----

7. A. Fine, *op. cit.*, pp. 35-36; Max Jammer, "The EPR Problem in Its Historical Setting," *Symposium on the Foundations of Modern Physics*, (World Scientific, 1985), pp. 142-146; and Don Howard, "Einstein on Locality and Separability," *Stud. Hist. Phil. Sci. 16*, p. 175-176.

8. Howard, *op. cit.*, p. 173.

> observation. But he did not just *assume* this. Instead, Einstein *grounded* his realism about physical systems and their properties in the deeper assumption of separability, which is important, because the latter assumption is susceptible to revealing kinds of physical and philosophical scrutiny that cannot touch unanalyzed postulates of physical realism.[9]

Howard argues that the best clue for understanding why Einstein held this assumption of "separability" to be a *necessary* condition for realism is found in his March 1948 letter to Max Born:

> ...if one renounces the assumption that what is present in different parts of space has an independent, real existence, then I do not at all see what physics is supposed to describe. For what is thought to be a 'system' is, after all, just conventional, and I do not see how one is supposed to divide up the world objectively so that one can make statements about the parts.[10]

As Howard notes, "what Einstein suggests here is that the separability principle is necessary because it provides the only imaginable objective principle for the individuation of physical systems. Needless to say this is an important and provocative claim, because quantum mechanics, interpreted as a theory about individual systems [which is certainly Bohr's view], denies the separability principle."[11] Nevertheless, it is still not obvious why Einstein regards the spatial separability of physical systems to be "the only imaginable objective principle for the individuation of physical systems." Here Howard puts forward the profound "guess" that Einstein's commitment reflects the need to separate the observer, *qua* physical system, from the observed in order to ground the "objectivity" of the resulting picture of physical reality:

> Like so many realists before him, Einstein speaks of the real world which physics aims to describe as the real '*external*' world, and he does so in such a way as to suggest that the *independence* of the real - its not being dependent in any significant way on ourselves as observers- is grounded in this 'externality'. For most other realists this talk of 'externality' is at best a suggestive metaphor. But for Einstein it is no metaphor. 'Externality' is a relation of spatial separation, and the separability principle, the principle of 'the mutually independent existence of spatially distant things', asserts that any two systems separated by so much as an infinitesimal spatial

----

9. Howard, *op. cit.*, p. 176.

10. M. Born (ed.), *Albert Einstein-Hedwig und Max Born Briefwechsel, 1916-1955* (Munich: Nymphenburger, 1969), pp. 223-224; as translated and cited by Howard, *Ibid.*, p. 191.

11. Howard, *Ibid.*, p. 191

> interval always possess separate states. Once we realize that observer and observed are themselves just previously interacting physical systems, we see that their independence is grounded by the separability principle, along with the independence of all other physical systems. Certain properties of the observed object can be changed by the observation interaction, but the locality principle guarantees that such changes cannot take place if the separation between observer and observed is space-like, as in the EPR *Gedankenexperiment*, and in any case the separability principle guarantees that the observed object always possesses its own cluster of properties, even if these might be changed by observation. The independence of the real consists not in any immunity to change, but in autonomy even through observation induced changes.[12]

We see in Howard's last remark that Einstein believed that the story physics tells about the natural world is the story of the changes of the properties possessed by ontologically independent substances. Einstein's commitment to the separability principle is not a matter of metaphysical dogmatism but in fact rests on the epistemological presuppositions canonized by the spectator theory of knowledge: the view that objective knowledge of a real external world requires a description of the careers of such substances from which the observer qua physical system is at least in principle excludable.

In contrast, Bohr's beliefs about the nature of physical reality were grounded in quantum theory itself, specifically in what he took to be that surprising empirical discovery distilled into the quantum postulate: the discovery that at the atomic level interaction has a granular rather than a smooth structure. This was for Bohr the reason why we must revise our concept of physical reality and therefore of how knowledge about reality is possible. Bohr repeatedly stressed that the break between classical and quantum physics, resides in the fact that "Planck's discovery of the *elementary quantum of action* ...revealed a feature of *wholeness* inherent in atomic processes, going far beyond the ancient idea of the limited divisibility of matter."[13] The consequence of adopting the quantization of action in physics is that in the fine structure of the physical world it is these *causal processes of interaction*, not bits of material substances, which are "atomized" - *i.e.* made "indivisible." The "atomicity" of an interaction" implies that the description of an "observation" of a microsystem which must subdivide the whole interaction into separate observed and observing systems is a conceptual idealization or "abstraction" employed for interpreting the

----

12. Howard, *Ibid.*, pp. 192-193.

13. N.Bohr, *Essays 1958-1962 on Atomic Physics and Human Knowledge*, New York: John Wiley & Sons, Inc., 1963, p. 2.

observation, but not a "picture" of an objective course of events.[14] On Bohr's concept of reality we individuate the parts of the world not *qua* spatio-temporally "separable" physical systems possessing properties apart from any interaction, but instead *qua* "individual" interactions each of which is a whole phenomenon the description and prediction of which must be the goal of a successful physical theory.

Thus, Einstein's realism is based upon the assumption of separability, whereas Bohr's realism is grounded in the quantum postulate which is effect is the very *denial* of Einstein's separability assumption.[15] A simple but suggestive way of expressing the philosophical reason for this difference between Bohr and Einstein is that for Einstein the ontological demands for an objective description precede epistemological demands, whereas for Bohr the epistemological precede the ontological. As we have seen, Einstein's rejection of Bohr's epistemological claim that we have in the quantum formalism a complete description of physical reality is based on his view that locality and separability must be essential conditions for any concept of physical reality that can be objectively described. Bohr does not deny that physics describes reality, but what he does hold is that -at least if we regard the effects of the quantum revolution as here to stay- physics cannot make these the necessary conditions for physical reality if that reality is to be objectively described. Thus Bohr's concept of reality is derivative from his epistemology, or as he would put it, from the "epistemological lesson" which the quantum revolution has taught us.

III. Atomic Physics and Human Knowledge: Bohr's Epistemological Lesson

Bohr's oft-repeated claim that the quantum revolution teaches an "epistemological lesson" is his way of making the point that we must revise our assumptions about the *knowledge* we acquire in the scientific description of nature. What we took to be a justified case of knowledge of reality in classical physics, he argued, rested on the presupposition that interaction takes place continuously. This belief expressed an "idealization" which was acceptable in the "special case" of

----

14. I have discussed the reasoning which led Bohr to this conclusion at length in H. Folse, *op. cit.*, 1985, pp. 108-167.

15. Howard. *op cit.*, p. 191.

ordinary experience, but could not be presupposed generally to hold beyond the range of our ordinary experience. Thus Bohr's lesson is the lesson that having knowledge of nature is not what we once that it to be. The separability between observer and observed which is necessary to support the spectator account of knowledge inherent in the framework of classical physics was erroneously believed to be characteristic of knowledge *generally*, but it is characteristic only of the "special case" of our knowledge of the objects of everyday experience. Thus with this lesson Bohr discards the spectator account of knowledge, the correspondence theory of truth that goes with it, and all of the paradoxes of reference associated with the "collapse of the wave-packet."

Bohr's frequent reference to the epistemological lesson of atomic physics as reminding us that we are both actors and spectators in the drama of existence tells us that we do not have knowledge of quantum mechanical objects by *picturing* them. To know reality at the microphysical level is not to be in the position of a spectator of a spatio-temporal history enacted by real particles and/or waves. How, then, do we gain knowledge of atomic systems? In precisely the way that Bohr indicates: by *interacting* with them and learning how to describe such interactions *unambiguously*. Each interaction is a whole phenomenon which can be described through quantum theory as the causal effect of the behavior of microsystems. Classically, in the phenomenon of "seeing a billiard ball" the interaction between the agency of observation, a human eyeball, and the object, the billiard ball, is such that each member of the interacting pair can be said to exist in a well-defined state. But, due to the quantization of interaction, at the microphysical level this interaction is such that it can be described adequately only by a formalism that characterizes both systems as in a single, non-decomposable state. There is no range of phenomena which cannot be statistically predicted by the quantum theoretical formalism, and there is no single phenomenon which can be interpreted as revealing more about the behavior of the microsystems than the theoretical formalism permits. As those phenomena which serve to confirm quantum theory are generically no different from all phenomena, they *can* be described unambiguously through the everyday concepts suitably refined by classical physics which are well-defined through reference to everyday phenomena. Nevertheless, because we must describe the phenomenon "objectively", by which Bohr means *unambiguously*, we must be able to distinguish between what is considered the observed object system and what is considered the observing system, and we can do this through the classical concepts. The necessity of distinguishing unambiguously between observing system and observed object in the theoretical "interpretation of the phenomenon" as an *observation* of a microsystem creates a conceptual distinction between the interacting systems where physically there is only one state. Consequently, while such descriptive concepts are perfectly well defined in the description of the observed object as it reveals itself phenomenally, they are *not* defined at all for reference to properties imagined as possessed by a substantial object existing *independently* of observation.

Hence, we know nature not by picturing it but by interacting with it. It is the wholeness of interactions at the microphysical level -the heart of the strangeness inherent in the quantum revolution- that demands altering our conception of the object of description in the philosophy of nature. It is not an object which "looks like" a phenomenal object when no one is looking at it.[16] Neither space-time accounts nor causal accounts using the dynamic conservation principles present a picture corresponding to the essential nature of the physical systems causing phenomena. Bohr's refusal to speculate about models of quantum mechanical reality follows from the fact that he sees such attempts at "description" of quantum mechanical entities as essentially attempts to *picture* that reality, not from some alleged positivist distrust of metaphysics. However, what should be obvious from Bohr's interpretation of quantum physics, is that such realities can be known only in terms of their potential for interaction. It is this fact which links the phenomenally observed to what can be known about the order of reality so revealed. Thus according to Bohr's concept of reality, real entities are entities that have the power to interact, and in interacting, to produce the phenomena which comprise the natural world.

Since what we can say about the quantum mechanical objects which cause these phenomena is simply this, that it is in interactions with these objects that certain specific phenomena are produced, then "to have knowledge" of such objects *means* to be able to predict (albeit only probabilistically) what specific sorts of phenomena such objects will cause in any particular given observational interaction. If "description" is assumed to mean the attribution of properties to substantial

----

16. Berkeley's lesson has finally caught up with physics. The old rationalist gambit of the primary/secondary quality distinction has been flushed out. Bohr would agree with the Kantian prohibition against the application of the substance property schema to objects outside of experience. However, he bases the prohibition on an empirical discovery of a physical fact, the quantization of action; Kant gives it a transcendental anchor in the nature of the human understanding.

entities, then *description* of objects is possible only for *phenomenal* objects, *i.e.* objects as we experience them. Complementarity is in accord with the empiricism of science in that it holds that what we describe in the scientific "description of nature" is, and can only be, nature as we experience it. Hence the definition of the "quantum state" of an isolated system in terms of *quantum* parameters is not a "description" in this sense, for unlike the classical state parameters, these parameters cannot be correlated with property terms that can be given direct empirical meaning. However, through defining the quantum mechanical state we are able to predict the likelihood of the occurrence of phenomena in specified conditions. Thus even though we cannot "describe" the quantum mechanical object in the way which we had presupposed in classical physics, because the prediction of the observational phenomena which form the empirical basis of quantum theory requires treating these phenomena as the causal consequences of the behavior of microsystems, it provides us with *knowledge* about the entities which produce these phenomena. By acquiring the quantum mechanical formalism surely we have advanced in our knowledge over the time prior to the quantum revolution when we could not even have dreamed of providing any such prediction.

Consequently complementarity supports a realist understanding of science in the two senses considered above. *First*, what never should have been doubted, Bohr defends *entity* realism with respect to the atomic systems whose behavior causes the phenomena described by quantum physics. *Second*, Bohr defends an *epistemic* realism which is in basic accord with realist values in that he holds that through the scientific description of nature we gain *knowledge* about the real entities which produce the phenomena we experience. It is true that this epistemic realism is not in accord with the assumptions of the dominant consensus, but, that should hardly be surprising, for Bohr reiterated time and again that quantum physics taught an epistemological lesson, that it forced on us *a change in what we considered knowledge to be*. Thus complementarity presents a revisionary definition of what it is "to have knowledge of an object". The *descriptive* statements quantum theory makes are restricted to statements about phenomenal objects. Moreover, considered as a formalism for predicting these phenomena the theory as a whole contains all the knowledge about atomic systems which can be confirmed by observation. Surely Bohr never intended to be understood as saying that in the advance of physics that constituted the quantum revolution we have learned nothing about the real atomic systems which produce the phenomena that confirm quantum theory.

IV. Epilogue: Why Bohr's Realism is Important

Let us review the course of these reflections about the role of reality in complementarity. Our basic question has been this: in the discovery that we can describe nature in the way it is described in contemporary atomic physics, have we discovered anything about the *real* inner structure of matter which causes the phenomena we point to as evidence of this mode of description, or is it the case that all we have done is contrived an intricate scheme for the correlation and prediction of such phenomena. I have argued that Bohr's doctrine of complementarity, and indeed his whole career, makes no sense unless we adopt the former alternative. However, Bohr's refusal to speculate on the nature of that underlying reality which complementarity must presuppose suggests that he may be in the uncomfortable position of presupposing the existence of real entities which, it would appear, we have no way to describe. However, this appearance is mistaken, and it is mistaken for the very same reason that much of Bohr's philosophy has been misunderstood, namely, that the criticism is launched from epistemological presuppositions which were perhaps plausible given the classical description of reality, but by the implications of the quantum postulate, denied in the quantum theoretical account. Bohr spoke of the recognition of this change of presuppositions for the description of nature as an "epistemological lesson" and he believed that lesson to have very far reaching consequences, at least in some respects parallel to the revolution which the description of nature underwent in the move from Copernicus through Newton. The reason Bohr *seems to be saying* that we can have no knowledge of the reality that produces the phenomena that count as evidence for the theory lies in our strong tendency to retain the classical presuppositions; we have not truly gone over to his complementaristic viewpoint. However, from the vantage point of his presuppositions, there is no mystery in Bohr's silence on the question of the nature of reality at the quantum level; indeed, it would appear instead that he has in fact said a good deal, perhaps even everything one can say, on the subject.

Now it may seem that to "save" the claim that realism can be defended in quantum physics only at the cost of a major overthrow of widely accepted epistemological presuppositions is a price to high to pay. Why not keep the basic epistemological presuppositions which have guided much of philosophy since the Enlightenment, and relinquish the foolish conceit of realism. After all if Bohr's view is realism, it is certainly not realism in the sense of the old correspondence/spectator/picturing account. Why, then, do I maintain it is wrong to call Bohr an anti-realist? After all, the labels hardly mean that much any more. In the current literature sometimes it seems you can call just about anything "realism" or "anti-realism" without it making much difference.[17] What this state of affairs indicates, of course, is that there are many definitions of "realism" clamoring for acceptance. Indeed, it has become fashionable to catalogue and develop typologies of the varieties and species of the genus *realisticae*. This, then, is the story of that species of realism which I have called "Bohr's realism." Insofar as Bohr's realism is a realism about quantum physics, and inasmuch as quantum physics is the present point to which the ancient, complex story of atomism has led, this could be said to be the latest chapter in the long story of atomism. So I think that it is very important to recognize that Bohr's viewpoint puts forward a kind of realism which endorses a radically new concept of reality, for without such a recognition, it is impossible to assess just where the long historical dialogue on atomism as a fundamental ontological hypothesis about the physical world now stands.

----

17. For example, Cartwright calls herself an anti-realist, but is welcomed as an ally by Hacking who claims to be a realist. Ellis argues realism must abandon correspondence while Fine takes it to be the essence of realism. As for Fine, who proclaims the death of realism, his candidate NOA seems to many to be exactly what realism has meant all along.

COMPLEMENTARITY AND UNCERTAINTY

Some measurement-theoretical and information-theoretical ascepts

PEKKA J. LAHTI
Department of Physical Sciences
University of Turku
SF-20500 Turku, Finland

ABSTRACT

In this lecture the usual Hilbert space formulation of quantum mechanics will be applied to study the properties of pairs of physical quantities which are complementary or which satisfy the uncertainty relation. The Heisenberg-Bohr interpretation of the uncertainty relations will be outlined within the extended Hilbertian framework of quantum mechanics. Probabilistic, information-theoretical, and measurement-theoretical aspects of the notion of complementarity, the uncertainty relations, and the Heisenberg-Bohr interpretation of the uncertainty relations will deserve special attention here.

1. INTRODUCTION

Max Born's proposal for a probability interpretation of the Schrödinger ψ-function[1)], Werner Heisenberg's discovery of the uncertainty relations[2)], and Niels Bohr's introduction of the viewpoint of complementarity[3)] were decisive steps towards a clarification of the conceptual foundations of the quantum theory and they led to what became known as the Copenhagen interpretation of quantum mechanics[4,5,6)]. Though this interpretation of quantum mechanics remained rather vague and intuitive and though it was never developed in to a systematic theory, the main intuitive ideas of that interpretation have served as starting-points or guide-lines for almost any subsequent systematic development of the theory, see e.g. references 4,6,7,8).

Indeed, the notion of probability is central in almost any formulation of quantum mechanics.[9)] The irreducability of the notion of probability[10] and the impossibility of justifying a subjective ignorance interpretation for quantum mechanical probabilities[11] are characteristics of the Hilbert space formulation of quantum mechanics, but they can also be used as building stones in some axiomatic reconstructions of the theory[12]. The notion of complementarity can be used to distinguish quantum mechanics from more general theories[13] or it can be used, for example, to characterize some of the main features of the usual (Hilbert space) formulation of quantum mechanics[14]. The existence of pairs of physical quantities which satisfy the uncertainty relation exhibits again a special feature of quantum mechanics.[15] In addition to this, the Heisenber-Bohr interpretation of the uncertainty relations [as the accuracy limits within which the values of complementary physical quantities can be simultaneously determined] requires an extension of the usual formulation of quantum mechanics[16] and it has led to the concept of unsharp reality[17] which is an important step towards a positive characterization of the concept of reality based on quantum physics[18].

In this paper some recent theoretical notions will be applied to study probabilistic, information-theoretical, and measurement-theoretical aspects of the notion of complementarity and the uncertainty relations.

Throughout the paper the usual Hilbert space formulation of quantum mechanics will be applied. To fix the notations and terminology I shall first (Sec. 2.1) recall some of its relevant ingredients. The notion of measurement is most important in the development of the theme of my paper. Thus, in Sec. 2.2, I shall briefly refer to some relevant results in the quantum theory of measurement. Sections 3 and 4 deal, respectively, with pairs of quantities which are complementary and which satisfy the uncertainty relations. Main properties of such pairs will be made explicit, the emphasis being on their measurement theoretical content. Different interpretations of the uncertainty relations will briefly be discussed (Sec. 4.2) in order to appreciate fully the novelty of the Heisenber-Bohr interpretation of those relations (Sec. 4.3). Section 5 discusses the information-theoretical aspects of the properties of the pairs of quantities which are complementary or which satisfy the uncertainty relations. The notion of informational completeness will be used in Sec. 5.1 to analyse the state distinction property of the relevant measurements. In Sec. 5.2 a special local information functional will be applied to explain the value determination property of those measurements. The final Section 6 summarizes in a verbal form the results presented in this paper.

2. GENERAL FRAMEWORK

2.1. Basic Notations

In the Hilbert space formulation of quantum mechanics (QM) the description of a physical system $\underline{S}$ is based on a (complex, separable, generally infinite dimensional) Hilbert space $\underline{H}$, with the inner product $\langle\cdot|\cdot\rangle$. Any <u>physical quantity</u> of the system is represented as (and identified with) a self-adjoint operator A in $\underline{H}$. The spectral measure of A is denoted by $E^A:\underline{B}(\mathbb{R})\to\underline{L}(\underline{H})$, where $\underline{B}(\mathbb{R})$ is the Borel σ-algebra of the real line $\mathbb{R}$ ($\underline{B}(\mathbb{R}^2)$ that of $\mathbb{R}^2$) and $\underline{L}(\underline{H})$ is the set of bounded linear operators on $\underline{H}$. Any <u>state</u> of the system is represented as (and identified with) an element T of $\underline{T}_s(\underline{H})_1^+$ of positive normalized trace class operators on $\underline{H}$. In this representation the pure states of the system appear as the (normalized) one-dimensional projection operators $|\varphi\rangle\langle\varphi|$ ($|\varphi\rangle\langle\varphi|\psi = \langle\varphi|\psi\rangle\varphi$), or $P[\varphi]$, on $\underline{H}, \varphi\in\underline{H}$, so that they may be identified, modulo a phase factor, with the unit vectors φ of $\underline{H}$. $\underline{H}_1$ denotes the set of unit vectors of $\underline{H}$.

Any pair (A,T) of a physical quantity A and a state T defines a real normalized measure on the Borel σ-algebra $\underline{B}(\mathbb{R})$ of $\mathbb{R}$, i.e. a probability measure

$$E_T^A:\underline{B}(\mathbb{R})\to[0,1],\ X\mapsto E_T^A(X) := \mathrm{tr}(TE^A(X)) \tag{1}$$

According to the <u>Born interpretation</u>, E_T^A is the probability measure (or distribution) of the possible values of the quantity A in the state T, i.e. $E^A_T(X)$ is the probability that the value of A is in the set $X \in \underline{B}(\mathbb{R})$ when the system is in the state T.

In studying the properties of pairs of physical quantities A and B their <u>commutativity domain</u> $com(A,B) := \{\varphi \in \underline{H} \mid E^A(X)E^B(Y)\varphi = E^B(Y)E^A(X)\varphi \text{ for all } X,Y \in \underline{B}(\mathbb{R})\}$ is very useful, as will soon be seen. com(A,B) is a closed subspace of $\underline{H}$, which is invariant under the spectral projections of A and B. Clearly A and B commute [in the sense that all their spectral projections commute] if and only if $com(A,B) = \underline{H}$. If $com(A,B) = \{0\}$, then A and B are <u>totally noncommutative</u>. The notion of commutativity domain admits, it particular, a probabilistic characterization: com(A,B) consists exactly of those (unit) vectors $\varphi \in \underline{H}$ for which the map $X \times Y \mapsto \langle\varphi| E^A(X) \wedge E^B(Y)\varphi\rangle$, $X,Y \in \underline{B}(\mathbb{R})$, extends to a probability measure $\underline{B}(\mathbb{R}^2) \to [0,1]$, $Z \to E^{A,B}_{|\varphi\rangle\langle\varphi|}(Z)$.[19-21] $E^{A,B}_{|\varphi\rangle\langle\varphi|}$ is the type-1 joint probability of A and B in the state $|\varphi\rangle\langle\varphi|$. As shown in ref. 21, this all extends immediately to arbitrary states T, as well.

The concept of a physical quantity as a (real) spectral measure E^A (or self-adjoint operator A) is unnecessarily restricted. In fact, it is insufficient e.g. for the Heisenberg-Bohr interpretation of the uncertainty relations. A natural generalization is a <u>generalized physical quantity</u> as a semispectral measure $E:\underline{B}(\mathbb{R}) \to \underline{L}(\underline{H})$, i.e. a positive operator valued measure for which $(0 \leq E(X) \leq)$ $E(\mathbb{R}) = I$. Spectral measures $E^A:\underline{B}(\mathbb{R}) \to \underline{L}(\underline{H})$ are exactly those semispectral measures $E:\underline{B}(\mathbb{R}) \to \underline{L}(\underline{H})$ for which $E(X) = E(X)^2$ for all $X \in \underline{B}(\mathbb{R})$. Clearly any pair (E,T) of a generalized physical quantity E and a state T defines again a probability measure E_T for which the Born interpretation may be adopted. The commutativity domain is also well-defined for the generalized quantities. However, for lack of the above-mentioned invariance property, no stright-forward generalization of the probabilistic characterization of the commutativity domain is known.

2.2. Measurement

The testability requirement of the basic QM-probabilities $E^A_T(X)$ requires, at least, the following two points. Firstly, there should exist a measuring apparatus $\underline{M}$, say, with which the values of a quantity A on the (object) system $\underline{S}$ in a state T can be determined. In effect, a suitable interaction between $\underline{S}$ and $\underline{M}$ should be established, after which the value of A on $\underline{S}$ will be read off the apparatus $\underline{M}$. Secondly, the numbers $E_T^A(X)$ should be given a measurement-theoretical content: $E_T^A(X)$ is the probability that the measurement of A on $\underline{S}$ with the apparatus $\underline{M}$ yields a result in X when the system $\underline{S}$ is initially (i.e. prior to the measurement) in the state T. Clearly such a measurement-theoretical interpretation (or rather reading) of the numbers $E_T^A(X)$ does not imply any restrictions on the notion of measurement, especially $\underline{M}$ being redundant here.

In the quantum theory of measurement (QTM) one restricts oneself to considering only such measurements which preserve the identity of the object system $\underline{S}$. If $\underline{M}$ realizes such a measurement of a quantity A of $\underline{S}$, then after the measurement, i.e. when the measuring interaction has ceased, $\underline{S}$ and $\underline{M}$ can be separated from each other. The other basic assumption of QTM is that the measuring apparatus $\underline{M}$, or at least a relevant part of it, is a physical system in the domain of (generalized) quantum mechanics. This assumption has no implication e.g. on the 'size' of $\underline{M}$ [microscopic or macroscopic] nor on the 'nature' of $\underline{M}$ [quantum or classical]. E.g. the number of the degrees of freedom of $\underline{M}$ may be big or $\underline{M}$ may have classical properties [which can be described within the generalized quantum mechanics e.g. in terms of the superselection rules].

With the above assumptions QTM is but a part of the theory of compound systems, and it remains to be analysed which kind of evolutions of the compound system $\underline{S}+\underline{M}$ may serve as measurements of a quantity A of the subsystem $\underline{S}$, and which kind of properties of $\underline{M}$ would allow one to determine in a unique way the value of A on $\underline{S}$ with $\underline{M}$. This is, of course, standard since the pioneering work of John von Neumann[22] and Wolfgang Pauli[23]. For our subsequent needs, it is, however, useful to condense the above ideas in the form of a definition. For this purpose, let $\underline{H}_M$ be the Hilbert space of the measuring apparatus $\underline{M}$, and let A_M be the so-called pointer observable of the measuring apparatus $\underline{M}$, i.e. a quantity of $\underline{M}$ which corresponds to the measured quantity A of the object system $\underline{S}$ [We use subindex M to refer to the quantities associated with $\underline{M}$.] Let T and T_M be the initial, or premeasurement, states of $\underline{S}$ and $\underline{M}$, respectively, so that $T \otimes T_M$ is the initial state of $\underline{S}+\underline{M}$. If $V: \underline{T}_s(\underline{H} \otimes \underline{H}_M) \rightarrow \underline{T}_s(\underline{H} \otimes \underline{H}_M)$ is the (measurement) evolution of $\underline{S}+\underline{M}$ [i.e. a trace preserving positive linear mapping], then the partial traces $T_V := tr_{\underline{H}_M}(V(T \otimes T_M))$ and $T_{M,V} := tr_{\underline{H}}(V(T \otimes T_M))$ are the final (or postmeasurement) states of $\underline{S}$ and $\underline{M}$, respectively.

Definition 2.2.1. With the above notations, a measurement (or a measurementprocess) of A on $\underline{S}$ is a quadruple $\langle \underline{H}_M, A_M, T_M, V \rangle$ which

(a) reproduces the statistics (A,T) through $(A_M, T_{M,V})$, i.e. $E_T^A = E_{T_{M,V}}^{A_M}$, for any $T \in \underline{T}(\underline{H})_1^+$, and

(b) leads to the objectification of the measurement result.

The second requirement, the objectification of the measurement result, is connected with the problem of justifying the subjective ignorance interpretation for the postmeasurement state T_V of the object system $\underline{S}$ through that of the postmeasurement state $T_{M,V}$ of the measuring apparatus $\underline{M}$ This is a well-defined though highly difficult problem, see e.g. 24-26). This question does not concern us here so that we shall now omit it from the above definition. The first requirement is a minimal condition for a measurement and it leads, in some cases, to the strong correlation between the values of A and A_M.[26)] The existence of a measurement process for each physical quantity A of the object system $\underline{S}$ is one of the main results in QTM. Indeed, we have:

Theorem 2.2.2. For any physical quantity of the object system $\underline{S}$ there is a quadruple $\langle \underline{H}_M, A_M, T_M, V \rangle$ for which $E_T^A = E_{T_{M,V}}^{A_M}$ for all $T \in \underline{T}_s(\underline{H})_1^+$.

An explicit construction of such a measurement can be given in the case of discrete quantities e.g. in terms of the so-called ideal first kind measurements[11,26]. This is, in fact, well known since the work of von Neumann[22]. That the theorem holds also for continuous as well as for generalized quantities was shown by Ozawa[27].

Any measurement process $<\underline{H}_M, A_M, T_M, V>$ of a physical quantity A (generalized or not) determines an <u>instrument</u> (i.e. an operation or state-transformation valued measure) $I_M^A : \underline{B}(\mathbb{R}) \to \underline{L}(\underline{T}_s(\underline{H}))$ through the relation $I_M^A(X)T := tr_{\underline{H}_M}(V(T \otimes T_M) \cdot I \otimes E^{A_M}(X))$ for all $X \in B(\mathbb{R})$, $T \in \underline{T}_s(\underline{H})_1^+$. The instrument I_M^A contains all the information on the measurement process $<\underline{H}_M, A_M, T_M, V>$ which is relevant to the object system $\underline{S}$: firstly, it gives the correct probabilities as $E_T^A(X) = tr(I_M^A(X)T)$ for all $X \in \underline{B}(\mathbb{R})$, $T \in \underline{T}_s(\underline{H})_1^+$; secondly, it gives the (nonnormalized) postmeasurement state of $\underline{S}$, $I_M^A(X)T$ being the state of $\underline{S}$ after the measurement with the condition that the measurement yielded a result in X. The properties of a measurement, like its ideality, repeatability, or first kindness, can then most directly be studied through the properties of their instruments.

The notion of instrument is, from the outset, more general.[28] In fact, any (generalized) physical quantity $E : \underline{B}(\mathbb{R}) \to \underline{L}(\underline{H})$ is associated with a unique family of instruments $I_i^E : \underline{B}(\mathbb{R}) \to \underline{L}(\underline{T}_s(\underline{H}))$, $i \in \underline{I}(E)$, $\underline{I}(E)$ a suitable index set, for which $E_T(X) = tr(I_i^E(X)T)$ for all $X \in \underline{B}(\mathbb{R})$, $T \in \underline{T}_s(H)_1^+$, $i \in \underline{I}(E)$. As we have seen above, any measurement process of A determines an instrument I_M^A of A. This instrument is completely positive (for this notion, see e.g. ref. 28) whenever the measurement evolution V is such. On the other hand, any completely positive instrument of A has a measurement-theoretical interpretation through a unitary measurement process, i.e. through a measurement process with a unitary evolution.[27] It is an open question whether all instruments have measurement-theoretical interpretations.

With these preliminaries we are now ready to analyse the problems posed in this paper.

3. COMPLEMENTARITY - What cannot be done

According to Wolfgang Pauli[23] modern quantum theory might have been called "The Theory of Complementarity". This is one of the many expressions indicating the importance of the idea of complementarity in the foundations of quantum mechanics - at least as they were understood within the circles close to Niels Bohr.

I shall not try to analyse here Bohr's general viewpoint of complementarity. I shall restrict myself to discuss complementary physical quantities only. In fact, it is this part of the notion of complementarity which serves to explain most directly some characteristic features of quantum mechanics and which thus also serves to distinguish quantum mechanics from more general theories[12,13).]

There seem to be two intuitive formulations of the notion of complementary physical quantities as discussed e.g. by Bohr[3,29)] and Pauli[23)]. One of these formulations refers to the mutual exclusiveness of the experimental arrangements which permit the unambiguous [operational] definitions, or value determinations or observations, of certain physical quantities, the other refers to the mutual exclusiveness of the certainly-yes predictions concerning the possible values of these quantities. Both of these formulations have been discussed and analysed in detail in the literature.[3,29,23,30,31,7,32,11,14)] I shall now give the proper formal definitions and discuss briefly their mutual inter-relations.

Definition 3.1. Instruments I_1 and I_2 are mutually exclusive if $I_1(X) \wedge I_2(Y) = 0$ for any bounded $X, Y \in \underline{B}(\mathbb{R})$ for which neither $I_1(X)$ nor $I_2(Y)$ is maximal.

Definition 3.2. Physical quantities A and B are complementary if any two of their instruments I_i^A, $i \in \underline{I}(A)$, I_j^B, $j \in \underline{I}(B)$, are mutually exclusive. (Cp. Bohr[29)], Pauli[23)].)

This formulation of complementarity has a direct connection to quantum measurement theory, cf. Sec. 2.2 above.

Theorem 3.3. Physical quantities A and B are complementary if and only if $E^A(X) \wedge E^B(Y) = 0$ for any bounded $X, Y \in \underline{B}(\mathbb{R})$ for which $E^A(X) \neq I \neq E^B(Y)$. (Cp. Ludwig[31,7], von Weizsäcker[32].)

<u>Definition 3.4</u>. Physical quantities A and B are <u>probabilistically complementary</u> if they share the following property:

$$\text{if } E^A_T(X) = 1, \text{ then } E^B_T(Y) < 1, \text{ or}$$

$$\text{if } E^B_T(Y) = 1, \text{ then } E^A_T(X) < 1,$$

for any $T \in \underline{T}_s(\underline{H})^+_1$ and for any bounded $X, Y \in \underline{B}(\mathbb{R})$ for which $E^A(X) \neq I \neq E^B(Y)$. (Cp. Pauli[23], Fock[30], Beltrametti and Cassinelli[11].)

<u>Remark 3.5</u>. Definitions 3.2 and 3.4 can be applied to generalized quantities E and F (semispectral measures) as well. Also in that case Theorem 3.3 gives an equivalent formulation of Definition 3.2. Definitions 3.2 and 3.4 are equivalent in case of standard physical quantities (spectral measures). In general, complementary physical quantities (Def. 3.2) are also probabilistically complementary (Def. 3.4) but not necessarily the other way round. This is very important, as shall be demonstrated later with unsharp position and momentum observables Q_f and P_g.

Our previous results[12-14] together with the results of Sec. 2.2 now show the following:

<u>Statement 3.6</u>. Complementary physical quantities A and B have no order independent sequential measurements, i.e., the result of a sequential measurement of A and B depends, in general, on the order in which A- and B-measurements will be performed.

<u>Statement 3.7</u>. Complementary physical quantities have no joint measurements, i.e., no A- and B-measurements can be combined to produce a joint measurement of A and B.

Statement 3.8. Probabilistically complementary physical quantities have no first kind, or preparatory, joint measurements.

Note that 3.6-3.8 hold true for generalized quantities as well. Furthermore, we have:

Statement 3.9. Complementary physical quantities A and B are totally noncommutative (i.e. $\mathrm{com}(A,B) = \{\varphi \in \underline{H} \mid E^A(X)E^B(Y)\varphi = E^B(Y)E^A(X)\varphi$ for all $X,Y \in \underline{B}(\mathbb{R})\} = \{0\}$).[15)]

With respect to the probabilistic interpretation of the commutativity domain (cf. sec. 2.1) this statement has the corollary:

Statement 3.10. Complementary physical quantities A and B have no type-1 joint probability [i.e. the mapping $X \times Y \to \mathrm{tr}(TE^A(X) \wedge E^B(Y))$ extends to a probability measure on $\underline{B}(\mathbb{R}^2)$ for no state $T \in \underline{T}_s(\underline{H})_1^+$].

It is unknown whether 3.9 and 3.10 hold true also for generalized quantities.

Example 3.11. The usual position Q and momentum P (as Fourier-related quantities) are complementary as e.g. $E^Q(X) \wedge E^P(Y) = 0$ for any bounded $X,Y \in \underline{B}(\mathbb{R})$. To define unsharp position Q_f and momentum P_g we specify the above (Q,P) pair to be defined in $\underline{H} = L_2(\mathbb{R},dx)$. Then e.g. Q_f is defined as the semispectral measure $X \mapsto E^{Q_f}(X) := (\chi_X * f)(Q)$, where $\chi_X * f$ denotes the convolution of the characteristic function χ_X, $X \in \underline{B}(\mathbb{R})$, and the confidence [or probability density] function f. If the confidence functions f and g are Fourier-related, i.e. $f = |\varphi|^2$ and $g = |\hat{\varphi}|^2$ for some unit vector φ and its Fourier transform $\hat{\varphi}$ in $\underline{H}$, then Q_f and P_g are not complementary, but they are always, i.e. for any f and g, probabilistically complementary.[14)]

With respect to the above example it is interesting to note that realistic position and momentum measurements need not be mutually exclusive (cf. Sec. 4.3). We also note that the (Q,P) pair shows up all the distinctive features one usually ascribes to the notion of complementarity. Q and P are <u>mutually exclusive</u> (in the sense of Def. 3.2), but they are <u>not contradictory</u> (as the disjoint properties $E^Q(X)$ and $E^P(Y)$ are not orthogonal). In a probabilistic sense, Q and P also <u>complete each other</u> (cf. Def. 3.4). However, it is most important to note that Q and P are <u>not informationally complete</u> (see Sec. 5.1.8).

4. UNCERTAINTY - What can be done

The existence of pairs of physical quantities which satisfy the uncertainty relations exhibits another special feature of quantum mechanics. Moreover, the Heisenberg-Bohr interpretation of the uncertainty relations is an important extension of the Born interpretation of these relations and it goes beyond the statistical interpretation of the uncertainty relations. As is already well known, the Heisenberg-Bohr interpretation of the uncertainty relations [as the accuracy limits within which the values of complementary physical quantities can simultaneously be determined] requires an extension of the usual formulation of quantum mechanics. This extension has led to the concept of unsharp reality which is an important step towards a positive characterization of the concept of reality based on quantum physics.

Here I shall first recall the general formulation of the uncertainty relations and show some of the main features of pairs of physical quantities which satisfy such relations. Then I will briefly mention the probabilistic and statistical interpretations of the uncertainty relations in order to do justice to the original Heisenberg interpretation as an important extension of the Born [probabilistic] interpretation.

4.1. General Formulation

Let A and B be any two physical quantities. We shall now study the product $Var(A,T) \cdot Var(B,T)$ of their variances in a state $T \in \underline{T}_s(\underline{H})_1^+$. [Here, e.g., $Var(A,T)$ denotes the variance of the probability measure E_T^A, i.e. $Var(A,T) := \int_{\mathbb{R}}(i - \int_{\mathbb{R}} i dE_T^A)^2 dE_T^A$.] We denote $h(A,B) = \inf\{Var(A,T) \cdot Var(B,T) \mid T \in \underline{T}_s(\underline{H})_1^+\}$ so that $h(A,B) = 0$ or $h(A,B) > 0$.

Definition 4.1.1. Physical quantities A and B satisfy the uncertainty relation if $h(A,B) > 0$.

If A and B satisfy the uncertainty relation, then the "uncertainty product" $Var(A,T) \cdot Var(B,T)$ of A and B has a positive lower bound which is independent of the state of the system. This means that the probability measures E_T^A and E_T^B of A and B are always, i.e. in any state T, so correlated that whenever e.g. the variance of A in a state T is "small" (i.e. $Var(A,T) < \varepsilon$) then the variance of B in this same state is "big" (i.e. $Var(B,T) \geq h(A,B)/\varepsilon$). This is the direct probabilistic content of the uncertainty relation. It can still be dramatized by noting that for any A there is a T such that $Var(A,T)$ is smaller than any given positive number ε. The following result leads, however, to a deeper understanding of the probabilistic content of the uncertainty relations.

Theorem 4.1.2. If A and B satisfy the uncertainty relation, then they are totally noncommutative [i.e. $h(A,B) > 0$ implies $com(A,B) = \{0\}$].[15)]

Consequently, if A and B satisfy the uncertainty relation, then they have no type-1 joint probability (cf. 3.10) and they have no order independent sequential measurements (cf. 3.6, see also item 6 in ref. 14).

The above results explain the probabilistic content of the uncertainty relations constituting thus the Born interpretation of the uncertainty relations. Though important and clarifying they do not, however, exhaust the interpretations of these relations.

4.2. Some Probabilistic Interpretations

The basic probability assertions of quantum mechanics take the form "$E_T^A(X) = p$" where $E_T^A:\underline{B}(\mathbb{R}) \to [0,1]$ is a usual, i.e. Kolmogorovian probability measure and $0 \leq p \leq 1$. Thus any interpretation of the notion of probability, like a relative frequency or a propensity interpretation, also gives an interpretation of the QM-probabilities, and, in particular, of the QM-variances. I shall not follow directly such an approach here. Rather, I shall first refer to the quantum theory of measurement which gives a measurement-theoretical interpretation of the QM-probabilities E_T^A. According to Theorem 2.2.2, for any pair (A,T) of a physical quantity A and a state T there is a measuring process $<\underline{H}_M, A_M, T_M, V>$ (even a unitary one) such that $E_T^A = E_{T_{M,V}}^{A_M}$. Accordingly, the basic QM-probabilities E_T^A can be interpreted [in a strict sense of a formal model] through the probabilities $E_{T_{M,V}}^{A_M}$ referring explicitly to a measurement process $<\underline{H}_M, A_M, T_M, V>$ of A. The number $E_T^A(X)$, $X \in \underline{B}(\mathbb{R})$, may then consistently be interpreted as the probability that a measurement of A, with the measuring process $<\underline{H}_M, A_M, T_M, V>$, on the system $\underline{S}$ in the state T leads to a result in the set X. This <u>measurement-theoretical interpretation</u> of the QM-probabilities will also lead to a measurement-theoretical interpretation of the QM-variances Var(A,T), and thus of the uncertainty relations.

When the above measurement-theoretical interpretation is combined with a relative frequency interpretation of probability, e.g. in the sense of van Fraassen[33], on obtains a <u>measurement-statistics interpretation</u> of QM.[34] In this interpretation Var(A,T) is identified with the variance $Var(\Gamma_T^A)$ of the measuring results of A in the sequence $\Gamma_T^A = \{a_1, a_2, \ldots, a_k, \ldots\}$ obtained in repeating the same measuring process of A on $\underline{S}$ in the state T many times, in fact infinitely many times. If A and B satisfy the uncertainty relation, then their measurement statistics are always so correlated that $Var(\Gamma_T^A) \cdot Var(\Gamma_T^B) \geq h(A,B)$. This statement is often rephrased as follows: Only such states T can be prepared for which the measurement statistics of A and B fulfil the requirement $Var(\Gamma_T^A) \cdot Var(\Gamma_T^B) \geq h(A,B)$. It is interesting to ask whether also the converse statement holds true, i.e. if the measurement statistics of two quantities A and B are always so correlated that $\mathrm{Var}(\Gamma^A_T)\cdot\mathrm{Var}(\Gamma^B_T) \geq h(A,B) \neq 0$ whether the statistics determine the state [initial preparation] uniquely. We shall return to this question in Sec. 5.1.

4.3. The Heisenberg-Bohr Interpretation

As already mentioned, the Heisenberg-Bohr interpretation of the uncertainty relations is an important extension of the probabilistic interpretation of QM. In that interpretation the quantity $\Delta(A,T) = \sqrt{\mathrm{Var}(A,T)}$ is identified with a characteristic accuracy $\delta(A,T)$, say, in determining the value of A in the state T. Then $\mathrm{Var}(A,T)\cdot\mathrm{Var}(B,T) \geq h(A,B)$ is interpreted as providing the accuracy limits within which A and B can simultaneously be determined.

In addition to this, there is also a measurement-theoretical aspect of the Heisenberg-Bohr interpretation. Namely, it is argued that "the characteristic limitations in the form of the uncertainty relations can be traced back to the interaction of the measuring apparatus and the system".[35)]

The intuitive Heisenberg-Bohr interpretation of the uncertainty relations can systematically be worked out by replacing a physical quantity A with a pair (A,f), where $f:\mathbb{R}\to\mathbb{R}$ is a suitable probability density function describing the relevant properties, as finite resolution, of the measuring apparatus employed to determine the values of A. Formally, the pair (A,f) defines a semispectral measure obtained as a suitable smearing of the original spectral measure (cf. Ex. 3.11, above). It is only after introducing this <u>new element</u> [formally replacing a spectral measure with a semispectral measure representing the same quantity] into the theory that the Heisenberg-Bohr interpretation of the uncertainty relations can systematically be given; the $\delta(A,T)$ above can then be identified with the standard deviation Δf of the "confidence function" f. (For details and for further references, see e.g. ref. 14.2.) <u>For example</u> (no general method known), in the important case of (complementary) position Q and momentum P, with $h(Q,P) = 1/4$ [taking $h/2\pi = 1$], one finds that if Q and P are replaced with the pairs (Q,f) and (P,g) such that f and g are Fourier-related, i.e. $f = |\varphi|^2$ and $g = |\hat{\varphi}|^2$ for some $\varphi \in \underline{H}_1$, then Q and P can simultaneously be measured within the accuracies characterized by f and g: $\Delta f \cdot \Delta g \gtrsim 1/2$. Indeed, in that case the unsharp position Q_f and the unsharp momentum P_g are obtained as the marginal observables of the joint observable $E:\underline{B}(\mathbb{R}^2) \to \underline{L}(L_2(\mathbb{R},dq))^+$, $Z \to E(Z) := (2\pi)^{-1} \int_Z |U_{qp}\varphi><U_{qp}\varphi| dqdp$, $U_{qp} = e^{i(pQ-qP)}$, $\varphi \in \underline{H}_1$ (with $f = |\varphi|^2$ and $g = |\hat{\varphi}|^2$).[28)]

The existence of the joint observable E quarantees the simultaneous measurability of Q_f and P_g both in the probabilistic sense as well as in the measurement-theoretical sense.

We emphasize that Q_f and P_g are <u>not complementary</u> so that they possess instruments and thus measurements which are not mutually exclusive. However, Q_f and P_g <u>are probabilistically complementary</u> so that none of their joint measurements are preparatory.

Following QTM one may again give measurement-theoretical interpretations for any of the probabilities $E_T^{Q_f}$, $E_T^{P_g}$, and E_T, $T \in \underline{T}_s(\underline{H})_1^+$. These interpretations do not, however, show that the "characteristic inaccuracies" Δf and Δg could be traced back to the measurement interactions. The question of reducing Δf and Δg to the measuring interactions between the apparata and the system is more delicate and it has been discussed only with respect to some models for Q- and P -measurements. As this is most important here, we shall briefly outline such models below.

Already von Neumann[22)] argued that the following realistic-type interaction (coupling)

$$H_{int}^{(1)} = a\, Q \otimes P_1$$

can be used to measure the position Q of the object system $\underline{S}$ by a measuring apparatus $\underline{M}_1$ initially in a state $T_1 = P[\Phi_1]$ employing its position Q_1 as the pointer observable, i.e. the quadruple

$<\underline{H}_1, Q_1, T_1, U_1 = \exp(-iaQ \otimes P_1)>$ is a realistic measurement for the position Q of the system $\underline{S}$. But, as it has been worked out in detail in Ref. 16, this measurement process defines an unsharp position Q_f with $f = f(|\Phi_1|^2, a)$. Similarly, a realistic measurement process $<\underline{H}_2, Q_2, P[\Phi_2], U_2 = \exp(-ibP \otimes P_2)>$ defines not the momentum P but the unsharp momentum P_g of the system, with the density g being again defined by the initial preparation $T_2 = P[\Phi_2]$ of the measuring apparatus $\underline{M}_2$ and the coupling parameter b, i.e. $g = g(|\Phi_2|^2, b)$.

The unsharp position Q_f and momentum P_g defined through the two measurement processes are probabilistically complementary, but the instruments defined by them need not be mutually exclusive. In fact, one may always choose the initial preparations of $\underline{M}_1$ and $\underline{M}_2$ such that this is the case. In any case, the combined measurement process $<\underline{H}_{1+2}, Q_1 \otimes Q_2, P[\Phi_1 \otimes \Phi_2], U_{1+2}>$, with $\underline{H}_{1+2} = \underline{H}_1 \otimes \underline{H}_2$ and $U_{1+2} = \exp[-i(aQ \otimes P_1 \otimes I_2 + bP \otimes I_1 \otimes P_2)]$, defines a joint observable E whose marginals are unsharp position $Q_{\bar{f}}$ and unsharp momentum $P_{\bar{g}}$ with Fourier-related confidence functions $\bar{f}$ and $\bar{g}$. This demonstrates that the coupling of the two measurement processes modifies the original (not necessarily Fourier-related) confidence functions f and g to a Fourier-related pair $(\bar{f}, \bar{g})$. (For further details, see ref. 14, and P. Busch, elsewhere in this volume.) The detailed properties of the resulting joint measurements of position and momentum depend on the initial preparations T_1 and T_2 of the measuring apparata $\underline{M}_1$ and $\underline{M}_2$. Due to the probabilistic complementarity of Q_f and P_g, none of these joint measurements are (strictly) preparatory.

It may be appropriate to close this section with the remark that von Neumann[22] could not come to the above conclusions because, instead of Q he considered a discrete position of the object system with respect to a given partition of the real line $\mathbb{R}$, the spectrum of Q. Now it is known that there is a principal dinstinction between the discrete quantities and the continuous ones; discrete quantities are characterized as those which admit ideal first-kind measurements.[27]

5. INFORMATION THEORETICAL CONSIDERATIONS

Any two physical quantities which are complementary or which satisfy the uncertainty relation are also totally noncommutative. In particular, such quantities have a joint probability in no state of the system. Then it becomes important, and nontrivial, to ask the following two questions:

(a) Under what conditions do the measurement statistics of such quantities determine the initial state of the system?

(b) To what degree of certainty can the values of such quantities be determined or known together?

These questions refer to the significance of the measurement results to the system before and after the measurement, respectively. With respect to the QTM-representation of an A-measurement these questions refer to the probabilities $E_T^A(X)$ and the initial state $T_i = T$ and the probabilities $E_T^A(X) = tr(I_M^A(X)T)$ and the postmeasurement states $T_f^X = I_M^A(X)T/E_T^A(X)^{-1}$, respectively. In view of the Heisenberg-Bohr interpretation of the uncertainty relations it is also very important to know what is the effect of replacing a sharp quantity with an unsharp one in these questions.

The first of the two questions can be studied in terms of the notion of informational completeness (Sec. 5.1). On the other hand, a quantitative analysis of the second question requires that an appropriate information functional $INFO(E_T^A, E_T^B)$ could be defined. For lack of any general theory I shall indicate in Sec. 5.2 a natural candidate for such a functional and discuss briefly its implications on the question (b).

5.1. Information Completeness - State Distinction Property of Measurements

The problem of the informational completeness of the complementary position and momentum observables was raised already by Wolfgang Pauli.[23] However, the "Pauli-problem" was seriously attacked only rather recently.[36] Though the notion of informational completeness was already introduced by E. Prugovecki[37] in 1977, the general theory waits still to be developed. I shall outline here some of the main concepts and results relevant for the present discussion.

<u>Definition 5.1.1</u>. A collection of physical quantities $\{A_1,A_2,\ldots,A_n\}$ is <u>informationally complete with respect to a set of states</u> $\underline{T}\subset\underline{T}_s(\underline{H})^+_1$ if for any T_1 and T_2 in $\underline{T}$ the condition

$$E^{A_i}_{T_1} = E^{A_i}_{T_2} \quad \text{for all } i = 1,\ldots,n \tag{2}$$

implies that $T_1 = T_2$. The collection $\{A_1,A_2,\ldots,A_n\}$ is <u>informationally (INFO-) complete</u> if it is informationally complete with respect to the whole set of states $\underline{T}_s(\underline{H})^+_1$. Similar definitions are given for generalized physical quantities $E_1,\ldots,E_n$, too.

Informationally complete physical quantities, and thus their measurements [represented by the relevant instruments] can distinguish between all the states of the system. The statistics of the measurement results $\Gamma^{A_i}_T$, $i = 1,\ldots,n$ of INFO-complete quantities allow one to infer the premeasurement or initial state T of the system.

Informational completeness is a property of a physical quantity [or a collection of physical quantities] and not a property of particular measurements or instruments as they all lead to the same measurement statistics.

<u>Statement 5.1.2</u>. No physical quantity A [with the spectral measure E^A] is INFO-complete whenever $\dim(\underline{H}) \geq 2$.

Proof. If A is not constant [i.e. a real multiple of the identity] then there is a $\varphi \in \underline{H}_1$ such that $\varphi' \equiv e^{iA}\varphi \neq e^{i\alpha}\varphi$ for any $\alpha \in \mathbb{R}$. Then $|\varphi'\rangle\langle\varphi'| \neq |\varphi\rangle\langle\varphi|$ but $E^A_{|\varphi'\rangle\langle\varphi'|} = E^A_{|\varphi\rangle\langle\varphi|}$ as $E^A(X)e^{iA} = e^{iA}E^A(X)$ for all $X \in \underline{B}(\mathbb{R})$.

Corollary 5.1.3. No set of commuting quantities $A_1,\ldots,A_n$ is INFO-complete. In particular, no complete set of commuting quantities nor any maximal quantity is INFO-complete.

Theorem 5.1.4. If the pair (A,B) is INFO-complete, then $\dim[\mathrm{com}(A,B)] \leq 1$.

Proof. If $\mathrm{com}(A,B) = \underline{H}$ (and $\dim\underline{H} \geq 2$), then by 5.1.3 they are not INFO-complete. If $\mathrm{com}(A,B) = \underline{K} \neq \{0\}$ and $\dim\underline{K} \geq 2$, then $\tilde{E}^A$ and $\tilde{E}^B$ commute on $\underline{K}$[21] and are not INFO-complete. But for any $\varphi \in \underline{K}_1 \subset \underline{H}_1$ $E^A_{|\varphi\rangle\langle\varphi|} = \tilde{E}^A_{|\varphi\rangle\langle\varphi|}$ and $E^B_{|\varphi\rangle\langle\varphi|} = \tilde{E}^B_{|\varphi\rangle\langle\varphi|}$. (Here e.g. $\tilde{E}^A(X) = E^A(X)|_{\underline{K}}$ for any $X \in \underline{B}(\mathbb{R})$.)

Example 5.1.5. The condition $\dim[\mathrm{com}(A,B)] \leq 1$ is not sufficient for (A,B) being INFO-complete. Indeed, Q and P are totally noncommutative but (Q,P) is not INFO-complete (see 5.1.6). The same holds true for any two spin components of a spin-$\frac{1}{2}$ system. In the spin-1 case we have e.g. $\mathrm{com}(S_x,S_y) = E^{S_z}(\{0\})$, $\underline{H} \triangleq \mathbb{C}^3$, so that $\dim[\mathrm{com}(S_x,S_y)] = 1$. But again (S_x,S_y) is not INFO-complete as can be demonstrated e.g. by the two different vector states $\varphi = (3)^{-\frac{1}{2}}(|1\rangle + i|0\rangle + |-1\rangle)$ and $\psi = (3)^{-\frac{1}{2}}(i|1\rangle + |0\rangle + i|-1\rangle)$, where $|1\rangle, |0\rangle, |-1\rangle$ are the eigenvectors of S_z in the standard representation.

Example 5.1.6 (The Pauli problem) In a footnote on p. 17 to his 1933 'Wellenmechanik' Pauli remarked that the question under which conditions the position and momentum distributions $|\varphi|^2$ and $|\hat{\varphi}|^2$ define the state function φ uniquely (modulo a phase factor) "has still not been investigated in all its generality". To some extent the problem is still open, though there is now a good number of important results on it. We shall collect them here.

1) Neither Q nor P is INFO-complete with respect to any state, i.e. neither Γ_T^Q nor Γ_T^P ever determines T. This is due to the fact that the point spectra of Q and P are empty.

2) Q and P are complementary, and thus totally noncommutative. Thus the pair (Q,P) fulfils a necessary condition (5.1.4) for being INFO-complete. But as already remarked, this pair is not INFO-complete. Indeed, if $\varphi \in L_2(\mathbb{R},dx)$ is a unit vector for which $\varphi(x) = |\varphi(x)|e^{i\theta(x)}$, $0 \leq \theta(x) < 2\pi$, $|\varphi(x)| = |\varphi(-x)|$ a.e. $x \in \mathbb{R}$, $\theta(x) + \theta(-x) \neq$ constant (mod 2π) then $\varphi_1(x) := |\varphi(x)|e^{-i\theta(-x)} = \bar{\varphi}(-x)$ represents another state than $\varphi(x)$, i.e. $|\varphi_1\rangle\langle\varphi_1| \neq |\varphi\rangle\langle\varphi|$, but $|\varphi_1(x)|^2 = |\varphi(x)|^2$ and $|\hat{\varphi}_1(y)|^2 = |\hat{\varphi}(y)|^2$, i.e. Q and P do not distinguish between these states. (This example is due to Prugovecki[37)], ref. 36 contains a wider class of similar examples.) Thus, in general, (Γ_T^Q, Γ_T^P) does not suffice to infer T.

3) (Q,P) is INFO-complete with respect to any family $\{P[U_{qp}\varphi] \mid (q,p) \in \mathbb{R}^2\}$, where φ is a unit vector in the common domain of Q and P, dom(Q)∩dom(P). [The condition $\varphi \in \mathrm{dom}(Q) \cap \mathrm{dom}(P)$ guarantees that the expectations $\mathrm{Exp}(Q, |\varphi\rangle\langle\varphi|)$ and $\mathrm{Exp}(P, |\varphi\rangle\langle\varphi|)$ exist and are finite.] Here again $U_{qp} = \exp(i(pQ - qP))$. Indeed, if $E^Q_{P[U_{qp}\varphi]} = E^Q_{P[U_{q'p'}\varphi]}$, then the expectations $\mathrm{Exp}(Q,P[U_{qp}\varphi])$ and $\mathrm{Exp}(Q,P[U_{q'p'}\varphi])$ are the same so that $q = q'$. Similarly, the condition $E^P_{P[U_{qp}\varphi]} = E^P_{P[U_{q'p'}\varphi]}$ implies that $p = p'$. In particular, the minimal uncertainty states (or coherent states) provide a set with respect to which the pair (Q,P) is INFO-complete. Thus, if the pair (Γ_T^Q, Γ_T^P) satisfies the condition $\mathrm{Var}(\Gamma_T^Q)\cdot\mathrm{Var}(\Gamma_T^P) = h(Q,P)$, then, in particular, the state T is uniquely determined by this statistics.

4) Q and Q_f as well as P and P_g are informationally equivalent, i.e. for any unit vectors $\varphi, \psi \in L_2(\mathbb{R},dx)$ $E^Q_{P[\varphi]} = E^Q_{P[\psi]}$ if and only if $E^{Q_f}_{P[\varphi]} = E^{Q_f}_{P[\psi]}$.[38)]

The informational equivalence of sharp und unsharp quantities means that there is no loss in the state distinction property of an observable when it is replaced by some of its unsharp counterparts. Clearly it follows that the pairs (Q_f, P_g), with confidence functions f and g, are INFO-complete with respect to the same sets of states than the pair (Q,P) is.

One of the important advantages of replacing the sharp pair (Q,P) with an unsharp pair (Q_f, P_g), with Fourier related confidence functions f and g, is the following:

5) The joint observable $E: \underline{B}(\mathbb{R}^2) \to \underline{L}(\underline{H})^+$, $Z \mapsto (2\pi)^{-1} \int_Z |U_{qp}\varphi\rangle\langle U_{qp}\varphi| \, dqdp$, $\varphi \in \mathrm{dom}(Q) \cap \mathrm{dom}(P)$ $\langle\varphi| U_{qp}\varphi\rangle \neq 0$ a.e. $(q,p) \in \mathbb{R}^2$, is INFO-complete.[37)] As already noted above, the marginal observables of E are the unsharp position Q_f and the unsharp momentum P_g, with $f = |\varphi|^2$ and $g = |\hat{\varphi}|^2$. Thus, though the statistics $\{(E_T(X \times \mathbb{R}), E_T(\mathbb{R} \times Y)) \mid X, Y \in \underline{B}(\mathbb{R})\}$, which is informationally equivalent with the statistics $\{(E_T^{Q}(X), E_T^{\underline{P}}(Y)) \mid X, Y \in \underline{B}(\mathbb{R})\}$, does not, in general, determine in the initial preparation T, the richer statistics $\{E_T(X \times Y) \mid X, Y \in \underline{B}(\mathbb{R})\}$ does always suffice to determine T uniquely. This is most remarkable, as the joint observable E can be taken to describe the approximate path of e.g. an electron.[16)]

I shall now summarize this important example. The canonical pair (Q,P) is complementary and satisfies the uncertainty relation. Though this pair is informationally incomplete, it is INFO-complete e.g. with respect to the set of minimal uncertainty states. Applying the Heisenberg-Bohr interpretation for the position-momentum uncertainty relation one may destroy the complementarity of Q and P. The probabilistically complementary unsharp pair (Q_f, P_g) is informationally equivalent to the sharp complementary pair (Q,P) so that the state distinction properties of these pairs are the same. However, the joint observable of Q_f and P_g, with Fourier related confidence functions f and g, is informationally complete.

5.2. Information on the Values of (Probabilistically) Complementary Quantities

Starting with the probability measures E_T^A and E_T^B, $T \in \underline{T}_s(\underline{H})_1^+$, one may define several information functionals $INFO(E_T^A, E_T^B)$ which could suit the study of the question (b) above. With respect to a given functional $\underline{T}_s(\underline{H})_1^+ \to \mathbb{R}$, $T \mapsto INFO(E_T^A, E_T^B)$ a state of maximal information is defined as a state which maximizes (or minimizes, depeding on the sign) the value of $INFO(E_T^A, E_T^B)$, $T \in \underline{T}_s(\underline{H})_1^+$. The choice of the information functional depends on whether we ask information about the observables as a whole (global information) or information on their values (local information). Here I shall consider only local information, so that an appropriate information functional is simply the sum of the probabilities of the values of the quantities:

$$G_{X,Y}^{A,B}(T) := E_T^A(X) + E_T^B(Y) \quad \text{for any } X, Y \in \underline{B}(\mathbb{R}). \tag{3}$$

A state $T_o \in \underline{T}_s(\underline{H})_1^+$ for which $G_{X,Y}^{A,B}(T_o) = \sup\{G_{X,Y}^{A,B}(T) \mid T \in \underline{T}_s(\underline{H})_1^+\}$ is then a state of maximal information of A and B associated with the value sets X and Y. Such states are characterized as those for which $supp(T_o) \subset kern(E^A(X) + E^B(Y) - aI)$, where $0 \leq a \leq 2$ is the largest eigenvalue of $E^A(X) + E^B(Y)$.[39] If $\varphi \in kern(E^A(X) + E^B(Y) - aI)$, $\varphi \neq 0$, $\|\varphi\| = 1$, then a positive outcome, the yes-answer, of an ideal first kind measurement of the simple observable $\{0, P[\varphi], P[\varphi]^\perp, I\}$ leaves the system in the state $P[\varphi]$ in which $G_{X,Y}^{A,B}(T)$ is maximal, i.e. the physical system is prepared $[T \to P[\varphi]TP[\varphi]/\langle\varphi|T\varphi\rangle = |\varphi\rangle\langle\varphi|]$ in the state $|\varphi\rangle\langle\varphi|$ in which we have maximal information on the values X and Y of A and B.

Everything that is said above on pairs of (sharp) quantites (A,B) applies to generalized quantities E_1 and E_2, as well.

I shall now refer to some results on the G-functional for the special pairs of complementary position and momentum (Q,P) and probabilistically complementary unsharp position and momentum (Q_f, P_g), respectively. Clearly in both cases the functional G cannot attain the value 2 for any bounded sets $X, Y \in \underline{B}(\mathbb{R})$. Moreover, it is known[40] that for any two bounded $X, Y \in \underline{B}(\mathbb{R})$ there is a maximal information state $|\varphi_o\rangle\langle\varphi_o|$ of Q and P:

$$G^{Q,P}_{X,Y}(|\varphi_o\rangle\langle\varphi_o|) = \sup\{ G^{Q,P}_{X,Y}(T) \mid T \in \underline{T}_s(\underline{H})^+_1\} = a < 2 ,$$

where a is the largest eigenvalue of $E^Q(X) + E^P(Y)$ (see Appendix 2 in ref. 39.1). Though there is no general result on the order of magnitude between $G^{Q,P}_{X,Y}(T)$ and $G^{Q_f,P_g}_{X,Y}(T)$, the following holds true:[39)]

$$G^{Q_f,P_g}_{X,Y}(T) \leq G^{Q,P}_{X,Y}(|\varphi_o\rangle\langle\varphi_o|) \tag{5}$$

for any bounded $X,Y \in \underline{B}(\mathbb{R})$, $T \in \underline{T}_s(\underline{H})^+_1$, and for any pair of confidence functions f and g. This show that the introduction of unsharpness cannot increase the maximal information on the values of complementary position and momentum quantities Q and P. We note, finally, that similar results hold true also for the global Shannon-type information functional for the pairs (Q,P) and (Q_f,P_g).[41,39)]

6. SUMMARY

The table below summarizes in a verbal form the state distinction and the value distinction properties of measurements of pairs of physical quantities which are commutative, complementary, or which satisfy the uncertainty relation. Whereas the state distinction property does increase when we get from commutativity to uncertainty, the value distinction property will, as a rule, decrease being optimal in the commutative case. This may be taken to reveal a kind of complementarity between different types of measurements.[42)] It could be a quantitative form of the complementarity between definition and observation as advanced by Bohr on several occasions.

Following the famous maxim of Einstein, "it is the theory which decides what can be measured", we may finally conclude the following: If we restrict ourselves to considering sharp (usual) quantities only, then we do not use the full power of the theory, and we voluntarily give up some measuring possibilities. *A fortiori*, the quantum theretical conception of reality then remains unnecessarily weak.

	SD-PROPERTY / DEFINITION	VD-PROPERTY / OBSERVATION
COMMUTATIVITY		
A single quantity or a commuting set. E.g. Q or P	Restricted to (common) eigenstates. No SD-property	Optimal: any possibility can be actualized (with an ideal first kind measurement).
COMPLEMENTARITY		
Complementary pairs E.g. (Q,P)	A necessary condition for SD-property fulfilled. SD-property extends e.g. to the set of minimal uncertainty states.	Mutually exclusive possibilities cannot be actualized simultaneously. No joint measurements; but maximal information states may exist, and they can be realized.
UNCERTAINTY		
Complementary pairs are replaced by probabilistically complementary pairs. E.g. $(Q,P) \rightsquigarrow (Q_f,P_g)$	Informational completeness, i.e. full SD-property may be obtained.	In general, information on value decreases; but a possibility for unsharp joint measurement is opened. No preparatory joint measurements, but ε-actualizability of probabilistically complementary unsharp properties.[14)]

Table: A comparison of the state distinction (SD-) and the value distinction (VD-) properties of measurements of pairs of quantities with different coupling properties.

ACKNOWLEDGEMENTS

Much of the material presented in this lecture arose from my collaboration with Dr. Paul Busch. The final form of this paper greatly benefited too, from my discussions with Paul Busch. I should also like to thank Drs. Kari Ylinen and Frank Schroeck for useful discussions on the commutativity problem as well as on the questions related to informational completeness.

7. FOOTNOTES AND REFERENCES

1) Born, M., Z. Physik 37, 863-867 (1926).

2) Heisenberg, W., Z. Physik 43, 172-198 (1927).

3) Bohr, N., Nature 121, 580-590 (1928).

4) Jammer, M., The Philosophy of Quantum Mechanics, John Wiley & Sons, 1976.

5) Mehra, J. and Rechenberg, H., The Historical Development of Quantum Theory, Springer 1982.

6) Primas, H., Chemistry, Quantum Mechanics and Reductionism, Springer 1981.

7) Ludwig, G., Foundations of Quantum Mechanics, Springer 1983.

8) Mittelstaedt, P., Sprache und Realität in der Modernen Physik, B.I. - Wissenschaftsverlag, 1986.

9) See e.g. 4,6,7,8,11).

10) Gleason, A., J. Math. Mech. 6, 885-893 (1957).

11) Beltrametti, E. and Cassinelli, G., The Logic of Quantum Mechanics, Addison-Wesley, 1981.

12) See e.g. Lahti, P. and Bugajski, S., Int. J. Theor. Phys. 24, 1051-1080 (1985).

13) See e.g. Lahti, P., Int. J. Theor. Phys. 19, 789-842 (1980), 22, 911-929 (1983), See also 12).

14) See e.g. Busch, P. and Lahti, P., Phys. Rev. D29, 1634-1646 (1984); Phil. Sci. 52, 64-77 (1985); J. Phys. A: Math. Gen 20, 899-906 (1987); Lahti, P., Foundations of Physics (eds. P.Weingartner and G.Dorn), hpt-verlag, 1986, pp. 117-137; Recent Developments in

Quantum Logic (eds. P.Mittelstaedt and E.-W.Stachow), Bibliographisches Institut, 1985, pp. 61-80; Report Series Turku-FTL-R104 1986; Lahti, P. and Ylinen, K., J. Math. Phys. (1987), in print.

15) See e.g. Pulmannova, S. and Dvurecenskij, A., Ann. Inst. H. Poincare Phys. Theor. 42, 253-265 (1985); and the items 13.1. and 14.7.

16) Busch, P., Int. J. Theor. Phys. 24, 63-92 (1985); and the items 14.1,2.

17) Busch, P., Symposium on the Foundations of Modern Physics (eds. P.Lahti and P.Mittelstaedt), World Scientific, 1985, pp. 343-357; Phys. Rev. D33, 2253-2261 (1986).

18) Mittelstaedt, P., Recent Developments in Quantum Logic (eds. P.Mittelstaedt and E.-W.Stachow), Bibliographishes Institut, 1985, pp. 215-234; Symposium on the Foundations of Modern Physics (eds. P.Lahti and P.Mittelstaedt), World Scientific, 1985, pp. 171-186; see also 17) and P.Busch, elsewhere in this volume.

19) Gudder, S., J. Math. Mech. 18, 325-335 (1968).

20) Pulmannova, S., Foundations of Physics 10, 641-655 (1980).

21) Ylinen, K., Symposium on the Foundations of Modern Physics (eds. P.Lahti and P.Mittelstaedt), World Scientific, 1985, pp. 691-694.

22) von Neumann, J., Mathematical Foundations of Quantum Mechanics, Princeton U.P., 1955. The original German text in 1932.

23) Pauli, W., Principles of Wave Mechanics, Springer, 1980. The original German text in 1933.

24) Mittelstaedt, P., Philosphische Probleme der Modernen Physik, Bibliographisches Institut, 1981, 6. edition.

25) Primas, H., elsewhere in this volume. See also 6.

26) Beltrametti, E., Gassinelli, G., and Lahti, P.J., "Unitary measurements of discrete quantities", manuscript.

27) Ozawa, M., J. Math. Phys. 25, 79-87 (1984).

28) See e.g. Davies, E.B., Quantum Theory of Open Systems, Academic Press, 1976.

29) Bohr, N., Phys. Rev. 48, 696-702 (1935).

30) Fock, V., Fundamentals of Quantum Mechanics, MIR Publishers, 1978.

31) Ludwig, G., Die Grundlagen der Quantenmechanik, Springer, 1954.

32) von Weizsäcker, C.F., Die Naturwissenschaften 42, No 19/20 (1955).

33) van Fraassen, B.C., Problems in the Foundations of Physics (ed. G.Toraldo di Francia), North-Holland, 1979, pp. 344-394.

34) Lahti, P., Report Series Turku-FTL-R105, 1986.

35) See e.g. Jauch, M., Foundations of Quantum Mechanics, Addison-Wesley, 1969, p. 160.

36) Corbett, J.V. and Hurst, C.A., J. Austral. Math. Soc. 20B, 182-201 (1978).

37) Prugovecki, E., Int. J. Theor. Phys. 16, 321-331 (1977).

38) Ali, S.T. and Doebner, H.D., J. Math. Phys. 17, 1105-1111 (1976).

39) Busch, P. and Lahti, P., J. Phys. A: Math. Gen. 20, 899-906 (1987); "States of maximal information and pairs of unsharp observables", manuscript.

40) Lahti, P., Rep. Math. Phys. 23, 289-297 (1986).

41) Grabowski, M., Int. J. Theor. Phys. 17, 635-641 (1978).

42) Busch, P., a private communication, 1987. See also Busch, P., Found. Phys. 17, 905-937 (1987).

WOLFGANG PAULI'S CONCEPTION OF REALITY

K.V.Laurikainen

Research Institute for Theoretical Physics,
University of Helsinki
Siltavuorenpenger 20 C, SF-00170 Helsinki, FINLAND

The conception of reality underlying the Copenhagen Interpretation is still a problem.[1] The reason for this is the *repression of the irrational* characteristic of Western thought. Wolfgang Pauli was the most radical but also the most consistent among the founding fathers of quantum theory. His conception of reality opens up a new perspective for science: a view into an *irrational world*. If we take Pauli's remarks seriously, we must evaluate many important questions in a new light.

Since 1976 Pauli's philosophical thought has been my main point of interest. I have published a book about it[2] in Finnish which has now been translated into English under the title "Beyond the Atom: The Philosophical Thought of Wolfgang Pauli".[3] This lecture is based on the foreword of the English manuscript.

The main source for my study of Pauli's philosophy has been the correspondence between Pauli and Markus Fierz stored in the Pauli Letter Collection (PLC) at CERN in Geneva.[4] It is a rare collection of documents which guides us to the deepest sources of these physicists' thought. In this way the fundamental philosophy of modern physics attains quite new enlightenment. Here physics joins the universal currents of contemporary thought. Pauli's letters have something to say to the philosopher, idea historian, theologian, biologist and psychologist--and to the average person of our time who is searching for answers to the basic questions of existence.

The correspondence between Fierz and Pauli is quite extensive--despite the fact that a great portion of their letters have been destroyed. They have been written during 1947-1958, over the last 12 years before Pauli's unexpected death. They deal with many different kinds of questions, for which reason Pauli's philosophical attitude can be understood much better in their light than by reading his very concise publications. A collection of Pauli's epistemological articles has been recently republished by Karl v.Meyenn.[5]

I am grateful to the late Mrs. Franca Pauli and to Professor Markus Fierz, as well as to the Pauli Committee which is responsible for the use of the Pauli Letter Collection at CERN, for the permission to publish rather long quotations from the Pauli letters in my book. Without this possibility of letting Pauli speak for himself, particularly in matters of a sensitive nature, it would not be possible to give a trustworthy picture of Pauli's thought.

It is time to open a discussion concerning Pauli's philosophy. Quite a bit has been written about the philosophical thought of Einstein and of Bohr, while people do not seem to know at all that Pauli had deep philosophical views which differ essentially from those of standard scientists and philosophers today. However, in the future Pauli may be found to have been the deepest philosopher of those three. His philosophy is close to Bohr's but Pauli is clearer and more consistent.

In order to understand Pauli's thought, it is necessary to pay regard to his *quaternarian attitude*, as he himself characterized it. Especially after World War II, Pauli tended more and more towards a "quaternarian" philosophy instead of a "trinitarian" one. To give some idea of the nature of the quaternarian attitude, this can be said to mean that instead of a thorough analysis of details Pauli put more and more emphasis in his later years on the intuitive comprehension of wholeness. Among the creators of quantum theory, Niels Bohr was a quaternarian thinker while Dirac, in parti-

inanimate (non-organic) matter should show "weak psychic parallel components".

I found this conclusion to be reasonable...[7]

In another letter to Fierz Pauli writes that when, in the process of the chemical evolution, an organic molecule which was able to duplicate itself emerged for the first time, how could one say that psyche was not present at that moment.[8]

Pauli describes the relations between matter and spirit, and between physics and psychology respectively, as complementary. It was a very fateful separation which Descartes made in the 17th century between spirit *(res cogitans)* and matter *(res extensa)*, although it was understandable at the time. This misconception still dominates Western thought. In physics it appears in the idea of the "detached observer": it is assumed that the influence of the observer on the observed phenomenon can always be eliminated, in principle. Niels Bohr also accepted this idea in his later years, while Pauli saw in quantum mechanics a "road sign" in another direction; one should not neglect the influence of the human psyche on observations.

The difference between Bohr and Pauli in their philosophical attitudes was discussed in my lecture here in Joensuu two years ago.[9] The question of the "detached observer" is characteristic in this respect. In general Bohr and Pauli understood the questions of the interpretation in quite the same way, *viz.*, in the sense of the complementarity philosophy of Bohr.

The idea of the "detached observer" is characteristic of the materialistic world view which subordinates the spiritual component of reality to the material one. The idea of the *anima mundi*, the world soul, has been completely expelled from the scientific world view. The indeterminism of the atomic world gave Pauli reason to speak of the return of the anima mundi (in an new sense of the word).[10]

When Niels Bohr chose the Taoist Yin-Yang symbol for his coat of arms in the Danish Elephant Order, he was no doubt aware of the essential

cular, represented trinitarian thought. It is exactly this quaternarian attitude which makes Pauli so interesting and timely today. (Ch.IX in ref.3 deals with *quaternity.*)

One characteristic of the quaternarian attitude of Pauli is that he considers drawing a sharp border between science and religion or between science and metaphysics to be very dangerous. Pauli often emphasizes the objective significance of the attitude "complementary to that in natural science". This I understand as an emphasis on the objective role of religion. Pauli's relation to religion remains, however, an open question for me. In connection with a deep spiritual crisis, when he was about thirty years old, Pauli left the Roman Catholic Church and remained a free-thinker in a legal sense for the rest of his life.[6]

Science was Pauli's starting point in these questions also. The message of particular importance which he has for modern man is that an unbiased delving into the foundations of science can guide a researcher to an understanding of the significance of religion.

* * * * *

The general trend in the investigation of the foundations of quantum mechanics after the death of its creators has been towards *realism.* There are differences in the more detailed characterization of this trend but the common aim has been to describe the "external world" as being independent of the "observer". One thinks of a universe where human beings and the human psyche exist just *as results of the evolution* which follows its "objective" laws quite independently of the existence of any "observer".

Pauli was critical with respect to such ideas. Basic to his world view is the inseparable *union of spirit and matter.* In a letter to Fierz Pauli mentions his agreement with some ideas of the German zoologist *B.Rensch:*

> R. means especially that it would be impossible, however, for the "psychic parallel components" to have "suddenly emerged" in the otherwise continuous ontogeny. Therefore even the so-called

role of spirit in the Taoist conception of reality. Any kind of mysticism was, however, strongly against his basic attitude.[11] For this reason Bohr's conception of reality became unclear, and his attitude has been generally interpreted as corresponding to positivism (or instrumentalism).

Pauli was more consistent. Acording to him it is wrong to imagine the existence of the material "external world" without including the "observer" and his psyche in the world view. The idea of a "detached observer" and the trend towards "realism"--in the usual sense of the word--has to be abandoned. According to Pauli, *spirit and matter must be considered as the two complementary basic elements of reality.*[12]

Heisenberg's view of the role of the observer was rather close to that of Pauli, but less definite. He gave a review of the interpretational problems of quantum mechanics, *e.g*, in the article "Die Entwicklung der Deutung der Quantentheorie" ("The Development of the Interpretation of Quantum Theory").[13] Heisenberg describes the role of the observer implicitly with the aid of the "Aristotelian" concepts of *potential* and *actual*. This difference is an expression of the limitations of our psyche when attempting to comprehend reality, and it is intimately connected with the fact that only statistical laws are possible in microphysics.

The mathematical description of the atomic world and especially its *wave function* relate to the *potential* attributes of the microworld. The *actual* properties of this world are expressed in the experimental data which are described by using the macrophysical language (*i.e.*, in a language which concerns an "objective" world where objects have "objective properties" independent of observations, quite as the idea of the "detached observer" presupposes). In the observation some of the potential properties of the object "become actualized", something formally expressed in the "reduction of the wave packet".

This difference between the potential and the actual level of our knowledge is an expression of the psychological limitations of the human observer. At the end of his article Heisenberg pointed out that we have certain "a priori" limitations in our comprehension of the world. Therefore the microworld can be described only in the sense of the "potential" level whence the wave functions, in particular, cannot be considered as "actual elements of reality" as materialistic philosophy would like to understand them.[14]

Heisenberg did not describe the implications of this situation with respect to our conception of reality as clearly as Pauli did. The former speaks, however, of the essential role of psyche in the quantum mechanical conception of reality, and in several articles Heisenberg has emphasized that the Copenhagen interpretation is not compatible with a materialistic world view.[15] Heisenberg interpreted the state function as "our knowledge" of the state of the system. Pauli still more clearly emphasized the indeterminism/statistical laws in quantum mechanics and the fact that this leads to the conception of the *irrationality of reality.*[16] He associated this with the unconscious functioning of our psyche.[17] For this reason empirical science cannot be isolated from psychology. The objectivity of human knowledge appears to be a utopian ideal, and scientific knowledge cannot be strictly separated from subjective beliefs. This is a consistent explication of the complementarity philosophy which Bohr himself did not dare to develop in all its implications because of his strong aversion to mysticism. In this respect Pauli has been the most critical and most consistent among the founding fathers of quantum theory.

If we are just interested in the logical structure of the theory, it is quite possible to avoid any talk of irrationality. This does not concern the logical structure but the empirical verification of the theory. Both the basic idea of complementarity by Bohr and the uncertainty principle by Heisenberg imply, however, that a rational theory does not attain "reality-in-itself" but represents only an approximate model of reality. Pauli used to describe this situation by emphasizing the statistical nature of the quantum mechanical laws. Statistical laws concern only expectation values and their accuracy (scatter) but do not describe individual events more than in a probabilistic sense. The characteristic scatter of individual experimental results around the mean value/expectation value is an expression of the basic irrationality of the world: a rational theory is not able to describe individual events in an unambiguous manner.[18]

Because of this situation Pauli considered reality as containing both rational and irrational elements which are complementary to each other.

This makes a Platonian and a Cartesian conception of reality only approximate. Pauli often compares his own conception of reality with Schopenhauer's philosophy where the world is considered to be a union of will and idea (Wille und Vorstellung).[19] *Idea* corresponds to the rational element of reality, *will* to the irrational. The rational description is bound to space and time, the irrational "will" is independent of space and time, in principle. The *creative* element of reality is irrational, the *conservative* element rational.

* * * * *

To give an example of an ontological view distinctly different from that of Pauli I shall try to describe, in short, *the indivisible wholeness of David Bohm.*

Bohm is a typical representative of those people who found the Copenhagen interpretation to be unclear. He developed a *hidden variable theory* in the fifties based on an idea which de Broglie also had proposed in the twenties. Characteristic of Bohm's theory is that it contains long-distance "influences" of a new kind. Because of them, the world becomes an *indivisible wholeness* in which everything more or less influences everything else.[20]

Bohm wishes to describe microphysical particles as "ordinary objects" with attributes (also kinematic attributes) independent of observations. This is in agreement with the idea of a detached observer.

Bohm accepts the use of the wave function (state function) of ordinary quantum mechanics and its derivation from the Schrödinger equation. This function, however, is interpreted as *a new kind of field* which influences the mechanical behaviour of the particles in the system with which the wave function is associated. It is then also possible to define exact position coordinates and momenta for these particles and speak of their definite orbits.

The accuracy of actual observations is limmited in the sense of the uncertainty relations of Heisenberg--at least in the methods we know today. Therefore the exact kinematic description of the motion as proposed by Bohm is only possible on a theoretical level, as if postulating a deeper level in reality, one which cannot be attained in observations. Only on this hypothetical level can particles be considered as "ordinary objects" with definite kinematic attributes (definite orbits).

The quantum mechanical wave function has a very remarkable feature of "wholeness" which Bohr has also emphasized: it does not only describe the system which is investigated but takes into account its *neighbourhood*--in principle everything that can have an influence on the system's behaviour in the case of a certain "phenomenon". In the act of measurement the *experimental arrangement as a whole* has an influence on the wave function to be used. It determines the "boundary conditions" for the solution of the Schrödinger equation which has to be used; this function describes the system *in the given neighbourhood.* This does not imply any influence due to the neighbourhood in the sense of the traditional causal description. It is a new kind of influence where space and time do not play the normal role as the basis of a causal description of phenomena.

By interpreting the wave function as a new kind of "field"--which in a many-particle system is not a field in the three-dimensional space but in the multidimensional "configuration space" of the system--Bohm thus introduces a new kind of "influence" which is independent of space and time and refers to the "wholeness" of the system and its neighbourhood.

Due to the limitations characteristic of the microphysical observations, only statistical predictions are possible with respect to observations. These predictions in Bohm's theory are exactly the same as those in ordinary quantum mechanics. In this sense he just gives a new interpretation for quantum mechanics. Bohm, however, speaks of his hypothetical orbits and kinematical attributes as belonging to a new level of reality where it is possible to describe the change in the state of the system in a strictly causal manner, not just in a statistical sense as in ordinary quantum mechanics. In this sense he has a *new conception of reality*, and in fact he is developing a new theory which could even lead to new experimental predictions, if the hypothetical level of definite orbits would in the future become attainable through experiment.

Bohm, in fact, emphasizes that his conception of reality concerns both the microphysical and the macrophysical levels. These levels are related to one another in the sense of Bohr's correspondence principle, as in ordinary quantum mechanics, the macrophysical level being a certain limiting case of the microphysical basic theory. Bohm interprets this unified description of both micro- and macrophysics as applicable to the measuring instruments too, and even to human beings and their brains! On this basis he develops a materialistic view of the universe which does not contain any basically non-material elements.

The psychological problems associated with observations are quite neglected by Bohm, not to speak of the manifestations of the unconscious! In his causal-deterministic universe Bohm does not see any "quantum mechanical paradoxes"--but there is no place for free will either, at least not without postulating a still deeper level of reality. Theory of measurements becomes a causal play between the robot-observer, his causal instruments and the observed system without any "collapse of the wave function". Even the universe as a whole is understood as a quantum mechanical system.

The problems concerning human knowledge which make the Copenhagen interpretation so revolutionary--especially if it is understood in the sense of Pauli's conception of reality--have in Bohm's reality disappeared. This is no solution to these problems but a way to avoid seeing them. Bohm constructs a world view which avoids discussing the basic problems of empirical knowledge and the role of psyche in the world.

It is interesting that Bohm, however, has also described human psyche in recent publications, and has, in fact, found parallels to Eastern mysticism in his "quantum reality". The dualism of matter and spirit is in his world view related to the dualism of particles and waves, which are considered to be the basic constituents of equal standing in Bohm's world. There are also other important features in Bohm's world view which bear similarities to Pauli's conception of reality. For instance, Bohm also emphasizes *wholeness*, not only in microphysical situations but more generally, and he also finds that *"influences" which are independent of space and time* are needed in microphysics. Further, he finds it necessary to `reconsider the *psycho-physical problem* which Pauli emphasizes as the most important problem of our time. Because Bohm comes to similar conclusions to Pauli concerning these important questions, despite the fact that Bohm is searching for a new conception of reality in quite another way than Pauli, it may be concluded that quantum mechanics is in these questions a road sign in a definite direction.

* * * * *

Comparison should still be made between Pauli's philosophy and some other views which are rather close to his thought. In this sense I would like to consider C.F.v.Weizsäcker and B.d'Espagnat.

C.F.v.Weizsäcker is a highly qualified expert in the Copenhagen interpretation. He has illuminated Bohr's and Heisenberg's philosophical attitudes in several articles and books, and has done it very well indeed. As compared to Pauli's thought there is, however, a characteristic difference in the question of the "detached observer" and, *a fortiori*, in their views concerning the role of spirit in the world view.[21]

v.Weizsäcker whole-heartedly accepts Bohr's emphasis on the objectivity of science, and therefore the idea of the "detached observer". The view that spirit and matter are complementary elements of reality continuously "influencing" each other seems to be foreign to v.Weizsäcker's thought. He seems, instead, to think of the psycho-physical question on the basis of a strict Kantian separation of the realms of spirit and matter.

In an (unpublished) letter to Fierz[22] Pauli strongly criticizes v.Weizsäcker's attitude in this respect, stating that it leads to an overly strict separation between science and religion. Pauli writes that for him the sources of scientific understanding and of religious feeling are one in the same.[23] He continues that if we go deeply enough into the depths of our psyche, the contrast between science and religion must in the end--"on the shore of the torrent of life (Styx) itself"--fade away. In this respect there also seems to be a difference between v.Weizsäcker and Heisenberg but Heisenberg has not expressed his views as clearly as Pauli who wrote[22] that Heisenberg often emphasized the objectivity of natural science too strongly.

The difference between v.Weizsäcker and Pauli may be further illuminated by quoting the end of Pauli's letter to v.Weizsäcker probably written in September 1956:

> In the end this conception, which is expressed in the Chinese symbol Tai-Gi [the Yin-Yang symbol; KVL] and more generally in the circularity, aims at the *conformity between internal and external.* This view is still quite unfamiliar to our present physics as well as to Bohr. This is what I mean by the "motives which have remained unconscious" as well as by the remark in my previous card that Bohr is *not* "circular". He will, *with* our present physics, "completely capsulize" the observer, in order that all that is psychic would remain outside physics. I feel, on the contrary, very strongly that this way of viewing things has timebound restrictions, in spite of all ideals of "objectivity". In this way, precisely, the desire for the unity of inner and outer natures remains namely an unconscious one (*cf.* also alchemy + Goethe), because it is there, *nevertheless.* The differences between Bohr and myself are in this non-physical area, not within the interpretation of quantum mechanics, and, I think, that is where your actual problems are also.[24]

Thus, the difference in attitude between v.Weizsäcker and Pauli concerns philosophy, not physics. But philosophy can influence the direction of our scienctific endeavours. In this case v.Weizsäcker seems to be interested now in problems which had scarcely interested Pauli. By this I mean interest in the theory of measurements and in the description of the universe in its entirety on the basis of quantum mechanics. The question of describing human psyche with the aid of quantum theory is certainly something that Pauli would judge as a materialistic misunderstanding of the role of the observer and of psyche in the world.

In his new book, "Aufbau der Physik,"[25] v.Weizsäcker stated concerning the observer:

> He will not be described when an experiment is described. He is, on the contrary, the one who does the describing.

v.Weizsäcker then emphasizes the intersubjectivity of the scientific description of the phenomena where "observer" does not mean any definite person, but rather, a representative of "human observers". Up to this point his view is exactly the same as that of Pauli. He then continues:

> An observer "objectivized" in this way, however, is not any "pure spirit". He must see with his eyes, work with his hands. He is an observer only because he--the dualists would say carefully: because his body--is a part of the world of phenomena. He can see his own hand, touch his own eye, he is in a physical interaction with the objects which he describes. The question of how the observer would be able to describe himself remains open.[26]

The difference between v.Weizsäcker and Pauli begins to appear here, and it concerns exactly the psycho-physical problem which Pauli found so fundamental. v.Weizsäcker comes back to this question some pages later under the title "Quantentheorie des Subjekts" and defines his view as follows:

> The hope of a unified picture of the world, however, and the progress of biology and of the cybernetic way of thought achieved during the six decades since the creation of quantum mechanics force us to think of the question of whether the description of the observer as an object must, *in principle*, remain incomprehensible or whether it would be possible, in principle. We support the second possibility.[27]

This is the crucial point. Pauli considers the psyche of a human observer as *complementary* to his body. It is *not* possible, then, to describe psychic phenomena on the basis of a theory concerning matter. Psyche and body together form a "whole" in which these two elements are in a similar relation to each other as that between the position coordinate x and its conjugate momentum p_x in quantum mechanics. Both are necessary as fundamentally different elements of the "whole" ("reality").

If we think of the role of the *unconscious* in an observation and its interpretation, it is impossible to think that quantum theory or any rational theory, in quite general terms, would be able to describe the functioning of the unconscious and, a fortiori, the human psyche in general. Here I cannot understand v.Weizsäcker: as far as empirical science is concerned he seems to accept an attitude which leads to a materialistic view of the universe while, when speaking of religious things, he seems to speak of another, independent spiritual world. This attitude seems to be characteristic of Protestant theology. This is exactly the point where Pauli finds a new conception of reality to be necessary. Pauli does not accept the Cartesian dualism and even less the "unitarian" materialistic conception of the world. He was searching for a new conception of reality on the basis of a "complementarity" relation between spirit and matter. This is intimately connected with his emphasis on the *irrationality* of reality and the essential *role of the unconscious* when forming a picture of the world.[28]

Bernard d'Espagnat has published many articles as well as several books on the epistemological and ontological questions in atomic theory. Earlier he has participated in phenomenological research in elementary particle physics, being in close contact with the work at CERN. Thus, he knows problems and methods both in experimental and theoretical atomic research very well. His books concerning the philosophy of quantum theory are well-known, especially his book "In Search of Reality"[29] which has been published in several languages.

d'Espagnat seems to be one of the persons nowadays who best understand the spirit of the Copenhagen interpretation. He describes the philosophical implications of quantum theory in a very similar way to Pauli, although the way of approach is different.

d'Espagnat describes the ontology characteristic of the atomic world by using the term *veiled reality*. He shows with the aid of a detailed analysis that the concepts used in everyday language and in classical physics are not suitable for describing the "reality-in-itself" or *independent reality*--independent of the observer and of the process of observation. Because of this situation physicists are in general inclined to accept the positivistic or instrumentalistic (d'Espagnat uses the term "operationalistic") attitude which avoids the concept of "reality" as an undefinable term. d'Espagnat, however, shows that such an uncritical attitude is difficult to defend consistently because there would then no longer exist any objective basis for judgment between correct and incorrect theories. He prefers the hypothesis of the existence of an independent reality, even if it cannot be exactly defined or characterized. He describes this independent reality as being "veiled", in principle.

This conception of a veiled reality differs from the Kantian concept of the world of "things-in-themselves" in that we get some knowledge of the veiled reality with the aid of empirical methods, even if we cannot describe it exactly with the aid of our concepts. Our theories are models which approximate reality, without ever being able to describe "reality-in-itself". Reality itself remains "always behind a veil."

The reason for this situation is that our senses have only a limited accuracy and our concepts, to which we are bound, are not suitable to describe the structure of "reality-in-itself". We can only describe *empirical reality* which contains experimentally verifiable things. Quantum mechanics, however, also contains "objects" which do not have any definite properties without an observation, as if the observation would "create" these properties. Therefore we have reason to make a distinction between an objective "reality-in-itself" as a world which we cannot describe by our concepts, and on the other hand, an "empirical reality" which is only an approximation of the "reality-in-itself" (independent reality) which remains "veiled".

This is a very similar picture of reality and of human knowledge to that which Pauli gives. Our psychological limitations have the effect, in both cases, that an objective reality is just an idealization of the empirical reality which our rational theories are able to describe. Instead of a "veiled reality" Pauli speaks of the *irrationality* of reality, and, for Pauli, the concept of the *unconscious* is fundamental for the understanding of the epistemological situation of the human observer. For d'Espagnat, psychological limitations of our knowledge are also essential but he bases the distinction between empirical reality and independent (veiled) reality on a formal analysis of our epistemological situation in atomic physics.

In the recent article "Empirical Reality, Empirical Causality and the Measurement Problem"[30] d'Espagnat defines exactly what he means by empirical reality. d'Espagnat is a good representative of the Cartesian clarity characteristic of French physics and philosophy, and therefore his articles should be understandable for the researchers presently most active in the foundations of quantum mechanics, as the trend is towards a logical exactness and clarity.

Personally, I find an intuitive understanding of the whole to be more important for the comprehension of the ontological problems of atomic physics than a logical analysis would be. In fact, d'Espagnat shows that people who develop very exact theories of the basic questions of quantum theory quite generally smuggle into their analysis certain views concerning "nature" or "reality" which are incompatible with the basic thoughts in quantum theory. People in general do not realize the difference between empirical reality which can be reached by scientific methods (through rational models) and independent reality which is more or less an object of belief--something just as necessary for the scientist as the man-on-the-street.

If the difference between these two conceptions of reality is not well understood, this can lead to two generally held one-sided attitudes. One is that of *scientism* which does not acknowledge any conception of reality independent of empirical reality. The other is that of *mysticism* which finds empirical knowledge in general untenable in consideration of "reality-in-itself".

Scientism is very generally accepted as a kind of modern religion in the academic world. In Finland, for instance, it represses all attempts to discuss seriously questions related to the distinction between empirical and independent reality.

The attitude characterized here as mysticism includes all religious world views which strictly separate questions concerning "reality-in-itself" or "final truths" from empirical knowledge. In this sense religion can be accepted as a historical phenomenon, also by those with a scientistic attitude, even though they do not grant religion any real value in questions concerning reality or truth. Such an attitude is, however, quite unacceptable for a natural scientist because he believes, for good reasons, that empirical science is able to clarify some questions concerning reality.

In his recent article mentioned above d'Espagnat wrote:

> The human means of apprehension and the human means of data processing on which this human experience rests cannot be kept out of consideration and science should not try to do so. Although such a conception was a part of Kant's philosophical doctrine it is considerably less detailed and specific than the latter was, so that it is not necessary to be a "Kantian" to subscribe to it. In fact it constitutes the--explicit or implicit--viewpoint of a great many thinkers and physicists of our times, most of whom have hardly heard of--and at any rate never took any interest in--the Kantian philosophy.[31]

Niels Bohr considered his philosophical attitude to be anti-Kantian because he did not consider the psychological limitations of our knowledge to be *a priori* given--as Kant did in his doctrine of the categories of understanding and in the forms of perception (space and time)--while Bohr considered these concepts to be objects of empirical research. Bohr wished here to emphasize that space, time, and causality do not have any meaning given *a priori* but are subjected to the investigation of physics.[32] However, the very idea that these concepts are fundamental for our comprehension of reality and characteristic of human knowledge means an acceptance of the basic idea of the Kantian epistemology, although not in the stiff form of Kant's categories. In this sense d'Espagnat finds that the quantum mechanical picture of the world necessarily corresponds to the Kantian conception of human knowledge, and in this respect he also interprets the situation in the same way as Pauli--and in fact, Bohr himself.

One can describe the role of these basic concepts by saying that space, time, and causality are not properties of reality-in-itself but are characteristic of our *knowledge* of reality. They belong to empirical reality but not (necessarily) to the veiled independent reality.

Characteristic of the human conception of reality, further, is that it consists of objects which have more or less invariant properties. This is not true with respect to the microworld because "objects" there have "properties" which depend on the manner of observation which the observer ("subject") can freely choose. In this sense the observer "creates" the objects and their properties.

This way of describing the ontological questions implied by quantum mechanics is quite in line with Pauli's thought. I would like to say that Pauli and d'Espagnat have the same view concerning the ontological implications of quantum theory. The "irrationality" of reality emphasized by Pauli creates the "veil" which hides "reality-in-itself" from the human observer. This "veil" is, according to Pauli, essentially associated with the unconscious functioning of our psyche which cannot be described with the aid of any explicit rational models.

The criticism which d'Espagnat addresses to the theories of measurements in atomic physics[33] would also be subscribed to by Pauli, as far as I understand Pauli's thought.

* * * * *

Perhaps this comparison of Pauli's views with those of some others makes it possible to get an idea of Pauli's philosophical thought. It must be emphasized that Pauli's message is revolutionary. He had the distinct opinion that the general trend of Western culture after the 17th century has been one-sided and dangerous. One can characterize this trend as a vision of a *clock-work* universe--a deterministic world where everything is, in principle, predestined. This vision has created a materialistic culture where the influence of religion has continuously diminished, and of which a strict separation between science and religion is characteristic. Pauli wishes to give us a vision where *spirit has again been acknowledged as a basic element* of the world along with matter. The universe should again be seen as an *organism*, not a clock.

The Copenhagen interpretation of quantum mechanics was, for Pauli, a "road sign" into this new world which is characterized by the *complementarity of spirit and matter*--and of religion and science. The view which Pauli opens is not bound to Eastern religions--although he also mentions the East-West problem in this connection. The West should find a Western way for putting in concrete terms the vision of a new world which Pauli opens. There is a very strong opposition in the Western academic world against allowing any *irrationality* in the picture of the world. This is the case both in science and in religion. This *repression of the irrational* is what we have to overcome in order to see the perspective which Pauli tries to open to us, and in this respect Eastern philosophies can help us find a new way.

REFERENCES

1. An account of different ontological views concerning quantum physics is given in Herbert, N., Quantum Reality: Beyond the New Physics (Anchor Press / Doubleday, Garden City, New York, 1985).
2. Laurikainen, K.V., Atomien tuolla puolen: Wolfgang Paulin ajatuksia hengestä ja aineesta, todellisuuden luonteesta ja pahan asemasta maailmassa (Kirjapaja, Mänttä, 1985).
3. Laurikainen, K.V., Beyond the Atom: The Philosophical Thought of Wolfgang Pauli (to be published by Springer-Verlag, Heidelberg). A preprint has been published in the preprint series of the Research Institute for Theoretical Physics, University of Helsinki, HU-TFT-87-6.
4. References to the Pauli letters are identified by using the numbering of the Pauli Letter Collection (PLC) at CERN (European Organization for Nuclear Research, CH-1211 Geneva 23, SWITZERLAND).
5. Pauli, W., Physik und Erkenntnistheorie in "Facetten der Physik" (Vieweg & Sohn, Braunschweig / Wiesbaden, 1984). First edition 1961.
6. For biographical references see ref.3, ref.5, and Pauli, W., Wissenschaftlicher Briefwechsel mit Bohr, Einstein, Heisenberg u.a. (Springer-Verlag, Berlin, Heidelberg, New York, Tokio; eds. Meyenn,K.von , Hermann,A., and Weisskopf, V.F.), Vol.I: 1919-1929 (1979),

Vol.II: 1930-1939 (1985); to be continued.
7. PLC 0092.200; a letter by Pauli to Fierz on March 5, 1957.
8. PLC 0092.107; a letter by Pauli to Fierz, no date.
9. Laurikainen, K.V., "Wolfgang Pauli and the Copenhagen Philosophy", Symposium on the Foundations of Modern Physics: 50 years of the Einstein-Podolsky-Rosen Gedankenexperiment, Joensuu, Finland, June 16-20, 1985 (eds. Lahti, P. and Mittelstaedt, P.; World Scientific, Singapore, 1985) pp.273-287.
10. See, *e.g.*, PLC 0092.078; a letter by Pauli to Fierz on Oct. 13, 1951 (*cf.* ref.3, Ch.III).
11. *Cf.* ref.3, Ch.V as well as Folse, H.J., The Philosophy of Niels Bohr (North-Holland, Amsterdam, 1985).
12. See, *e.g.*, ref.3., Ch.V.
13. Heisenberg, W., "Die Entwicklung der Deutung der Quantentheorie", Physikalische Blätter 12, 289-304 (1956).
14. *Cf.* ref.3, Ch.V. (Acad. V. Fock's discussions with Bohr.)
15. See, *e.g.*, Heisenberg, W. "Die Plancksche Entdeckung und die philosophischen Probleme der Atomphysik", Universitas 14, pp.135-148, 1959.
16. See, *e.g.*, ref.3, Ch.VI.
17. *Ibid.*
18. See, *e.g.*, PLC 0092.078 (part 3); a letter by Pauli to Fierz on October 13, 1951 (ref.3, Ch.III).
19. Schopenhauer, A., Die Welt als Wille und Vorstellung I-II (Brockhaus, Altenburg, 1818).
20. David Bohm has developed his ontological view in several publications of which we shall just mention two monographs and a recent article:

Bohm, D., Wholeness and the Implicate Order (Routledge & Kegan Paul, London, Boston and Henley, 1980). Idem, Unfolding Meaning: a Weekend of Dialogue with David Bohm, ed. by Factor, D. (Foundation House Publ., Mickleton and Loveland, 1985).

Bohm, D., Hiley, B.J., and Koloyerou, P.N., "An Ontological Basis for the Quantum Theory", Phys. Reports 144 no.6, pp.323-375, 1987.
21. See, *e.g.*, Weizsäcker, C.F.v., "Quantum Theory and Space-Time", Symposium on the Foundations of Modern Physics, Joensuu, June 16-20, 1985 (see ref.9), pp.223-237, and ref. 25.
22. PLC 0092.051; a letter by Pauli to Fierz on July 15, 1948.
23. PLC 0092.165; a letter by Pauli to Fierz on January 27, 1956.
24. PLC 0495.71; a letter by Pauli to v.Weizsäcker in September 1956 (also appendix in PLC 0092.176; a letter by Pauli to Fierz on October 2, 1956).
25. Weizsäcker, C.F.v., Aufbau der Physik (Hanser, München, Wien, 1985), p.530.
26. Ibid., p. 530.
27. Ibid, p. 535.
28. Pauli accepted the concept of the "unconscious" in the sense in which C.G.Jung used it and which introduces a fundamentally irrational element.
29. d'Espagnat, B., A la Recherche du Réel (BORDAS, Paris, 1979); in English: In Search of Reality.

30. Idem, "Empirical Reality, Empirical Causality and the Measurement Problem" (Foundations of Physics, 17, 1987) pp.507-529.
31. Ibid, p.2.
32. See, *e.g.*, ref.11, Folse, p.217 *ff*.
33. Ref.30, p.24.

LANGUAGE AND REALITY IN QUANTUM PHYSICS

PETER MITTELSTAEDT

Institut für Theoretische Physik
der Universität zu Köln
5000 Köln, Federal Republic of Germany

ABSTRACT

The relations between language and reality are investigated here for the domain of quantum physics. It is shown that an object language of quantum physics can be obtained by constituting it exclusively on the basis of operations which are compatible with the laws of quantum mechanics. This language provides a description of single object systems which is realistic in the sense of Einstein's criterium of reality and which is complete with respect to the quantum physical reality. The relation between this language and the reality is, however, less direct than in the case of the language of classical physics, since the Tarski semantics must be replaced by a weaker "process semantics".

INTRODUCTION

In the domain of classical physics there is a direct one to one relation between the physical reality and the formal language which is used for its description. According to Aristotle the truth of a proposition of this language corresponds to a matter of fact in the physical reality. In the 20th century this realistic semantics was further elaborated by Tarski[1)] with respect to formal languages of higher order. Tarski illustrated the meaning of this semantics by the famous example: The sentence "it snows" is true if and only if it snows. The same semantics was used by Wittgenstein[2)] and applied to a language of classical physics by Carnap[3]. Apart from the limitations of this semantics in higher order languages, which have been investigated by Tarski, there are no other reasons to restrict in any way the application of this "Aristotle-Tarski semantics" within the framework of classical physics.

Moreover in classical physics it is possible in general to relate properties to an object system and predicates to a referent respectively. There are of course well known necessary requirements which must be fulfilled, if propositions can be related to a referent corresponding to an individual system, but in many cases it is merely a question of convention, whether one speaks only about properties or about properties of an object system. It may happen also in classical physics that the use of a convenient referent to which various predictions can be related, is not sufficiently justified. But even in these cases one could imagine that there is an object system which can be used as referent without thereby getting into conflict with any general law of nature or of logic.

1. BOHR'S INTERPRETATION

a) Complementarity: If the domain of reality is extended such that also quantum mechanical phenomena are taken into account, the simple adaequatio-relation between language and reality does not hold any longer. Nils Bohr[4)] made the important discovery that for dealing with the paradoxical features of quantum physics it is not necessary to use new physical concepts and laws of nature, - but to apply a new methodological aspect which he called "complementarity". Since the experimental arrangements which correspond to the measurement of different physical quantities, e.g. position and momentum, are in general mutually exclusive, the measuring results cannot simply be related to a physical object as its properties. Instead one can merely describe the preparation of the object, the measuring procedure and the experimental outcomes.

This meaning of the complementarity as a relation between two observable quantitites, the measuring instruments of which are mutually exclusive was later confirmed by Lahti[5)] within the framework of Hilbert space quantum mechanics. By means of a convenient definition of an instrument as a family of operations and a relation of mutual exclusiveness of instruments, two observables A and B can be defined to be complementary if and only if the corresponding instruments are mutually exclusive. In a more intuitive way this concept of complementarity can be illustrated by a photon split beam experiment. In a Mach-Zehnder-interferometer the incoming photon is split by a beam splitter into two components B and ¬B, which are reflected at two mirrors and brought back together with a phase difference at a beam recombiner. In this experiment there are two mutually exclusive measuring arrangements: Either one observes which way (B or ¬B) the photon came, or one observes the interference pattern.

b) Incompleteness: According to Bohr the notion of complementarity is not completely described by the mutual exclusiveness of measuring instruments. Mutually exclusive instruments correspond to different aspects under which the object system can be observed, but only the totality of various aspects leads to a complete description of the system. Bohr emphasizes that "evidence obtained under different experimental conditions cannot be comprehended within a single picture, but must be regarded as complementary in the sense that only the totality of the phenomena exhausts the possible information about objects".[6)]

This property can, however, not be confirmed within the framework of Hilbert space quantum mechanics. Instruments in the sense of Lahti[5)] are in general not incomplete in the described sense. On the contrary, if the instrument corresponds to an ideal measurement of the first kind the information which can be obtained by this instrument about the system is maximal, i.e. it cannot be increased by further measurements of the same kind. The reason is that results of subsequent measurements can in general not be related to the same system, since every single measuring process possibly destroys the result of the preceding one. Hence the "totality of phenomena" has no obvious meaning.

On the other hand complementarity of two observables does not mean that the different aspects under which the object is investigated are completely independent. Although the results of subsequent maximal measurements cannot be accumulated, they are correlated by probability laws. These probabilistic correlations have important consequences for the possible interpretations of quantum mechanics.

c) Nonobjectifiability: In the language of classical physics it is merely conventional whether measuring results which were obtained by different experimental arrangements are interpreted as properties of the same physical system. In particular, if a given system S is known to have the property A, then it follows for any other property B that the system S either possesses B or $\neg B$ (not B). In logical terms this means that the relation $A = (A \wedge B) \vee (A \wedge \neg B)$ holds in the language of classical physics. - If in a given situation it is at least known that the system in question has the property B or $\neg B$ this property B will be called "to be <u>objectified</u>" in the system.

One could guess that a similar situation also happens in quantum physics. Even if the complementarity of two observables A and B forbids any joint measurement, it could happen that for a system S with property A the complementary property B could be objectified in the system in the described sense. In this case complementarity would merely describe the impossibility of joint measurements for observables, the measuring instruments of which are mutually exclusive. However, the object system could still be considered as an entity which possesses all possible properties (or their contraries) - even if they cannot be measured simultaneously.

It is a most important result that quantum physics does not allow for this interpretation. If on a system $S(\varphi)$ with preparation φ the property A was measured and if one applies the logical equivalence $A = (A \wedge B) \vee (A \wedge \neg B)$ which means that the complementary property B is assumed to be objectified, one is led to a contradiction. Namely from the logical equivalence one obtains the relation

$$p_\varphi(A) \leq p_\varphi(B \sqcap A) + p_\varphi(\neg B \sqcap A)$$

for the probabilities $p_\varphi(A)$, $p_\varphi(B \sqcap A) = p_\varphi(B)p_B(A)$ and $p_\varphi(\neg B \sqcap A) = p_\varphi(\neg B)p_{\neg B}(A)$, where $p_\varphi(A)$ is the probability to measure A and $p_\varphi(B \sqcap A)$, say, is the sequential probability first to measure B and then to measure A. However quantum mechanical probabilities do not fulfill this relation. Instead we find that in quantum physics the probability law

$$p_\varphi(A) = p_\varphi(B \sqcap A) + p_\varphi(\neg B \sqcap A) + p_\varphi^{int}$$

holds, where p_φ^{int} is the interference term, which only vanishes for observables which are not complementary.+)

+) The classical probability relation $p_\varphi(A) \leq p_\varphi(B \sqcap A) + p_\varphi(\neg B \sqcap A)$ contradicts the quantum mechanical equation for $p_\varphi^{int} > 0$.

This impossibility to objectify an observable B if another complementary observable A was already measured, can again be illustrated by means of the photon split-beam experiment mentioned above. If the interference pattern, i.e. observable A, was measured, then it is not allowed to say that the photon came either on the B-way or on the ¬B-way. Instead we must assume that the photon came on both ways since otherwise the interference pattern could not be explained.

§2 PHILOSOPHICAL INVESTIGATIONS

a) Consequences of Bohr's analysis: The results of the preceding section are most important for the relation between language and reality in quantum physics. The nonobjectifiability of complementary propositions shows that it is not possible to speak about quantum object systems and their properties in terms of classical language and logic. In order to avoid these difficulties, Bohr[4)] used a description of quantum mechanical phenomena which deals merely with the preparation of the system, the measuring process and the registration of the results, but not with objects and their properties. Since preparation as well as registration are performed by macroscopic instruments, there are no problems to apply classical language and classical logic to this "positivistic" description of quantum physics. For this reason Bohr refuted any attempts to change the language in which quantum phenomena are formulated. "... the recourse to three-valued logic, sometimes proposed as means for dealing with the paradoxical features of quantum theory, is not suited to give a clearer account of the situation, since all well defined experimental evidence, even if it cannot be analysed in terms of classical physics, must be expressed in ordinary language making use of common logic".[7)]

This argument is correct but not really striking. It is obvious that macroscopic bodies and in particular measuring instruments can be described in terms of classical language. Hence for Bohr there is no need to discuss systems of non-classical logic.[+)] It is, however, still

+) It should be mentioned that the non-classical system of quantum logic, which is investigated in the next section, is not a three-valued logic.

an open question, why quantum physical systems cannot consistently be described by means of classical language and logic and whether a consistent description is possible within the framework of an object language which differs distinctly from the language of classical physics. This problem will be investigated in the next section.

b) Positivistic versus realistic description: For a better understanding of the result that speaking about objects and their properties is contradictory in quantum physics, one should recall some well-known arguments from the traditional philosophy. In his "Treatise of Human Nature" (1739) David Hume described the important fact that we never observe objects but only qualities and that it is nothing but imagination if we consider the observed qualities as properties of an object. Hence any scientific cognition begins with the observation of qualities and it seems to be merely a question of interpretation whether in addition to the observed phenomena a ficticious object, "an unknown something" is used for the description of the experimental results. Within the positivism of the 19th century, Ernst Mach denied for similar reasons the existence of atoms, since - at this time - atoms could not directly be observed but only by their effects on thermodynamical quantities. Although Boltzmann defended the atomic viewpoint throughout his career, it is obvious that in the 19th century the Mach-Boltzmann debate could not finally be decided. It was rather a question of convention whether atoms were considered as real entities or as a convenient way of speaking about the results of thermodynamics.+)

The same problem was treated by Kant in his "Critique of Pure Reason" (1787). However in contrast to Hume, Kant emphasized that "objects of experience" are not only arbitary imaginations but that objects must be constituted from the observational data by means of some well defined conceptual prescriptions, the categories of substance and causality. Hence the interpretation of the empirical data as properties of an object can only be justified if the object as a carrier of pro-

---

+) The whole problem is treated from a historical point of view by S.G. Brush [8] -

perties was constituted by the categories mentioned, i.e. by some laws of nature. In particular Kant formulated necessary conditions which must be fulfilled by the observational data, if the measuring results can be considered as properties of an "object of experience". Hence a description of the physical world which deals merely with observations, sensations, phenomena etc., can be enriched by introducing new entities, things or objects, if these objects can be constituted by means of physical laws from the experimental data. (Fig. 1)

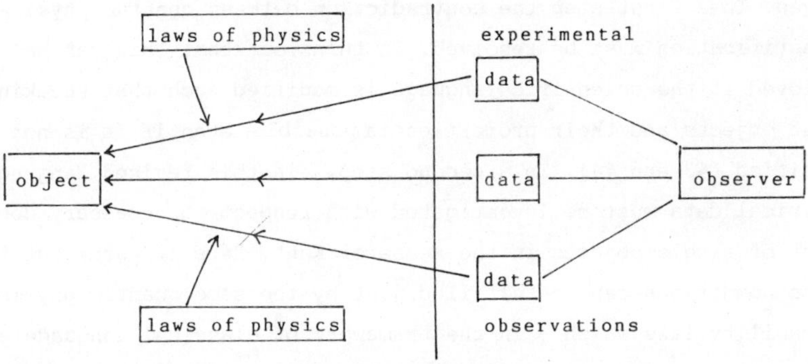

Fig. 1: The constitution of objects in physics

c) Quantum physics and reality: This method of constituting objects is used in many fields of physics, e.g. in astrophysics. On the basis of the astronomical observations the astronomical object, a small planet, say, is constituted by means of the observed data and the Newtonian law of motion. In this way one can extend the physical theory which deals with the astronomical observations by the incorporation of physical objects. Hence it is obvious that also quantum physics begins with observations and measuring results in accordance with Bohr's interpretation. However in contrast to classical physics the constitution of object systems by means of physical laws fails. Moreover it turned out that the laws of quantum probability plainly contradict the most general properties of an object system, which correspond to some laws of classical logic. Hence the necessary conditions for objects of experience seems to be violated by the laws of quantum physics.

Compared with the situation in classical physics, the reality problem in quantum physics is much more difficult. Whereas in classical physics the constitution of objects sometimes fails, since the physical laws which are needed for the constitution are not available, in quantum physics the principle of complementarity is considerably strengthened by the proabability laws which contradict the necessary requirements of object systems. These considerations show that for a realistic description of quantum physical phenomena two separate problems must be solved. In a first step the contradiction between quantum physics and objectification must be resolved. It turns out that this can be achieved if the scientific language is modified such that speaking about objects and their properties is possible, even if it is not yet justified (§3 and §4). In a second step, in this revised language the empirical data must be investigated with respect to necessary conditions of single objects in the sense of Kant. Here it turns out that these conditions can be fulfilled just by the same quantum physical probability laws which - in the framework of classical language and logic - led to inconsistencies (§5). In this way one arrives at a formal language of quantum physics which allows for a consistent speaking about single physical objects and their properties.

§3 CLASSICAL LANGUAGE AND QUANTUM LANGUAGE

a) Critique of classical language: In the preceding sections we have seen that a careless application of classical language and common logic to quantum objects and their properties generally leads to contradictions. In this situation Bohr's relinquishment of a realistic description of quantum phenomena seems for the present to be unavoidable. However one could also ask whether the use of classical language and logic is actually justified in the domain of quantum physical phenomena. In order to answer this question, it is necessary to go back to the foundation of classical language and logic and to formulate explicitly the preconditions and hypotheses on which the language of classical physics is based.

The propositional object language $\mathcal{S}_C$ of classical physics (C-language) is based among other assumptions on the ontological hypothesis that it is decided objectively whether a given system S possesses arbitrary elementary properties (or their contrary) and moreover irrespective of the possibilities of observing these properties. Furthermore it is presupposed that elementary properties are independent and can be tested independent of each other by finite testing procedures. The most general possibilities of the speaker of this language (who is at the same time an observer in the sense of physics) to prove or to disprove propositions $P_i(S)$ about a system S follow from these ontological premises. Furthermore the possibilities to prove (or to disprove) elementary propositions of $\mathcal{S}_C$, which are summarized in the "pragmattics" (C-pragmatics) have some influence on the semantics and the syntactical structure of the language $\mathcal{S}_C$ which is built up of these propositions.

According to the strong preconditions formulated in the ontology and pragmatics, the language $\mathcal{S}_C$ can be equipped with a semantics which is in the spirit of <u>Aristotle</u> and <u>Tarski</u>. Since all possible properties P_A of an object S are assumed to be objectively decided and since these properties can be tested independently by finite means, for propositions $A(S) \in \mathcal{S}_C$ one can use the following criterion of truth: The proposition A(S) is true if and only if the system S possesses the property P_A and A(S) is false if and only if S possesses the counterproperty $P_{\neg A}$ of P_A. In this way the <u>truth</u> of a proposition is uniquely related to a corresponding situation in the physical <u>reality</u>. Hence this kind of semantics will also be called "realistic".

The syntactical structure of the language $\mathcal{S}_C$ follows from the pragmatical possibilities and from the assumed semantics. In particular it turns out that the most general syntactical rules, i.e. the laws of formal logic, are strongly dependent on the pragmatic possibilities of the "speaker-observer" of the language $\mathcal{S}_C$. - On the other hand the ontological hypotheses and the pragmatics of C-language are not in accordance with the basic principles of quantum physics, the <u>complementarity</u> and <u>non-objectifiability</u>. Hence one should be prepared for inconsistencies if the laws of classical logic are applied to quantum physical phenomena. The demonstration of the contradiction between objectifiability and quantum probability confirms this suspicion. - (Fig. 2)

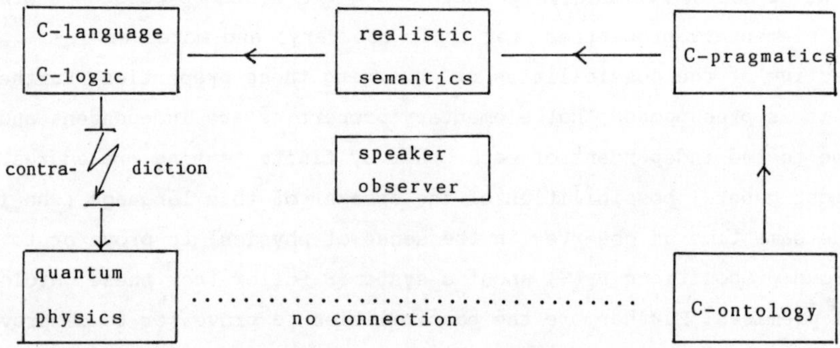

Fig. 2: Schematic representation of the relations between classical C-ontology, C-pragmatics, C-language and the inconsistency of these structures with Q-physics.

b) <u>Pragmatics and semantics of quantum language</u>: In order to construct an object language $\mathcal{S}_Q$ of quantum physics (Q-language) which allows for a consistent description of quantum systems we must dispense with those ontological presuppositions of $\mathcal{S}_C$ which are not in accordance with quantum mechanics. Hence we start from a new ontology (Q-ontology) and a weaker pragmatics (Q-pragmatics) which incorporates from the very beginning the restrictions of proving propositions which come from the principle of complementarity. The methods of proving or disproving propositions of a quantum object language $\mathcal{S}_Q$ which are treated in the Q-pragmatics are - from a physical point of view - the measuring procedures for quantum mechanical properties, and are thus restricted by the complementarity principle.

We will assume here that a proposition A(S) which states that the system S has the property P_A can always be proved or disproved. However if after the successful proof of A(S) another proposition B(S) has been shown to be true, it is not guaranteed that a new proof attempt of A(S) is again successful. Hence two propositions A(S) and B(S) are in general not simultaneously decidable. This is only the case if the simultaneous decidability of A and the subsequently tested proposition B has explicitly been demonstrated. In this case the propositions A and B are said to be "<u>commensurable</u>". Within the present formal language approach the commensurability of two propositions A and B is merely a contingent property which must be tested experimentally.

On the basis of the weak ontological premises of quantum physics (Q-ontology) and the corresponding Q-pragmatics, we can now formulate the basic principles of a semantics for the language $\mathcal{S}_Q$. A realistic semantics of truth as it was used successfully in C-language cannot be applied to Q-language. The reasons are the non-objectifiability of quantum physical properties and the peculiarities of the quantum mechanical measuring process. For a system S with a given preparation W an arbitrary property P_A is in general not objectified. However since any property is assumed to be measurable, this means that an experimental test of P_A consists of two steps: In the first step the initial state W of the system is tranformed into a new state W', such that P_A is objectified in W'. In the second step one merely observes by reading the measuring instrument, whether S(W') possesses the property P_A or its contrary.

It is obvious from this description of the measuring process that the truth of a proposition A(S) can be related to a property P_A of the system S only if A is objectified in the state W of the system. Hence the semantics must take into account the difficulties to prove a proposition A(S) by means of a quantum mechanical measuring process. For this reason a <u>proposition A(S) will be called to be true if and only if by means of a measuring process the system S was shown to possess the property P_A</u>. A semantics which in this way relates the truth

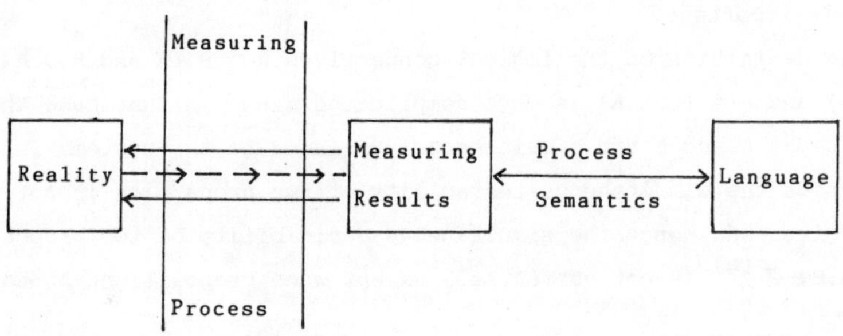

Fig. 3: The relations between language and reality with process semantics

of a proposition merely to the result of a measuring process and not to some real situation will be called "process semantics". (Fig. 3)

It is obvious that the process semantics is a relaxation of the realistic semantics, since it provides only a weaker connection between true propositions and facts of the physical reality than the Aristotle-Tarski semantics. If a proposition is true in the sense of the process semantics, it is also true in the sense of the realistic semantics, - but nothing is known about the inverse relation.+)

§4. SYNTAX AND LOGIC OF QUANTUM LANGUAGE

a) Compound propositions: On the basis of the weak Q-ontology and the process-semantics mentioned, the formal language of quantum physics can now be established. Starting from the set $\mathcal{S}_Q^{(e)}$ of elementary propositions, the language can be extended by incorporating compound propositions. In a first step we introduce the sequential conjunction $A \sqcap B$ (A and then B) which describes the temporal sequence of two propositions and which will be defined here by the possibilities to attack and to defend this connective. These possibilities and thus the meaning of the compound proposition can best be illustrated by a proof tree which describes the sequence of successive proof attempts. The proof tree for $A \sqcap B$ consists of a test of A and a succeeding test of B at a system S with preparation W. $A \sqcap B$ is defined to be true with respect to S(W) if the A-test as well as the subsequent B-test are successful. It should be emphasized that the temporal order of the two proof attempts is important.

The definition of the logical connectives $A \cap B$ (A and B), $A \cup B$ (A or B) and $\sim A$ (not A) is more complicated since in that case the propositions A and B are simultaneously related to the system. In quantum physics the simultaneous testability of two properties is not generally given and hence the simultaneous decidibility of two propositions $A, B \in \mathcal{S}_Q^{(e)}$ is not guaranteed, except when propositions A and B

+) Moreover according to a theorem by Giuntini[10)] under very general conditions there is no possibility to extend the process semantics such that any proposition of $\mathcal{S}_Q$ is directly related to certain facts of a hidden reality.

are commensurable. Hence we introduce "availability propositions" $k(A,B)$ and $\bar{k}(A,B)$ which state the commensurability and the incommensurability of A and B respectively and which can be tested by convenient series of measurements. In particular, $k(A,B)$ is considered to be true if and only if A and B can be tested in arbitrary sequence without thereby influencing the result of the respective trials.

By means of these avialability propositions one can now define the logical connectives by the possibilities to attack or to defend them and by the corresponding proof trees. E.g. in case of the logial conjunction $A \cap B$, it is stated that the propositions A and B both are simultaneously true for a system S. In order to make sure that after the test of B the result of the preceding A-test is still available, the proof tree for $A \cap B$ contains - compared with the proof tree for $A \sqcap B$ - an additional $k(A,B)$-test. According to this proof tree, the logical conjunction is considered to be true with respect to S(W) if and only if the three proof attempts for A,B and $k(A,B)$ are successful. - In a similar way the other logical connectives can be defined.

b <u>The language $\mathcal{S}_Q$</u>: For propositions $A \in \mathcal{S}_Q$ we have to distinguish two kinds of equivalence relations. Two propositions A and B are called <u>proof equivalent</u>, denoted by $A \equiv B$, if in each proof tree in which one of the propositions occurs, it may be replaced by the other one without thereby influencing the result of the whole proof process. On the other hand, A and B are called <u>value equivalent</u>, denoted by $A = B$, if the proof trees for A and B have the same outcome. It is obvious that the proof equivalence is the stronger relation, since the proofs of A and B may affect propositions which are not commensurable with A and B in different ways. Furthermore we make use of the binary relation "$\leq$", the implication, which can be defined by $A \leq B \leftrightarrow A \equiv A \cap B$.

We will assume here that to any logically connected proposition A, the elementary subpropositions of which are related to the same time instant, there exists and elementary proposition $A^{(e)}$ such that $A = A^{(e)}$. One can then define inductively new connectives $A \wedge B = (A \cap B)^e$, $A \vee B = (A \cup B)^e$, and $\neg A = (\sim A)^e$. The structure which is imposed by these connectives on the set of elementary propositions will be of particular interest for the following investigations.

A proposition $A \in \mathcal{S}_Q$ is called to be true with respect to the preparation W of S, denoted by $W \vdash A$, if a proof tree of A with the initial preparation W has only branches of success. If in particular the proof tree for a proposition A has only successful branches irrespective of the initial preparation W, we call A formally true and write $\vdash A$. If the negation $\neg A$ is formally true, i.e. $\vdash \neg A$, A is said to be formally false. Furthermore we introduce two special propositions, the true proposition (verum) V, such that $\vdash V$, and the false proposition (falsum) $\Lambda = \neg V$. By means of these definitions it follows that $\vdash A$ is equivalent to $V \leq A$.

The language $\mathcal{S}_Q$ can be used for a formal description of the quantum mechanical measuring process. If at a system S with preparation W the proposition A is measured, we have to distinguish two different steps. In step (I) the proposition A is <u>objectified</u>, which means that the preparation W is transformed into the <u>sequential Gemenge proposition</u> $W \sqcap (A \cup \neg A) = (W \sqcap A) \cup (W \sqcap \neg A)$ which describes an ensemble of two possible outcomes, $W \sqcap A$ and $W \sqcap \neg A$. In a second step (II) this ensemble can be reduced simply by reading to one of the possible components $W \sqcap A$ and $W \sqcap \neg A$ respectively.

c) <u>Quantum logic:</u> A compound proposition $C \in \mathcal{S}_Q$ is said to be formally true if it is true for any preparation W of the considered system, i.e. if it is true irrespective of the truth-values of the elementary propositions and availability propositions contained in the proof tree of C. By means of the true proposition V, the formal truth of C can be expressed by $V \leq C$. There are infinitely many formally true compound propositions in the language $\mathcal{S}_Q$. As examples we mention here: $V = C \vee \neg C$ (principle excluded middle), $C \wedge \neg C = \Lambda$ (law of contradiction). In contrast to classical logic the equivalence $A = (A \wedge B) \vee (A \wedge \neg B)$ does not generally hold except when A and B are commensurable. Instead of this equivalence only the weaker implication $(A \wedge B) \vee (A \wedge \neg B) \leq A$ is formally true in $\mathcal{S}_Q$.

The totality of formally true propositions of $\mathcal{S}_Q$ can be summarized in the calculus $\mathcal{L}_Q$ of quantum logic. It consists of "beginnings", i.e. implications $A \leq B$ which are formally true, and "rules" which allow for the construction of new formally true implications from given formally true implications. The calculus $\mathcal{L}_Q$ is known in the literature and need not be repeated here.[11] The Lindenbaum-Tarski algebra of the calculus $\mathcal{L}_Q$ of quantum logic is an orthomodular lattice L_Q which is atomic and which fufills the covering law.[11] Atomicity and the covering law follow from the requirement that the propositions of the language $\mathcal{S}_Q$ are related to a single physical system.[12]

The formal language $\mathcal{S}_Q$ of quantum physics can be applied to the objectifiability problem discussed in §1. Since in $\mathcal{S}_Q$ the restrictions due to possible incommensurabilities are already incorporated into the syntax and logic of the language $\mathcal{S}_Q$, no additional rules must be kept in mind if the language is applied to a physical situation. In particular it should be possible to speak about an object-system and its properties in terms of $\mathcal{S}_Q$ without thereby running into a contradiction with quantum probability. This can easily be demonstrated by considering again the above mentioned split-beam experiment. For the propositions A and B in quantum logic only the implication relation $(A \wedge B) \vee (A \wedge \neg B) \leq A$ holds instead of the classical equivalence relation. This means that at a system S for which the proposition A is true, another proposition B is not already objectified by the laws of logic. Furthermore within the framework of quantum probability theory[13,14] from the implication mentioned one obtains only the inequality $p_W(A \wedge B) + p_W(A \wedge \neg B) \leq p_W(A)$ which does not contradict any more the quantum mechanical probability law which contains an interference term.

Hence we find that the contradiction between objectification and quantum physics can be resolved if the language $\mathcal{S}_C$ is replaced by the weaker language $\mathcal{S}_Q$, the ontological premises of which are in accordance with quantum physics. Moreover this restriction is really necessary since according to the above mentioned result of Giuntini[10] there is no consistent extension of the new quantum language $\mathcal{S}_Q$ which could restore classical objectifiability without thereby leading to contradictions with quantum physics.

§5. THE CRITERION OF REALITY

a) Truth and Reality: In the language $\mathcal{S}_Q$ of quantum physics the concept of truth is determined by the process-semantics: A proposition A(S) is said to be true iff it is shown by a measuring process that the system S possesses the property P_A. According to this semantics the relation between truth and reality is weaker than in the Aristotle-Tarski semantics which can be used as a semantics for classical language. The measuring process separates the reality from the measuring results and thus also from the language. Only those properties of the system considered, which were actually shown to be there by measurement, correspond to true propositions of the language. Hence one could conjecture that the measuring process hides like a veil some features of the reality. In this case quantum physics as well as quantum language were incomplete.

The incompleteness conjecture was particularly advocated by Einstein. However within the framework of the formal language approach to quantum physics, this conjecture is no longer tenable. According to the above mentioned theorem[10)], there is no consistent extension of the process-semantics which would provide truth values to all propositions of the language $\mathcal{S}_Q$ and which extension could thus be considered as a realistic semantics. Hence within the formal language approach the incompleteness conjecture is no longer tenable. This argument will be further discussed in §5c.

b) Einstein's criterion: Under these conditions it is of particular interest to investigate the question whether there are at all situations in which the physical reality is directly related to elements of the formal language. That this is in general not the case, follows from the reasons for replacing the realistic semantics by the process semantics. The decisive argument in favour of the process-semantics is the non-objectifiability of arbitrary properties of a system. The measuring process of the property P_A transforms in general the preparation W of the system into a new (Gemenge) state in which either P_A or $P_{\neg A}$ is realized. Moreover since P_A is in general not objectifiable in W, nothing can be said about the reality of P_A before the measuring process was performed.

If once the reasons are known for replacing the Aristotle-Tarski semantics by the process-semantics one can try to characterize the reality of some property by a pragmatic situation in which the reasons mentioned are not relevant. This is the case if two (for the present independent) conditions are fulfilled: If (R1) the speaker-observer can predict with certainty that the system S with the preparation W posesses the property P_A and if (R2) he knows that P_A can be observed without thereby changing the system in any sense, then (R3) he may consider P_A as a real property of S = S(W). - By means of this sufficient criterium of reality one can describe situations in which the speaker-observer may speak about reality in the sense of classical physics - and that in spite of all the restrictions of quantum language.

A sufficient criterium for reality of this kind was first formulated by Einstein[15)]. Einstein hoped in this way to be able to circumvent Bohr's verdict that speaking about reality is impossible at all in quantum physics. Einstein's criterium should not be misunderstood as an attempt to restore classical reality in quantum physics. Instead it characterizes situations in which the classical concept of reality is still relevant and meaningful even in the framework of quantum physics. However, the example by means of which Einstein tried to illustrate his criterium turned out to be plainly false.

c) <u>The reality criterium in Q-language:</u> Within the framework of the formal language of quantum physics the reality criterium turns out to be a semantic principle already contained in $\mathcal{S}_Q$. In terms of the formal language, the first premise (R1) means that the preparation W and the proposition A are related by the implication $W \leq A$. The second premise (R2) assumes that A can be measured on the system S = S(W) with preparation W without thereby influencing the system in any way. Since in $\mathcal{S}_Q$ the successful test of A is described by $W \sqcap A$, the premise (R2) reads $W \equiv W \sqcap A$, which is equivalent in $\mathcal{S}_Q$ to the first premise $W \leq A$. - The conclusion (R3) of the reality criterium states that the property which corresponds to A pertains to the system S, which means in terms of $\mathcal{S}_Q$ that $S(W) \vdash A$ holds, i.e. that A is true with respect to W(S). Hence the reality criterium can be expressed by the implication $W \lessapprox A \rightarrow S(W) \vdash A$. It is obvious that this relation is not a new principle which must be added to the language $\mathcal{S}_Q$, since it is already contained in it as the well known semantic criterium for the truth of A with respect to S(W), i.e. for $S(W) \vdash A$. Hence the language $\mathcal{S}_Q$ fulfils the criterium of reality mentioned and can thus be considered as a realistic language[16)].

If the reality criterium can be applied to a given situation then the process semantics can be replaced by the Aristotle-Tarski-semantics which means that true propositions correspond to elements of the physical reality and vice versa. However since this happens only in very particular cases, one could ask whether quantum language and process-semantics really exhaust the quantum physical reality or whether there are veiled parts of the reality which are not covered by the formal language. In the latter case quantum language would be incomplete. According to the above mentioned theorem[10)] this incompleteness suspicion can be excluded. Hence the criterium of reality which is formulated in the present section is not only correct in those cases in which it is applicable but also exhaustive in the sense that it covers the full quantum physical reality.

The reality criterium which is formulated in this section is a <u>sufficient</u> criterium: <u>If</u> the premises $W \lessapprox A$ and $W \equiv W \sqcap A$ are fulfilled, <u>then</u> A is true with respect to W. However, in the case of a single physical system S = S(W) this inference can also be inverted. This means that <u>if</u> the system S with preparation W possesses a property P_A which corresponds to the truth of the proposition A, <u>then</u> the relation $W \lessapprox A$ holds. The reason for this somewhat surprising statement will be discussed in the following section. It turns out that the necessary conditions for the reality of properties of an individual system have important consequences for the structure of the formal language.

§6. THE LANGUAGE OF A SINGLE OBJECT

a) <u>Constitution of objects:</u> The formal language $\mathcal{S}_Q$ of quantum physics is for the present a language of predicates. According to the process-semantics the validity of a predicate corresponds to the experimental confirmation of an observable property. An object system as a carrier of these properties was up to now simply assumed to exist. What is meant by this assumption? Are there object systems irrespective of possible properties or are objects merely theoretical concepts which are constituted by means of the observable properties? These questions must be answered if object systems will be introduced as new linguistic terms into the formal language.

The problem of consituting objects was already briefly discussed within the framework of classical physics. Since objects do not appear primarily in the physical experience, they must be constituted as carriers of accidental properties. According to Kant's analysis, the constitution of objects is not purely conventional but must take into account some laws of nature which connect the various observations. Conversely it follows from these considerations that if an object with time-dependent accidental properties is given as an element of our experience, then the accidental properties of this object must fulfill some physical laws. Hence these laws of nature - the causality law in Kant's terminology - are necessary conditions for the cognition of objects of experience.

An object system as a carrier of time-dependent accidental properties is an element of the exterior physical reality. It must persist in time and it must be observable at different time instances and from different positions of the observer. Without the possibility of repeated observation, the observer would never be able to recognize the system as a persistent carrier of accidental properties which belongs to the objective and exterior reality. Since in nonrelativistic physics observers which are connected by a Galilei-transformation are equivalent in some sense, an object system should be invariant with respect to the full group of Galilei-transformations. In mathematical terms this means that an object-system is given by a convenient representation of the 10-parameter Galilei-group, the parameters of which characterize the permanent "essential" properties of the respective system type.[17,18)]

b) Objects in quantum language: In quantum physics the constitution of individual objects by means of physical laws fails if objects are incorporated as new elements into the language $\mathcal{S}_C$ of classical physics. The most general properties of object systems which correspond to some logical tautologies contradict the laws of quantum probability. However it could be shown in the preceding sections that the contradiction between objectification and quantum physics disappears if the language $\mathcal{S}_C$ of classical physics is replaced by the language $\mathcal{S}_Q$ of quantum physics which is based exclusively on ontological premises which are in accordance with quantum physics.

An object to which all predicates of the language can be related as their referent is generally characterized as a unitary representation of the full Galilei group with parameters which correspond to the essential properties of the system type (charge, mass, intrinsic spin, etc.). Conversely, if there exists a system with well defined essential properties, then some features of the language $\mathcal{S}_Q$ must be invariant with respect to Galilei transformations. Hence the existence of a symmetry group is a necessary condition for a language the propositions of which are related to objects at all.

Furthermore if all propositions of the language $\mathcal{S}_Q$ are related to an individual system S, this object must have been constituted by means of a "causality law", i.e. by some physical laws. Within the framework of the formal language there are at least two kinds of laws, the strict causal laws which can be expressed by an implication $A \leq B$, and the less rigorous probability laws $p(A;B) = p_o$. For the constitution of an individual object system causal implications such as $A \leq B$ are needed, whereas probability laws would provide only an approximate constitution of a single system. However if the totality of all probability laws between arbitrary propositions A,B,... of $\mathcal{S}_Q$ is taken into account, then probabilities can also be used for an exact constitution of an object.

c) Necessary conditions: If a measured proposition A can be related to a system S = S(W) with preparation W, then the propositions A and W are both related to the same object system S. It follows from the arguments mentioned above that this system S must have been constituted by means of accidental properties and physical laws. In the present case there are only two accidental properties W and A and the physical law $W \leq A$ which could have been used for the constitution of S. Hence if one states that A was measured at the system S with the preparation W, it follows that W and A are connected by the implication $W \leq A$. For this reason the criterium of reality mentioned above is not only sufficient but also a necessary criterium for the reality of A on S(W): If the property which corresponds to A pertains to the system S which originally had the preparation W, - then the relation $W \leq A$ holds, which means that A can be predicted with certainty for the system S = S(W) with preparation W[19,20].

If one assumes that all propositions $A \in \mathcal{S}_Q$ of the formal language are related to an individual system S, then the system S must have been constituted by means of physical laws which hold between arbitrary propositions of the language. However these laws must, in general, be probability laws since implications can only hold for commensurable propositions. Hence we conclude that the propositions A,B,C,... of $\mathcal{S}_Q(S)$ are at least connected by probability relations such as $p(A;B) = p_o$ by means of which the system has been constituted. It follows from the existence of probability laws of this kind that for the propositions A(S),B(S)... of $\mathcal{S}_Q(S)$ the law of atomicity and the covering law hold.[12] Hence atomicity and covering law are necessary conditions for a language the propositions of which are related to an individual object system.

Conclusion: The invesitigations of the present paper can be summarized in the following way: Quantum theory in its present form is neither a positivistic theory in the sense of Bohr, which cannot be described by a formal object language, - nor is it a realistic theory which can be formulated by means of classical object language and a realistic semantics in the sense of Aristotle and Tarski. It can, however, be formulated by means of an abstract object language of quantum physics $\mathcal{S}_Q$, which is based on the most general conditions of operations which are compatible with quantum physics. This quantum language is concerned with physical objects and their properties but it avoids any contradictions with quantum probability.

The relation between this formal quantum language and the physical reality is less direct than in the case of classical language. The realistic Aristotle-Tarski semantics must be replaced here by the weaker process semantics which relates only the results of a measuring process to true propositions and vice versa. There are, however, special situations such that language and reality are connected in the same way as in classical physics. These situations correspond exactly to the sufficient criterium of reality which was first formulated by Einstein[15] and which turns out to be also a necessary condition for the quantum reality if the language is related to a single object system.

REFERENCES

1) Tarski, A:, Studia philosohica I, 261-405 (1936).
2) Wittgenstein, L., Tractatus Logico-philosophicus, London (1922).
3) Carnap, R., Logical Syntax of Language, London and New York (1937).
4) Bohr, N., Nature 121, 580-90 (1928).
5) Lahti, P., Intern. Journ. of Theor. Phys. 22, 911-929 (1983).
6) Bohr, N., in: Schilpp, P.A., (ed.), Albert Einstein: Philosopher-Scientist, Library of Living Philosophers, Evanston, Ill. (1949).
7) Bohr, N., Dielectica 2, 312-319 (1948).
8) Brush, S.B., The Kind of Motion we call Heat, Amsterdam (1985).
9) Mittelstaedt, P., Sprache und Realität in der Modernen Physik, Mannheim (1986).
10) Giuntini, R., Studia Logica XLIX, 17-35 (1987)
11) Mittelstaedt, P., Quantum Logic, Dordrecht (1978).
12) Stachow, E.-W., in: Recent Developments in Quantum Logic, P. Mittelstaedt and E.-W. Stachow (eds.), Mannheim (1985) pp. 129-145.
13) Jauch, J.M., Synthese 29, 131-154 (1974).
14) Mittelstaedt, P., in: The Foundations of Probability Theory, Statistical Inference and Statistical Theories of Sciences III, W.L. Harper and C.A: Hooker (eds.), Dordrecht (1976) pp. 155-167.
15) Einstein, A:, Podolsky, B. and Rosen, N., Phs. Rev. 47, 777-780 (1935).
16) Mittelstaedt, P. and Stachow, E.W., J. Phil. Log. 7, 181-208 (1978).
17) Levy-Leblond, J.M., J. Math. Phys. 4, 776-788 (1963).
18) Piron, C., Foundations of Quantum Physics, New York (1976).
19) Mittelstaedt, P., in: Symposium on the Foundations of Modern Physics, Singapore (1985), pp. 171-186.
20) Busch, P., Phys. Rev. D 33, 2253-2261 (1986).

CONTEXTUAL QUANTUM OBJECTS AND THEIR ONTIC INTERPRETATION

Hans Primas

Laboratory of Physical Chemistry, ETH
CH-8092 Zürich , Switzerland

ABSTRACT

The main reason for the disagreement about the interpretation of quantum mechanics is due to the neglect of the influence of the environment. A mathematically rigorous description of open quantum systems together with the relevant environments, the classical aspects of measuring instruments, and symmetry breakings is possible in the framework of algebraic quantum mechanics. An object is defined as an open quantum system which in spite of its interaction with its environment is not Einstein-Podolsky-Rosen-correlated with the environment. Since quantum mechanics describes the world as a whole which is not made out of parts, quantal objects exist not in an absolute but in a contextual sense. An individual interpretation of a quantum system is possible if and only if this system is an object. Some consequences and examples are discussed.

1. TRADITIONAL QUANTUM MECHANICS HAS A VERY LIMITED RANGE OF VALIDITY

A modern reevaluation of the Copenhagen view has to consider very carefully the progress in the mathematical formulation of quantum mechanics in the last fifty years. The historical Copenhagen interpretation was based on an only rudimentarily developed formalism and systematically ignored all effects of the environment [1]. Even the Hilbert--space formalism as summarized in Johann von Neumann's book "Mathematische Grundlagen der Quantenmechanik" of 1932 [2] represents a theory of *very limited validity* since it is stringently restricted to systems without superselection rules and having only finitely many degrees of freedom. Many of the conceptual difficulties of traditional quantum mechanics can be traced back to the limited validity of the Hilbert-space formalism and to the fact that usually the influence of the environment of a quantum system is neglected or not taken into account properly.

The acceptance of the time-honored idea of perfectly isolated systems without much inquiry is definitely causing concern. We are under the obligation to check whether such an assumption is theoretically consistent and empirically correct. A proper theory of quantum systems has to fulfill a basic stability requirement in the sense that a small change in the environment should induce only a small change in the description of the object system.

If we single out a "system", we have divided the universe into two parts. All we have not singled out will be called the *environment* of the system. We say a mathematical model of a system is *robust* if the inclusion of the effects of its environment does not change the description in a *qualitative* way. Unfortunately, most Hamiltonian models with finitely many degrees of freedom are not robust. Therefore *the environment of an object system should never be left out of consideration*. If one disregards this elementary requirement, one can get easily absurd results as Schrödinger's famous cat example drastically demonstrates.

At first sight one may think this problem could be solved in a conceptually simple manner by treating the object system together with its environment as a single system. That is possible but there is a nontrivial difficulty. The Hilbert-space formalism of traditional quantum mechanics is valid only for *strictly closed Hamiltonian systems with finitely many degrees of freedom*. Sometimes it is argued that even "measuring apparatuses as well as human observers are finite systems" [3]. This remark is very misleading since these systems have no robust description in terms of finitely many bare elementary systems so that they have to be considered as open systems. Even a so-called "free atom" or "free molecule" is interacting and hence entangled with the rest of the world via the unavoidable electromagnetic radiation field. This radiation field is part of the environment of every material system, so that the inclusion of the environment leads to quantum systems with *infinitely* many degrees of freedom. For such infinite quantum systems both the Hilbert-space formalism of traditional quantum mechanics and the Fock-space formalism of quantum field theory lose their validity.

The Hilbert-space formalism of quantum mechanics depends in a crucial way on the Stone - von Neumann uniqueness theorem [4,5] for the canonical commutation relations which proves that for systems with *finitely* many degrees of freedom there is up to unitary equivalence a unique irreducible Hilbert-space representation of the algebra of observables. For infinite systems the Stone - von Neumann theorem fails and in fact there are infinitely many *inequivalent representations* of the canonical commutation relations [6,7]. These inequivalent representations are of great conceptual importance since they limit the validity of the superposition principle and can lead to *superselection rules, spontaneous symmetry breakings*, the emergence of *classical observables* and the possibility of truly *irreversible dissipative processes*. Such phenomena do not violate the first principles of quantum mechanics but they can neither be discussed in the framework of the Hilbert-space formalism of traditional quantum mechanics nor in terms of a Fock-space type quantum field theory. As we will discuss in section 5, spontaneous symmetry breaking can occur already with rather simple molecular systems, they are generic for complex quantum systems and by no means rare exceptions. Therefore it is not meaningful to draw on the basis of the Hilbert-space formalism of traditional quantum mechanics far-reaching conclusions of philosophical nature such as the nondescribability of the measurement process or the nonexistence of an objective reality. Before we can enter into a discussion of the possibilities and limitations of quantum mechanics, we have to develop rigorous mathematical methods for moderately complex systems. Only if we have attained a sufficiently strong foothold we may try to transfer ideas and methods to really complex systems like actual measuring instruments and cats.

2. QUANTAL OBJECTS AND THEIR ENVIRONMENTS

The fact that there are many inequivalent Hilbert-space representations of quantum systems with infinitely many degrees of freedom forces us to rephrase the first principles of quantum mechanics in a representation-free manner. This program has been realized with great success and we have nowadays a physically sound and mathematically rigorous formalism of generalized quantum mechanics, generally known under the name *algebraic quantum mechanics* [8].

Algebraic quantum mechanics is a straightforward generalization both of traditional quantum mechanics and of classical mechanics. In the domains proper of these theories, nothing is changed. Both classical mechanics and quantum mechanics are treated in the same formal framework, and the systems under discussion may possess either finitely or infinitely many degrees of freedom. In the framework of algebraic quantum mechanics one can discuss many problems which are outside the domain of traditional quantum mechanics, like

- molecules in interaction with the radiation field
- the interaction between quantal and classical systems
- the emergence of classical molecular structures
- the relaxation behavior of single quantum systems
- nonlinear feedback effects via the environment of an object system
- thermodynamic systems and phase transitions.

Above all, algebraic quantum mechanics gives us the tools to understand and to decide under which circumstances a quantum system can have *classical features*. As a rule, not only the dynamics but also the kinematics of an object system depends on the environment. That is, *the W*-algebraic structure of the observables is not given a priori*. Given the basic kinematical group, the quasilocal structure and the state of the environment, the methods of algebraic quantum mechanics allow (via the so-called GNS-construction) to construct in a unique way the relevant W*-algebra of observables. This algebra may have a center, so that one can have nontrivial observables which commute with *all* observables. Such observables are called *classical observables*, they represent the *classical properties* of the system. If *all* observables are classical (i.e. if the W*-algebra of observables is commutative), the corresponding system is called *classical*. If the center consists only of the multiples of the unit operator, it is called trivial, and since there are no nontrivial classical observables such a system is called *purely quantal*. These are the two extreme cases, in general one has to expect that there are classical observables but that not all observables are classical. Such systems are called *partly classical and partly quantal*.

Algebraic quantum mechanics gives us the conceptual framework to include the interactions of a quantum system with its environment. But the conceptually difficult problem of the nonseparability of the quantum world remains. That is, the main problem of a theory of open quantal systems is not the interaction as such but the *entanglement* of the open system with its environment.

In the Hilbert-space formalism a joint system is described by tensor product of the Hilbert spaces of the constituent systems. In algebraic quantum mechanics the algebra of observables of a joint system is given by the tensor product of the constituent algebras of observables. In the purely classical case, where all algebras of observables are commutative, *every* pure state of the joint system is a *product state*, that is, a tensor-product state of pure states of the constituent systems. For purely quantal systems this is no longer valid, in general a pure state of a joint quantal system is not a product state. Or in other words, the restriction of a pure state on a tensor-product algebra to one of its constituent noncommutative algebras is not a pure state. A pure state which is not a product state is called an *Einstein-Podolsky-Rosen-correlated state*, and we say with Schrödinger that the constituent systems are *entangled*. If the constituent systems of a joint system are all nonclassical, then most pure states of the joint system are Einstein-Podolsky-Rosen-correlated states. This fact often is expressed as a *principle of nonseparability:*

> If two closed finite quantum systems have interacted in the past, they are inextricably entangled, no matter how far apart they are in space, and it is impossible to assign a pure state to either of the two subsystems.

If this entanglement of quantal systems is generally valid, then nature is holistic and nonseparable in the sense that *the world is not made out of parts*, so that the system concept of classical natural science becomes problematic.

In order to have a precise language, we will distinguish between the concepts "system" and "object". By a *system* we just mean the referent of a theoretical discussion, without any ontological commitment. On the other hand we choose to use the word *"object"* as the most general ontological term for something having individuality and properties. We do not restrict the notion of an "object" to concrete things or to entities which are localized in space. Objects may be quite abstract individuals which not necessarily can be isolated from the rest of the world.

We consider objects as the most general referents of an individual interpretation of quantum mechanics. Therefore we adopt the following definition:

> *Definition*
> An object is an open quantum system which in spite of its interaction with the environment is not Einstein-Podolsky--Rosen-correlated with the environment.

It follows that objects are exactly those quantum systems for which at any time a maximal description in terms of a pure state is possible. Objects may change their actualized properties but they keep their identity in course of time. If a quantum system is not an object, it does in general not possess a maximal description in terms of a pure state so that no individual interpretation is possible.

An important theorem of algebraic quantum mechanics says that

(i) all classical systems are objects,
(ii) a nonclassical open system is an object if and only if its environment is classical [9].

Accordingly, nonclassical objects in nonclassical environments do not exist. Since classical systems have no natural place in the Hilbert--space formalism of traditional quantum mechanics, the only objects which can be discussed in the framework of traditional quantum mechanics are perfectly isolated systems, a quite unrealistic idealization. As a consequence, any attempt of an individual interpretation of *traditional* quantum mechanics is marred by the unavoidable but untenable fiction of a perfectly isolated system (compare also the discussion by Bohr [1]). In contrast to this, algebraic quantum mechanics allows the description of the interaction of quantal objects with their environment and so paves the way for an individual interpretation of quantum mechanics.

3. REGULATIVE PRINCIPLE OF REALISM AND THE ONTIC INTERPRETATION OF ALGEBRAIC QUANTUM MECHANICS

According to algebraic quantum mechanics we have to expect that open systems both have quantal and classical properties. This is in fact the typical situation we find in the *mesoworld*, the domain between the microworld and the macroworld. Typical *chemical systems* always are partly classical and partly quantal. In one and the same molecular system quantal and classical features coexist and interact with each other. A typical example is the material carrier of genetic information, the *DNA-molecule*. Molecules of this type possess many biologically important properties which definitely require a *quantum-theoretical* description, for example their photochemical reactivity. On the other hand, DNA-molecules have a tertiary structure which is manifestly *classical* and also of biological importance, for example for the mechanisms of genetic recombination. Nowadays electron micrographs are used routinely to investigate the tertiary structure of single DNA-molecules. For example, circular DNA-molecules may be knotted, and there are enzymes which can change the knot type of a circular DNA-molecule [10,11]. Hence the knot type of a circular DNA-molecule is an example of a classical property of biological importance. If one wants to understand the molecular basis of heredity, one needs a theory of matter which can describe both the quantal *and* the classical properties of individual objects.

There is a practically continuous transition from the microworld to the macroworld. For a chemist there is no fundamental difference between an atom and a very small molecule, or between small and large molecules, between large molecules and giant molecules or crystals. It would be entirely unreasonable to require that a molecular scientist should ever use different languages for describing small and large molecules, simply because there is no qualitative difference between these systems. If a biochemist describes the knot type of a single circular DNA-molecule, he adopts an *ontological description* which goes far beyond a phenomenalist view. No scientist adopts the view that the knot type of DNA is not real until it is observed but all molecular biologists claim that DNA-molecules really exist and that these molecules *have* well-defined tertiary structures. Whether we observe or know these structures is entirely irrelevant for their existence. Such a belief in an objectively existing reality is, of course, not provable. Nevertheless, it is the deepest motivation for all scientific research and the basis of the realistic interpretation of the classical natural sciences. A realistic interpretation has nothing to do with experimental facts but characterizes a way of thinking. Since in everyday life everybody is committed to some kind of realism, it is important to adopt also for quantum theories a realistic way of thinking.

Such an *ontological committment* is permissible provided we adopt an appropriate interpretation of quantum mechanics. By an *interpretation* of a formally consistent and experimentally corrobated theory we mean some regulative principles which neither can be deduced nor disproved on the basis of the mathematical formalism and the pragmatic working rules associated with the theory.

Classical realism in the sense that all attributes of an object are truth-definite at every instance, is inappropriate for quantum mechanics, but we can ask for an interpretation which is as close as possible to the classical realistic world view. We say that an interpretation of algebraic quantum mechanics fulfills the *regulative principle of realism* if its restriction to the classical part corresponds to the generally accepted realistic individual interpretation of the traditional classical physical theories. Here, the *classical part* of a quantum system with the algebra A of observables is defined to be the classical system with the center Z of A as commutative algebra of observables. This regulative principle requires the consistency of an interpretation of algebraic quantum mechanics with the realistic interpretation of classical physical theories. Since both classical mechanics and traditional quantum mechanics are special cases of algebraic quantum mechanics, this regulative principle restricts the admissible interpretations of traditional quantum mechanics.

In algebraic quantum mechanics it is logically consistent to extend the regulative principle of realism to the following *ontic interpretation :* The referents of the theory are individual objects whose *potential properties* are represented by the selfadjoint elements of an appropriate C*-algebra A of observables. The *ontic state* at time t of an object is defined by the set of all potential properties which are actualized at time t. It can be represented by an extremal, normalized, positive linear functional on the algebra A (usually called a "pure state"). If the ontic state at time t is given by the pure state $\rho_t \in A^*$, then an observable $A \in A$ corresponds to a property actualized at time t if and only if A has a dispersionfree value with respect to ρ_t, that is iff $\rho_t(A^2) = \{\rho_t(A)\}^2$. That is, while the *potential* properties characterize the *object*, the *actualized* properties characterize the *state* of the object. These rules fix the *ontology* of algebraic quantum theory. It is not radically different from the ontology traditionally accepted for the classical physical theories. We just have to distinguish carefully between potential and actualized properties. In classical theories this is not necessary since all classical potential properties are always truth-definite.

4. ALGEBRAIC QUANTUM MECHANICS AND THE COPENHAGEN VIEW

If we accept algebraic quantum mechanics, we have to modify some basic tenets of the Copenhagen view *) since they just reflect the limitations of the historical Hilbert-space *formalism* of quantum mechanics. The main points I propose to change can be summarized in five theses:

(i) *The distinction between microphysis and macrophysis has become obsolete*

It makes no sense to distinguish between microscopic and macroscopic systems, the proper terms are *quantal* systems and *classical* systems. This remark is important since there are *microscopic systems having classical properties*, and there are *macroscopic systems having quantal properties.*

(ii) *We do not need two different languages*

In contradistinction to Bohr's expectations we have now a consistent *common language* for the description both of microscopic and macroscopic systems, as well as for quantal and classical systems.

(iii) *Macrophysics is not a limiting case of microphysics in the sense of Bohr's correspondence principle*

Bohr's correspondence principle was a historically most important heuristic guiding rule which can only be justified by specifying an appropriate context. This context-dependence can be made explicit in the formalism of algebraic quantum mechanics which allows a mathematically consistent description of systems which are partly quantal and partly classical.

(iv) *The interaction between a microsystem and a macroscopic measuring system is not an "unanalyzable" link*

While in the framework of the Hilbert-space formalism of traditional quantum mechanics a detailed description of irreversible processes and of measuring processes is indeed impossible, these problems are well-posed in the framework of algebraic quantum mechanics and there are

---

*) Compare for example the 10 theses formulated by Laurikainen [12].

already relevant rigorous results proving the analyzability of these interactions.

(v) *A pure state is more than just a bridge between the preparation and registration*

According to Bohr, the state vector allows predicting the probability of the occurence of phenomena under specific conditions but does not refer to the properties possessed by an object system. This tenet is in contradiction to the regulative principle of realism. It is logically possible to adopt an ontic interpretation whereby pure states refer to the *actualized* properties of an object.

While I propose to regard theses aspects of the historical Copenhagen view as enforced by a not quite appropriate formalism, hence as of only historical importance, there are other aspects which still are relevant today. Of particular significance for modern science is the possibility of an *individual* interpretation of quantum mechanics.

The modern experimental techniques have been developed to a point where one can start to experiment with *single* microscopic systems, like spectroscopic experiments with single atoms or biological experiments with very few molecules. In order to describe such experiments in a theoretically transparent manner, an individual interpretation is almost indispensable. Of course, every individual description of a quantum phenomenon will contain an irreducible stochastic element but that does not imply the necessity of a purely statistical ensemble interpretation. A well-known example is given by the classical Brownian motion which has two different descriptions: a purely statistical ensemble description in terms of a Fokker-Planck-type diffusion equation and an individual description by a stochastic Langevin-type differential equation. Since every classical probability measure has a unique decomposition into pure states, the two descriptions of Brownian motion are essentially equivalent. This equivalence does not extend to quantum theories because in nonclassical theories a mixed state never has a unique decomposition into pure states. There is no difficulty to obtain an ensemble interpretation from a stochastic individual interpretation but the reverse is not true so that in quantum mechanics an individual description in terms of pure states is as a rule more general (and more difficult) than a mere statistical ensemble description.

Accordingly, I propose that from a modern viewpoint the following theses of the historical Copenhagen view should be considered as the essential points which should be retained in an individual interpretation of algebraic quantum mechanics:

(i) The referent of quantum mechanics is a *single* system.

(ii) Each observed phenomenon has to be considered as a *whole*. Different experimental arrangements can exclude each other so that the measurement of one property may be *complementary* to the measurement of another. Nevertheless, the experimenter has the free choice between mutually exclusive experimental arrangements.

(iii) Every measurement is terminated by an *irreversible* registration. Consequently, each observed phenomenon is associated with an irreversible process whose statistical laws are governed by *primary* probabilities.

(iv) Measuring instruments must have descriptions in terms of *classical* concepts (or more generally, in terms of a *Boolean* language), and the results of experiments are to be expressed in *realistic* terms.

The last point is in accordance with the present-day experimental methods of physical sciences. Indeed, *without any exception*, every experimental arrangement and all experimental results can be described in engineering terms, using only a few concepts from *classical* physics. In passing, we may mention that this fact does *not* imply that the *human* perception necessarily corresponds to the classical notion of measurement. There is a possibility that everyday experience goes *beyond* the classical domain and that there may be a more direct way to get into contact with quantum reality than via a Boolean projection. However, such a possibility - if it exists at all - has never been investigated in the physical sciences.

While Bohr insisted that classical concepts have to be used for the registration of measurements, he did not have the proper mathematical tools to prove that this requirement is consistent with the formalism of quantum mechanics. His correspondence principle does not help since the limits postulated in general do not exist for all observables. Therefore a generalization of traditional quantum mechanics is necessary to prove in a mathematically rigorous way the compatibility of Bohr's requirement. This desideratum is fulfilled by algebraic quantum mechanics.

The historical Copenhagen view is furthermore confronted with the apparent paradox that the observed phenomena are irreversible while the Schrödinger dynamics for a closed quantum system is strictly deterministic. However, there is no inconsistency. As we know from the theory of classical K-flows, strictly deterministic processes can be genuinely irreversible and dissipative. The uncomfortable situation in traditional quantum mechanics is enforced by Bohr's requirement that in the specification of the state of a quantum system all influences of the environment have to be excluded [1]. In algebraic quantum mechanics the influence of the environment *can* be included and it can be *shown* that the proper dynamics of a quantum system strongly interacting with a K-flow-type *classical* environment is not given by the usual Schrödinger equation but by a nonlinear stochastic differential equation. Accordingly, the probabilities of quantum mechanics are related to the interactions with an appropriate *classical* environment - a phenomenon which in Bohr's historical view is taken to be "unanalyzable".

Whether the interaction of a quantum system with its environment leads to measurement-type process or not, depends solely on the nature and the strength of the interaction but in no way on the existence of a human observer. Or, in the words of Wolfgang Pauli: "Hat der physikalische Beobachter einmal seine Versuchsanordnung gewählt, so hat er keinen Einfluss mehr auf das Resultat der Messung, das objektiv registriert allgemein zugänglich vorliegt. Subjektive Eigenschaften des Beobachters oder sein psychischer Zustand gehen in die Naturgesetze der Quantenmechanik ebensowenig ein wie in die der klassischen Physik" [13]. That is, in quantum physics the free will and the consciousness of the observer play exactly the same role as they play in *classical* physics.

In the following section we discuss an example of a measuring-type interaction of a small quantum system with its environment which is entirely independent of the existence of any observer but which is of crucial importance in the biochemistry of all living beings.

5. A SIMPLE EXAMPLE FOR THE EMERGENCE OF A SUPERSELECTION RULE

Some people ask whether the superposition principle of traditional quantum mechanics has an unrestricted validity, even on the macroscopic level [14]. For a theoretical chemist this is a curious question since we know empirically and can explain quantum-theoretically that the universal validity of the superposition principle breaks down already with rather small molecules like alanine.

Alanine is one of the simplest amino acids and has the chemical formula $CH_3CH(NH_2)COOH$. A single molecule of alanine consists of 13 nuclei and 48 electrons, a rather small molecule well in the reach of modern computational quantum chemistry. If we consider this molecule as a strictly closed Hamiltonian system, then the Hilbert-space formalism of traditional quantum mechanics is relevant and predicts a *nondegenerate parity-invariant* ground state. This prediction is in contradiction to all experimental results: *there exist no parity-invariant pure states for alanine*. The experimentally observed states of lowest energy are energetically *degenerated* and *chiral*, that is, the molecules exist only in a left- and righthanded form. The chirality of biologically important molecules like alanine is of fundamental significance for molecular biology.

The fact that the Hilbert-space formalism of traditional quantum mechanics leads to a wrong prediction for the ground state of alanine does not imply that quantum mechanics is wrong but only that we did not apply quantum mechanics properly. A detailed analysis has shown that one cannot treat molecules like alanine as closed systems. If one includes the quantized electromagnetic radiation field, in particular the interaction with the soft photons, and works out the problem correctly (what leads to tremendous mathematical difficulties), then the quantum-theoretical description is in agreement with the experimental facts. A mathematically correct discussion can neither be done in the framework of the Hilbert-space formalism of traditional quantum mechanics nor in a Fock-space type quantum electrodynamics but requires the more powerful tools of algebraic quantum mechanics. The final result is that alanine acquires by its interaction with the electromagnetic radiation field (an environment having infinitely many degrees of freedom) a *new classical property* : the chirality of alanine. This classical observable represents an objective, always truth-definite property : there is a strict superselection rule between L-alanine and D-alanine [15].

This result is in accordance with the fact that one can synthesize L-alanine and D-alanine, but *not* the parity-invariant superposition of L-alanine and D-alanine. The ground state expected from a naive discussion of alanine as a closed system would approximatively be the equal-weight superposition of the two enantiomeric forms which is, however, strictly forbidden and therefore does not exist as pure state.

6. CONTEXTUAL OBJECTS

If quantum mechanics is valid, atomism is dead. The world is not composed of minute things. Also the empirist claim that objects exist in an absolute sense proved to be inadequate. In a holistic theory, notions like objects and patterns have no a priori meaning. The human cognitive reality is not a representation of the external reality but is created by brain events, including perception and cognition (compare also Varela [16]). Pattern recognition turns out to be the same as pattern *creation*, it can be described as a map of the non-Boolean reality into a Boolean description of this reality, whereby different recognition criteria may use different methods of abstractions and so may lead to incommensurable but equally legitimate descriptions. The selection of a particular criterion creates some patterns but makes it impossible to ask complementary questions.

All concepts of empirical science refer to observations obtained by some pattern recognition methods which distinguish between relevant and irrelevant features. What has to be considered as relevant, and what as irrelevant, is neither god-given nor written down in the first principles of a fundamental theory but has to be introduced into this theory by making the very same abstractions as the pattern recognition methods of empirical science do. The relevant aspects of an experimental result are considered as *facts*, the irrelevant features are relegated to the category of *noise*. Note that there are no noise-free experiments and that noise always represents ignorance.

The abstraction from not directly relevant features is the *conditio sine qua non* for the objectivity of science, it does not falsify our description of reality but *creates the patterns of reality*. All objects we discuss in natural science are *contextual* objects, their fictitious existence depends both on the environment as well as on some well-chosen abstractions.

The abstractions necessary to institute a context break the holistic unity of nature. In algebraic quantum mechanics the contextual objects connected therewith are created *by breaking the logical symmetry* of fundamental quantum theory whereby the actually existing Einstein-Podolsky-Rosen correlations between the object system and the rest of the world are suppressed. In order to get a theoretically consistent abstraction, the logical symmetry of the resulting theory has to be given by a subgroup of the group of all logical symmetries of the fundamental theory, and for nonclassical objects the environment has to be described in purely classical terms. Note that this neglect of Einstein-Podolsky-Rosen *correlations* does not imply the neglect of *interactions* between the object and its environment. To summarize : *disentangling a system from its environment creates a contextual object whose patterns can be discovered by a pattern recognition method that rejects the Einstein-Podolsky-Rosen correlation between the object system and its environment as irrelevant* [17].

A contextual quantum object admits an individual description, one can even adopt the "ontic interpretation of algebraic quantum mechanics" (as sketched in section 3) for contextual objects. That is, it is consistent to adopt *within a fixed context* a way of speaking *as if* the contextual objects would really exist. While contextual objects depend on abstractions they are mind-independent in the sense that they do not depend on the mind of a biological individual.

Since quantum objects necessarily have classical environments, noncontextual absolute objects do not exist if quantum mechanics is universally valid and has only a few strict discrete superselection rules. This situation does not invalidate a realistic world view in the sense that the material world exists independently of the minds of mankind and that natural science enables us to gain some understanding of this reality. The patterns of a contextual object can be measured objectively and they definitely reflect some objective aspects of an independent reality.

However we cannot restrict us to one single context, for a comprehensive description the richness of reality requires the use of many complementary contexts.

7. EXAMPLE : MOLECULES AS CONTEXTUAL OBJECTS

In general, objects do not exist indepedently of their environment. The inclusion of the environment can have dramatic effects, even if the individual interaction constants are extremely small. In our example of chiral molecules, the actually realized molecular state, the *"dressed state"*, is accompanied by an infinite number of photons whose degrees of freedom are hidden, manifesting themselves only by a few parameters which can be described either by phenomenological concepts or the corresponding theoretical notions (like "classical observables", "molecular superselection rules"). From a fundamental point of view, the crucial phenomenon is a change of the equivalence class of the algebra of observables of the system including its environment. Further well-known examples are the concept of a "dressed electron" consisting of a bare electron accompanied by a cloud of infinitely many soft photons (which necessitates the use of non-Fock representations for the free asymptotic photon field), or the concept of "solvated ions". A dressed state may be well disentangled from its environment. That is, it is possible that the environment is responsible for the existence of objects. In this case the objects are well-adapted to the environment and structurally stable with respect to small changes in the environment.

To illustrate this situation, we consider again our example of the alanine molecule. For the *bare* molecule alanine in interaction with its radiation field there exists no stationary pure state, so that the bare molecule cannot be an object. The interaction of the bare molecule with the bare radiation field is so strong that the system undergoes a spontaneous symmetry breakdown so that the system has to be described in a representation *inequivalent* to the Fock representation. In this new representation the molecular system is accompanied in an inseparable manner by clouds of soft photons, and we speak of a *dressed molecule* which is in a parity-broken chiral state, that is, either left- or right-handed. Furthermore, there is a superselection rule between the two enantiomers. There is still a rather weak interaction of the dressed molecule with the quantized dressed radiation field, which is, for example, responsible for small radiation dampings and level shifts. The traditional method to deal with these effects is the Wigner-Weisskopf method which accounts for the most important part of the residual interaction but neglects the effects of the Einstein-Podolsky-Rosen correlations between the dressed molecule and the dressed radiation field. The Wigner-Weisskopf approximation is equivalent to the replacement of the quantized radiation field by a *classical* stochastic electromagnetic field. That is, in the Wigner-Weisskopf approximation the environment of the chiral dressed molecule becomes classical so that a single molecule of L- or D-alanine becomes an *object* in the sense of algebraic quantum mechanics.

If we accept quantum mechanics as a correct theory of reality, then neither atoms nor molecules exist in an absolute sense, hence we cannot say that matter is made out of quarks, electrons, atoms or molecules. However, we can *describe* matter in terms of these entities *provided* we are willing to abstract from some holistic entanglements and to replace them by classical interactions as we do in the Wigner-Weisskopf approximation. Of course, other kinds of abstractions are feasible and lead to different descriptions, but abstractions are unavoidable.

8. DYNAMICS OF OBJECTS AND MEASUREMENTS

Since objects are as a rule not independent from their environments, the description of their dynamical behavior becomes a very complex problem. One cannot expect that the time evolution is governed by a Hamiltonian dynamics. In the ontic interpretation an object is at every instant taken to be in a pure state, hence its effective dynamics is given by a one-parameter map which transforms pure states into pure states. This dynamics can be represented by a *stochastic differential equation*. Thereby the environmental effects are represented by fluctuating forces and damping terms. Furthermore, polarisations and reaction fields can lead to feedback effects which give rise to *nonlinear* dynamical equations. For a large class of purely quantal objects the state of the object can be represented by a state vector which then fulfills a *nonlinear stochastic Schrödinger-Langevin equation* in the sense of Itô. This equation has a dilation to a one-parameter group dynamics which is *deterministic* but *not determinable*. That is, the associated stochastic process is neither predictable nor controllable but can describe the irreversible behavior of a *single* quantal object and explain the dynamics of measurements in terms of an individual interpretation. Of course, such measurement-type evolutions occur always if an object is strongly interacting with an appropriate classical environment, and not only in measurements arranged by an experimenter.

A measurement process of the first kind is a special kind of time evolution of a joint system consisting of a quantum object together with a classical instrument which leads to the actualization of a given observable of the object system. Such measurements of the first kind are somewhat unrealistic idealisations, they correspond hardly to actual measurements but they can be realized at least in arbitrarily close approximation. For example, if the object can be described by a factor A of type I_2, represented by the algebra of all linear operators acting on a two-dimensional Hilbert-space H, an ideal measurement of an atomic projector $P \in A$ is described by a stochastic process governed by a bistable nonlinear stochastic differential equation. Its asymptotic solutions for $t \to \infty$ can be summarized as follows:

$$\omega = \omega_O \otimes \omega_M \to \begin{cases} \omega_O^+ \otimes \omega_M^+ & \text{with probability } \; p = \omega_O(P) \\ \omega_O^- \otimes \omega_M^- & \text{with probability } \; 1 - p \end{cases} \tag{1}$$

for all initial states ω_O of the object system. Here, ω_O, $\omega_O^\pm$ are pure states of the object, while ω_M, $\omega_M^{\pm}$ are pure states of the measuring instrument. In the final object states $\omega_O^{\pm}$ the atomic projector P is *actualized*, $\omega_O^{+}(P) = 1$, $\omega_O^{-}(P) = 0$, while the final states ω_M^{+} and ω_M^{-} of the classical measuring instrument are *disjoint*. The pure states of the object system can also be represented by state vectors Ψ, $\Psi^{\pm} \in H$, so that the reduced individual description of this ideal measurement can also be summarized as

$$\Psi \rightarrow \begin{cases} \Psi^{+} = P\Psi/\|P\Psi\| & \text{with probability } \; p = \langle\Psi|P\Psi\rangle \\ \Psi_{-} = (1-P)\Psi/\|(1-P)\Psi\| & \text{with probability } \; 1 - p \end{cases} \tag{2}$$

for all initial state vectors Ψ. This time evolution is governed by a nonlinear stochastic differential equation which maps pure states into pure states*). Due to an extremely sensitive dependence on the precise *pure* initial state of the measuring instrument, the final state of the bistable measurement process is only statistically predictable. Averaging over the initial pure states of the measuring instrument and over the fluctuations due to the classical environment, one gets a Fokker-Planck equation describing a diffusion in a bistable potential. This diffusion equation can be rewritten in terms of a density operator $D_t \in A$ which fulfills a *linear* equation of motion. In this well-defined statistical ensemble description, the measurement operation can be expressed by the usual projection postulate

$$|\Psi\rangle\langle\Psi| = D \rightarrow \lim_{t\rightarrow\infty} D_t = PDP + (1-P)D(1-P) \tag{3}$$

which describes just the transition from a superposition to a mixture. While this relation (3) is a necessary condition for a measurement of the first kind, it is not sufficient for validity of the individual description given by eq.(1).

---

*) An interesting nonlinear stochastic Schrödinger equation which reproduces measurements of the first kind has been postulated by Gisin [18]. In spite of some similarities, our description is different. Above all our nonlinear stochastic evolution equation is not postulated but *derived* from a physically sensible interaction of the object with a K--type classical system in the limiting case of a measurement of the first kind.

The traditional interpretation of quantum mechanics assumes two fundamentally different time evolutions : the deterministic time evolution of the unobserved system and the stochastic time evolution associated with the so-called "reduction of the wave packet". In contradistinction the time evolution in algebraic quantum mechanics is uniquely given, the ordinary Schrödinger equation and the dynamics of a measurement process being just extreme special cases. In this framework, the so-called "collapse of the state vector" is neither abrupt nor discontinuous but given by a multistable continuous dynamics. Moreover, the physically decisive interaction is between the quantum object and the classical instrument, the consciousness of an eventual observer plays (of course!) no role whatsoever.

The individual description of the measuring process by a nonlinear stochastic Schrödinger equation depends in a crucial way on the *classical* nature of the measuring instrument since interactions between purely quantal systems never give rise to *nonlinear* evolution equations for states. So the question arises : are measuring apparatuses really classical systems? However, this question is not quite properly posed. There can be no question that every laboratory apparatus also has quantal aspects. The proper question is : *do the experimentally relevant variables of a measuring apparatus admit a classical description?* Algebraic quantum mechanics can in principle give a positive answer but a rigorous derivation of the classical behavior of some collective modes from first principles as for the superselection rule associated with chiral molecules (sketched in section 5) is inconceivable for any realistic macroscopic measuring instrument. However, the conclusion that "macroscopic systems cannot be assumed to be in definite pure states" [19] is hasty. The fact that we do not have the tools to know in which pure state an object is does not imply that the object is not in a definite pure state.

If the essential variables of an instrument do not admit a classical description, this instrument does not qualify as a measuring instru-

ment. In this sense, we can palm responsibility off on the experimenter. In order to describe his experimental tools in terms of classical physics and engineering science, the experimenter makes many tacit abstractions. In order to get a correspondence between experiment and theory, we have to make the very same abstractions in the theoretical descriptions as the experimenter makes in the discussion of his tools. Every application of quantum mechanics *presupposes* the existence of classical domains where measuring instruments can be described as classical dynamical systems. To say that such classical domains do not exist is the same as to say quantal objects do not exist. If we speak of quantal objects and of classical instruments we tacitly mean *contextual* objects which exist only in virtue of abstractions which usually are taken for granted.

REFERENCES

[1] Bohr, N., "Atomic Theory and the Description of Nature", Cambridge University Press, Cambridge, 1934; p.54.

[2] Von Neumann, J., "Mathematische Grundlagen der Quantenmechanik", Springer, Berlin 1932.

[3] Ghirardi, G.C., Rimini, A., Weber, T., "Unified dynamics for microscopic and macroscopic systems", Phys.Rev. D 34, 470 - 491 (1986); footnote 4.

[4] Stone, M.H., "Linear transformations in Hilbert space. III. Operational methods and group theory", Proc.Nat.Acad.U.S. 16, 172 - 175 (1930).

[5] Von Neumann, J., "Die Eindeutigkeit der Schrödingerschen Operatoren", Mathematische Annalen 104, 570 - 578 (1931).

[6] Emch, G.G., "Algebraic Methods in Statistical Mechanics and Quantum Field Theory", Wiley-Interscience, New York 1972.

[7] Bratteli, O., Robinson, D.W., "Operator Algebras and Quantum Statistical Mechanics", Springer, New York; vol.1, 1979, (second edition 1987); vol.2, 1981.

[8] For a review, compare e.g. Primas, H., "Chemistry, Quantum Mechanics and Reductionism", Springer, Berlin 1981 (second edition 1983); Emch, G.G., "Mathematical and Conceptual Foundations of 20th-Century Physics", North-Holland, Amsterdam, 1984; Sewell, G.L., "Quantum Theory of Collective Phenomena", Clarendon Press, Oxford, 1986.

[9] Raggio, G.A.,"States and composite systems in W*-algebraic quantum mechanics". Thesis ETH Zürich, No.6824, ADAG Administration & Druck AG, Zürich, 1981.

[10] Spengler, S.J., Stasiak, A., Cozzarelli, N.R., "The structure of knots and catananes produced by phage λ integrative recombination: Implications for mechanism and DNA structure", Cell 42, 325 - 334 (1985).

[11] Stahl, F.W., "Genetic recombination". Scientific American 256, No.2. pp. 52 - 63 and 104, February 1987.

[12] Laurikainen, K.V., "Wolfgang Pauli and the Copenhagen philosophy", in: Symposium on the Foundations of Modern Physics", ed. by Lahti, P. and Mittelstaedt, P., World Scientific Publishing Co., 1985; pp. 273 - 287.

[13] Pauli, W., "Naturwissenschaftliche und erkenntnistheoretische Aspekte der Ideen vom Unbewussten", Dialectica 8, 283 - 301 (1954).

[14] Leggett, A.J., "Quantum mechanics at the macroscopic level". In: "The Lesson of Quantum Theory", ed. by de Boer, J., Dahl, E. and Ulfbeck, O., North-Holland, Amsterdam, 1986, pp. 35 - 57.

[15] Pfeifer, P., "Chiral molecules - a superselection rule induced by the radiation field". Thesis ETH Zürich, No. 6551, ok Gotthard S+D AG, Zürich, 1980.

[16] Varela, F.J., "Principles of Biological Autonomy", Elsevier - North Holland, New York, 1979.

[17] Primas, H., "Pattern recognition in molecular quantum mechanics", Theor.Chim.Acta 39, 127 - 148 (1975).

[18] Gisin, N., "Quantum measurement and stochastic processes", Phys.Rev. Lett. 52, 1657 - 1660 (1984).

[19] Zeh, H.D., "On the irreversibility of time and observation in quantum theory". In "Foundations of Quantum Mechanics", ed. by B.d'Espagnat; Academic Press, New York, 1971, pp. 263 - 273.

HEISENBERG'S PHILOSOPHY.

C.F. von WEIZSÄCKER
Arbeitsgruppe Afheldt in der Max-Planck-Gesellschaft
D 8130 Starnberg, Bahnhofplatz 4
FR Germany

CONTENT

1. What Do We Mean By "The Copenhagen Interpretation"?

> I think it is safe to say that no one understands quantum mechanics. R.Feynman (1967)
>
> The Copenhagen Interpretation _is_ quantum mechanics.
> R.Peierls (1985)

What do we mean by making a distinction between a theory and its interpretation? What is a theory? What is an interpretation?

These questions aim at a systematic philosophy of science. Yet a symposium on the Copenhagen Interpretation 60 years after the Como lecture puts itself by its very name into an explicitly historical context. So does the task assigned to me by asking me to speak on Heisenberg's philosophy. This title defines a philosophy not by a systematic term but by the name of a man who developed it.

This historical setting is natural since it has turned out that up to the present time there exists no systematic philosophy of science that would have gained general consensus on the interpretation of quantum theory. I shall try to explain this situation by beginning with Heisenberg's probably most important contribution to the philosophy of science, the concept of the progress of physics as a sequence of "closed theories" (abgeschlossene Theorien; Heisenberg 1948). This concept places the systematic achievements of physics (closed theories) into an historic concept (sequence). The concept originated in a reflection on what Heisenberg himself had achieved twenty years earlier when he was lucky enough to lay the foundation of the latest closed theory which so far has emerged in physics.

A "closed" theory is not an ultimate, all-embracing theory, but a theory which cannot be improved any more by small changes. Classical mechanics, thermodynamics, electrodynamics, special and general relativity, quantum mechanics are examples. A new closed theory differs from its predecessors not only by new assertions but by new concepts. The closed theories of the 20th century have confined the earlier ones - now called "classical" - to limited fields of application. Heisenberg did not either consider quantum mechanics as ultimate; he aimed at finding a closed theory beyond it.

It was by no means a priori evident that physics would develop along such a scheme. The systematic explanation of this historical phenomenon would be a great step forward in the philosophy of science. Our initial questions on the meaning of terms like "theory" and "interpretation" are aspects of this unsolved problem. I propose some descriptive steps.

We take "theory" in these questions to mean a closed theory; quantum mechanics is our actual example. Such a theory can be described as consisting of a mathematical structure and a physical semantics. In quantum mechanics the mathematical structure is a Hilbert space H, together with the set of representations R of the additive group of real numbers in this Hilbert space. The beginning of the semantics is indicated by statement like: The linear subspaces of H describe possible states of some physical object; the eigenvalues of a self-adjoint operator L in H are the possible values of some "observable" (measurable quantity) of that object; the representations R are possible laws of dynamics, the real number measuring time.

The "closed" feature of the theory is evidently in its well-defined mathematical structure. The semantics, however, is open-ended. Which observable quantity corresponds to some mathematically given operator? How to refer the mathematical structure to experience, or to some preconceived ideas on reality? "Semantics" means to tell the meaning of certain words; this meaning, however can only be told by other words. Bohr said "vi haenger i sproget", we are suspended or stuck in language. The semantics of a new closed theory must refer to a language which existed before the invention of the new theory and which was needed to make the new theory possible - let it be everyday talk in the contemporaneous culture, or the language developed in the earlier closed theories.

Now I maintain: The Copenhagen Interpretation (CI) was invented with the purpose of being the minimal semantics of quantum mechanics, i.e. that amount of semantics which was needed to speak understandably at all on the meaning of quantum mechanics. This statement explains the two motti on the head of this section. As far as CI was successful in its purpose it is quantum theory. As far as it needs further semantical elucidation of its own terms nobody has so far understood quantum theory.

It seems evident that a full semantical effort on a new closed theory must lead into the classical debates of philosophy. The founding father of critical philosophy, Socrates, asked anyone who made a strong assertion:"Please, tell me what you have meant to say", and he became unpopular by forcing his partners to admit that they had not known it themselves. In this respect, Bohr always reminded me of Socrates. My present task is to follow the way on which Heisenberg tried to answer the Socratic question.

The classical philosophical answers are best represented by great philosophers, rather than by abstract terms (like "realism", "positivism", "idealism")which stay themselves in need of further semantical clarification. Einstein, I think, described himself rightly as a follower of Spinoza who, in the great tradition of ontology, asked what is. Bohr, I feel, is rather a follower of Kant who asked what we can know; Bohr's questions are epistemological, Socratic. Heisenberg, especially in his later years, recognized Plato as the leading thinker, the poet of the "central order". Pauli referred to Schopenhauer, and he learnt most from the Psychology of C.G.Jung.

6o years after the Como lecture, 2400 years after Socrates - this is not a long time for the clarification of our ideas.

2. The Observer: Bohr and Heisenberg 1922-1935.

"Sommerfeld taught me optimism, the Göttingen school taught me mathematics, Bohr taught me physics" - so Heisenberg told me in later years. His autobiographical book "Der Teil und das Ganze"(1969), English translation "Physics and Beyond"(1971), describes the developement of his ideas through his whole life, written in the form of talks which I should compare to Platonic dialogues; "through them the author hopes to demonstrate that science is rooted in conversations" (G 9, E XVII; G means pages in the German, E in the English edition). In the present lecture I can only select three topics : the early emergence of the "observer" problem, his later physics, and the ideas of his mature philosophy.

Quantum mechanics was a "scientific revolution" in the sense of Th.S.Kuhn (1962), a change of paradigm. I shall not repeat the well-documented history of that revolution but present my personal interpretation of its central problem. When I had induced Heisenberg in the last years of his life to read Kuhn's book, he said:"Alright, but the new 'paradigm' is a new closed theory". In Kuhn's Description there is a "prerevolutionary phase" in which the earlier paradigm fails to solve some puzzles. The reaction in general is to make the theory more and more complicated. The revolution then consists in seeing that the puzzles could not possibly be solved by the old paradigm. The new paradigm brings a new simplicity. Only its success finally opens the eyes for the fact that the difficulties of the earlier phase admitted of no solution in the old frame.

The insoluble problem of classical physics became visible in the "ultraviolet catastrophe". Einstein (1905) saw that there can be no thermodynamical equilibrium in a classical continuum dynamics. The problem was already understood by Boltzmann and used as the decisive argument for the existence of atoms. It was rediscovered whenever a clear model of micro-dynamics was presented, so by Planck for the Maxwell field, and by Bohr in the Rutherford model of the atom. In the Bohr model the quantum conditions restricted the continuity of motion; the continuum reappeared in the shape of the transition probabilities between the discrete states. In the first talk between Bohr and Heisenberg, 1922, the inconsistencies of the remaining classical description of the permitted orbits were the decisive problem.

Heisenberg (1925) cut through the Gordian knot by completely abolishing the orbits. The knife for this cut was a version of Occam's razor: the methodological principle that the theory ought to connect observable quantities only. But then Einstein in a talk 1926 explained to him (see G 92, E 63) that "it is the theory which decides what we can observe". The uncertainty principle of 1927 is the result of Heisenberg's acceptance of Einstein's statement. The impossibility of simultaneously observing precise values of position and momentum does not mean the pseudo-positivistic nonsense "what cannot be observed does not exist" but the test of consistency: "what does not exist in the theory cannot be observed in an experiment describable by the same theory".

It is interesting to consider the mathematical background of the uncertainty principle. In his 1925 paper Heisenberg considered position q and momentum p like energy and time as observable. He used the results of dispersion theory (Kramers and Heisenberg 1925), i.e. of transition probabilities, for directly connecting q and p as noncommuting algebraic quantities. When Schrödinger had invented wave mechanics it was soon understood that Heisenberg's algebraic quantities (matrices) were operators in the Hilbert space of Schrödinger's wave functions. In this Hilbert space there are no common eigenvectors of the noncommuting p and q; this is the mathematical basis of the uncertainty priciple. The self-consistency of the theory made Born's statistical interpretation of the wave function inevitable. Yet a probability can only be given an operational meaning as a prediction on the outcome of an observation (more precisely: a prediction of a relative frequency of such events). Thus the concept of observation could not be eliminated if one wanted to give a semantics to the mathematical structure of the theory. The observer is a precondition of the <u>meaning</u> of quantum theory; but thereby he is not yet an <u>object</u> of quantum theory.

I have heard the story that Bohr soon after his Como lecture explained the new philosophical results to an old friend, and that he got the reply:" O yes, but you cannot deny, Bohr, that you have been saying precisely the same things already twenty years ago". Complementarity is basically a philosophical term, connected with what Bohr called "the limitations of our means of expression". Bohr's approach is epistemological from the outset. When Einstein said "God does not play at dice", Bohr replied:" The problem is not whether God plays at dice or not but what we are meaning when we say that God plays at dice or not". Thus Bohr always analyzed the role of the observer as precondition and not as object of the theory. Thus the objectivity of the observational result becomes all-important for Bohr. The observer is just not to be considered in physics as far as he is an individual with a private field of experience and emotions. Science presupposes that all observers can agree on the result of an observation. This is the reason why Bohr emphasizes the classical description of the measurement. The measuring result must be objectively available. It is essential to see that Bohr's later concept of phenomenon (1935) precisely means a situation which is available to all participant observers.

But why should the observer not also be an object of the theory? The measuring instrument, at least, consists of atoms like all other objects; thus it ought to obey the quantum laws if any macroscopic object does. Heisenberg never seriously considered the idea that quantum theory might be limited to microscopic objects; like nearly all presentday physicists, he considered quantum theory as (so far, to our knowledge) universally valid and classical physics as a macroscopic approximation. Then the classical description of the instrument just meant that only as far as it was an approximation would the instrument be of use <u>as</u> an instrument.Then it was our free choice whether we wanted to describe the instrument by a (imprecisely known) wave function or by classical terms; the cut between classical and quantum description was to some degree "mobile", i.e. arbitrary. The cut is then the place in which determinism breaks down: the wave function obeys the deterministic Schrödinger equation, the instrument in described by causal classical physics, but the translation between the two languages is done in the cut by the concept of probability.

Bohr, however, objected (I only remember an exchange of letters between Bohr and Heisenberg on this matter in the early thirties) that it was just a precondition for the meaning of the words in which we describe all experience that we should not draw the cut into the realm of directly observable phenomena. Heisenberg had taken this limitation of the mobility of the cut into account at least in the sentence:"The chain of cause and effect could be quantitatively verified only if the whole universe were considered as a single system - but then physics has vanished, and only a mathematical scheme remains". (1930, p.58). But this was less than Bohr craved. For him, it was not only the inclusion of the whole universe into a quantum description in which semantics, the meaning of quantum mechanics would break down; it was already the inclusion of the human observer. The limits of quantum theory were included for him within the limits of physics, and he considered complementarities between physics and an adequate description of organic life, between objectified knowledge of the human body and self-awareness, between scientific analysis and the ethical and emotional aspects of human life. For him, the great philosophical illumination by quantum theory was just that <u>even</u> in physics complementarity relations turned out to be fundamental.

I finish this section by two additional remarks: on the uncertainty priciple and on continuity. The weakness of the Planck-Bohr phase of quantum theory was that it excluded most of the states in classical physics. Quantum mechanics, however, contains more possible states of an object than its classical limiting case. This "surplus information" is protected by the exclusion of the classical orbits which is the purpose of the uncertainty principle. This purpose is just not to limit our possible knowledge but to open the door to an amazing widening of possible knowledge. This is why Heisenberg always considered the search for hidden variables which would re-establish classical physics as a wish to undo progress; I propose to apply to it the Freudian term "Trauerarbeit".

In this context a remark on continuity. I would bet that a description of a wave function not as an expression of probability, that is of a set of possibilities, but as an actuality (a "reality" in the vocabulary of "realism") would re-introduce the thermodynamical paradoxes inherent in all known classical dynamics of continuum. It is my personal philosophical supposition that even in mathematics continuity means possibilty; even Cantor said in defence against the Aristotelians that actual infinity means an infinity <u>of</u> possibilities. If this were true, quantum theory would just mean to lift physics to the level of a consistent theory of the continuum. But this remark goes beyond my present task.

3. Heisenberg's Later Physics.

Neither Bohr nor Heisenberg originally considered quantum mechanics (even including quantum field theory) as a final theory within physics - hence they were not too interested in the possible philosophical consequences of the idea of quantum theory either as a universal description of reality, or as an obstacle to another, e.g. more "realistic" universal theory. For Heisenberg, "closed" was not indentical with "final", but the sequence of closed theories rather hinted at the idea of physics as an open-ended enterprise. Yet what he actually achieved in his remaining forty years of work in physics was his participation in a magnificent exploration of the continent discovered in quantum mechanics, but no discovery of another continent. Quantum mechanics proved far more stable than he or Bohr had expected.

The atomic nucleus turned out to obey quantum theory after the discovery of the neutron. Heisenberg's attempt at introducing a "smallest length" into physics (1938) stayed inconsequential. In the years after 1945, molecular biology made it more and more probable that even the phenomena of life needed no other laws for their explanation than those of existing physics and chemistry; and chemistry had fully merged with the basic laws of quantum theory.

Heisenberg's last theory, the nonlinear spinor-field theory (1959) was in principle a not fully conventional quantum field theory.Still, if successful, it might have deserved the name of a new closed theory. While classical quantum mechanics is only an abstract frame for any possible objects and dynamical laws, the spinor-field theory intended to determine all possible particles and their dynamics from one basic field with a fairly simple symmetry group. It meant, so to speak, the transition from "abstract" quantum theory to a closed "concrete" quantum theory. That it was not accepted by the community of particle physicists was a profound grief to Heisenberg. Personally I think that its weakness was not in the corrigible neglect of more recent developements like quarks but that just by its bold approach to basic problems it ran into a difficulty which has so far not either been resolved in any of the competing theories as soon as they try to go beyond the phenomenological level.

I guess that the problem is an essential incompatibility between quantum theory and the model of a (relativistic) field. In their naive forms both classical and quantum field theories produce singularities. Renormalization is a successful trick; in principle its success only means that there are ("renormalizable") theories in which we can postpone the real solution of the problem by "assuming the problem to be solved" and calculating relativistically invariant consequences of this assumption. I think Heisenberg was right in not accepting this as a final position. He wanted a theory which would produce only finite interaction energies which would then not have to be subtracted but interpreted as empirically meaningful. His own assumptions, however, forced him into an indefinite Hilbert space metric which Pauli, I think rightly, rejected. Heisenberg's own attempts at a solution got stuck in the mathematical complications. I suppose that the assumption of an <u>actual</u> field in a pre-existing <u>actually</u> noncountably infinite space-time continuum violates the possible meaning of continuity and hence will always produce paradoxes. But this remark again goes beyond my present task.

4. Heisenberg's Later Philosophy.

Heisenberg's was a philosophical mind. Yet by natural gifts, inclination, and sportive ambition he was a theoretical physicist, with creative physical and mathematical intuition. His interest in mathematical rigor was, however, only limited; his instinctive criterium for accepting mathematical proposals in physics was rather their mathematical beauty and simplicity. Kramers liked to say: "Heisenberg has the great gift of inventing mathematical approximations and applying them in the only case in which they converge".

Correspondingly, his interest in philosophy was primarily neither ontological nor epistemological but rather of an aestetic nature. He could not believe that logical pedantry would reach the really important questions. One of his famous late philosophical lectures had the title "Die Bedeutung des Schönen in der exakten Naturwissenschaft" (1970). There he begins by telling of the fascination of beauty he found, still at high school, in the mathematical theory of numbers, and soon in mathematical physics; he discussed the Greek definition of beauty as harmony of the parts and, in Plotinus, as "the splendor of truth". The only professional alternative to mathematics and physics which he considered when leaving high school was music.

Heisenberg's philosophical reflection was awake even before his great early achievements in physics; it may have been instrumental for his finding the right way in physics. His first reaction, still at school, to a chemical textbook which represented atoms like little spheres "with hooks and eyes" was sheer contempt. What is the value of a mechanical model of an atom which presupposes the very same mechanical laws it is supposed to explain? Young Heisenberg, like, 25 years earlier, young Einstein, learnt to admire Ernst Mach. Mach's penetrating criticism of Newton's concept of absolute space paved the way for Einstein's relativity theories. Mach's criticism of the belief in unobservable "material qualities" inspired Heisenberg's early emphasis on observable quantities. But neither Einstein nor Heisenberg was ever impressed by the later school of Logical Positivism which unsuccessfully attempted to reduce physics to logical inferences from sense-data. What was lacking?

The history of philosophy teaches us that Plato and Spinoza, Leibniz, Hume and Kant, and, in our days, Popper, were aware that universal laws cannot be logically deduced from a collection of single observations. The philosophers differ only in the proposed solutions. Einstein chose ontology as a basis, in his concept of "reality". Bohr argued for the necessary classical description of experiments by saying that only in classical physics is the intuitive space-time description united with strict causality - this is an epistemological, strictly Kantian argument. Heisenberg had read Plato's Timaeus while still at school. He was then surprised and skeptical at the use of the regular, "Platonic" bodies as atom models, but impressed by the idea that mathematical unity was the foundation of physical laws. When, forty years later, he used symmetry groups as the basic principles of physics, he recognized the Platonic bodies as the first examples of group representations.

Heisenberg's philosophical vision was the unity of being. In his late book "Der Teil und das Ganze" the leading concept is the "central order". Already 1942 he wrote a philosophical manuscript which, being without an original title, has been printed in his Collected Works, vol. C I, under the appropriate title "Ordnung der Wirklichkeit". There he starts from a conceptual frame for the "regions of reality" given by Goethe (1823) as different levels of experience: "accidental, mechanical, physical, chemical, organic, psychological ("seelisch"), ethical, religious, genius (genial)". This is an ascension which shows Goethe's and, consequently, Heisenberg's kinship with Plato. I like to call Goethe "the Plato of the senses".

Indeed, Plato is absolutely misrepresented when he is called a dualist. The awakening to philosophy, to be sure, begins by learning to distinguish true being from the sensual appearances, but the completed philosophy, the philosophy of descent from the highest vision back to the world of time and senses, reveals the world of appearences as an optimal representation of the highest forms (ideai). The simile of the cave, representing this linkage of ascent and descent, admits of three interpretations: physical, political, and mystical. Physically seen, the appearances are sense-perceptions, the higher forms are mathematical models and laws of nature, then pure mathematical forms, created by the Pythagorean numbers, highest the One. Politically, the appearances are social success and failure, the higher forms are the strict ethical norms, highest the Form of the Good. Mystically, the appearances are mental states, the higher forms are the levels of self-awareness, highest in Unio Mystica. An adequate interpretation of Plato's philosophy would create an awareness that the three highest steps - "One, "Good", "Unio Mystica" - are identical.

Heisenberg was never a Platonist in the sense of the school. He read Plato as the poet who presented the inscrutable central order in great parables. So Heisenberg also saw religion on which he gave a late lecture (1973), as one of the time-honoured high steps on the ladder towards an expression of the central order. This is visible both in Plato's and in Goethe's ascents.

The early problem of the "observer" fits into this view of reality. The Cartesian dualism of matter and mind was never accepted by Heisenberg but rather considered as an unsolved puzzle. The destruction of the world view of classical physics was the gate into the freedom of an unfinished philosophical quest. To say it in Platonic terms: As far as we understand nature mathematically we understand its participation (methexis) in the true reality which is mirrored in the human mind and is essentially form (idea), i.e. spiritual.

5. A Remark on Heisenberg and Pauli.

Prof Laurikainen has presented us with a most impressive analysis of Pauli's philosophy. It is, I think, not too far, but still clearly distinct from Heisenberg's view.

Heisenberg and Pauli met as students of Sommerfeld 1920. Mathematically, Pauli was definitely superior to Heisenberg, perhaps not in mathematical intuition, but in his absolute mastery of the breadth of knowledge, especially of the logical connections, and in his relentless critical power. Far later Heisenberg told me:"I have never publishd a paper in physics without having first presented it to Pauli for criticism. If Pauli agreed, the paper was correct. If Pauli did not agree, it might still be essentially correct, but then caution and, perhaps, revision was certainly needed". As a creative physicist, Heisenberg was superior.

In their philosophies, both adhered firmly to the Copenhagen school in contradistinction to Einstein and Schrödinger, and both dared in later years to express their basic intuitions in a metaphysical imagery which Bohr's epistemological self-critique forbade him to use. Their difference might perhaps in a first step be expressed in psychological terms. Heisenberg's strength was an ever-resurrecting optimism which made it natural for him to believe in the unity of being under the image of a central order. Pauli had an ever-present awareness of evil.This permitted his role as a critical "conscience of the Copenhagen school".It is mirrored in the many stories on the "Pauli effect", the destruction of instrument by his mere physical presence, of which I might tell some examples which I have seen myself.It also led him into a profound psychological crisis at the age of thirty years. But when the crisis was overcome, not without the help of C.G.Jung, whose patient, then friend and finally critical philosophical partner he became, it opened his eyes also for what he called the irrational element in reality. This brought him closer to Schopenhauer than to the Platonic tradition. In Jung's language, he described science and Platonism as "trinitarian", but a philosophy which included evil as an indispensable element of a complete reality as "quaternarian".I cannot persue this difference further in the present paper.

Pauli's difference with Bohr, as presented by Laurikainen, concerns the idea of the "detached observer". Personally I feel that both authors were fully consistent in their respective ways of inquiry. Bohr's epistemological analysis must make a clear distinction between the knower and the known. Thus the observer must be methodologically "detached" from the object of observation; if this, according to quantum theory, is only an approximation, it is, according to Bohr, that approximation in which observation can be described as an objective event at all. Pauli's interest, on the contrary, lies just in the fact that this approximation is never strictly true, and that, if taken as absolute, it describes precisely the presupposition of the "trinitarian", rationalistic misjudgement of reality. "Consciousness is an unconscious act" (Eduard v Hartmann, William James - to mention two philosophical precessors of Freud, Jung and Pauli).

6. A Modern Evaluation .

In the sixty years since the Como lecture the perception of the philosophical problem of quantum theory has considerably changed through changes in the analysis of the theory itself, and, I should like to add, through changes in biology.

In the theory itself, the attempts at local hidden variables have finally suffered their downfall which we in the Copenhagen school have been justified in quietly and convincedly expecting. On more recent attempts to re-interpret the theory without changing it mathematically, Th.Görnitz will present a paper by him an myself in this symposium; see also references Kochen 1955. Deutsch 1985, Cramer

1986, Görnitz and Weizsäcker 1987[1,2,3].

In the present paper I shall confine myself to a remark on the modern aspect of the observer problem. All three new interpretations quoted above contain as more or less self-evident the postulate that the observer should be included into the quantum description. Görnitz and I believe that this is a completely natural generalisation of the Copenhagen Interpretation if we take into account modern biology on the one hand, an abstract reconstruction of quantum mechanics on the other hand. Modern biology sees no difficulty in accepting quantum theory as basic and sufficient in the description of organic bodies includindg humans. In the problem of matter and mind we agree with Heisenberg and Pauli that it is a "Scheinproblem", a misunderstanding created by Descartes' mixing up of epistemology and ontology ("what is differently defined must be different substances"). We (Weizsäcker 1985, ch.8; Drieschner, Görnitz, Weizsäcker 1987) have tried an abstract reconstruction of quantum theory in which classical theory is not taken as a starting point of "quantization", but probabilistic predictions on any empirically decidable alternatives are described by simple postulates. Then a prediction on my own mental state is as natural a subject-matter for quantum theory as any prediction on objects known in space. In this view it would be permitted to consider my brain as a "surface description" of the - mainly unconscious - mind in that experimental arrangement in which "my" mind is observed from outside, i.e. in that approximation in which it can be considered as a separable object, thereby making an aspect of a super-personal spiritual reality appear as a separate person, as "my mind".

For the last time I have to conclude a section by the sentence that this is a problem going beyond the task assigned to my present paper.

References.

Bohr, N. (1935), Phys.Rev. 48, 696

Cramer, J.G. (1986), Rev.Mod.Pys. 58, 647

Deutsch, D. (1985), Int.Jour. Theor.Phys. 24 , 1

Drieschner, M.,Görnitz,Th., Weizsäcker,C.F.v. (1987), to be submitted to Int.Jour.Theor.Phys.

Einstein, A. (1905), Ann.d.Phys. 17 ,132

Feynman, R.P. (1967), The Character of Physical Law, MIT, Cambridge MA, p.129

Goethe, J.W. (1823), Nachträge zur Farbenlehre, Verschiedene Nachträge, Nr. 31

Görnitz, Th., Weizsäcker, C.F.v. (1987[1]), this volume

id.(1987[2]), Int.Jour. Theor.Phys., Oct.1987

id. (1987[3]), Loyola Conference 1987

Heisenberg, W. (1925), Z.Phys. 33 , 879

id.(1930), Die physikalischen Prinzipien der Quantentheorie, Leipzig, Hirzel; engl. The Physical Principles of Quantum Theory, Univ. of Chicago Press (1930)

id. (1938), Z.Physik 110, 251

id. (1942),Ordnung der Wirklichkeit.Ges.Werke C I(1984),217

id. (1959), Z.Naturforschung 14a,441 (with H.P.Dürr, H. Mitter, S. Schiedar, K.Yamasaki)

id. (1969), Der Teil und das Ganze, München, Piper engl. Physics and Beyond, Harper & Row (1971)

id. (1970), Die Bedeutung des Schönen in der exakten Naturwissenschaft. Ges.Werke C III, 369

id. (1973), Naturwissenschaftliche und religiöse Wahrheit, Ges.Werke C III, 422

Kramers, H.A., Heisenberg, W.(1925), Z.Phys.31 , 681

Kochen, S. (1985), Symp.Found.Mod.Phys., World Scientific, Singapore, 151

Kuhn, Th.S. (1962), The Structure of Scientific Revolutions, Chicago Univ. Press

Peierls, R. (1985), oral remark, Joensuu Conf.

Weizsäcker, C.F.v. (1985), Aufbau der Physik, München, Hanser

Part III: CONTRIBUTED PAPERS

THE BREAKDOWN OF THE COMMON SENSE

F. BORG

Palo 15, SF-67400 Karleby

FINLAND

0. INTRODUCTION

Any lecture delivered at this symposium, following the conventional style of symposial rethoric, should open with the frank denial there ever existed anything to be named the Copenhagen Interpretation, or Philosophy ; like writers opening their meeting with the statement "the Litterature is dead", or the liberal theologicans who beging their debates declaring God is dead. The other trick of the trade is to end whatever one says with a questionmark. Sympathizing with the latter tradition we could ask whether there existed or still exists an Orthodox Copenhagen Philosophy ? - If such an Orthodox Philosophy existed it did, however, never exist in the minds of Pauli, Bohr and Heisenberg. We do not enter the inquiry into exactly how such an Orthodox Philosophy may have become established (cf. note 1). Instead we arrest our attention at some key-problems which did - or, did not - occupy the Copenhagen People, and whose philosophical nature will be brought out by focusing on these matters through the frames of Analytical Philosophy (Wittgenstein) and semiotics (structuralism). But first we may give a characterization of the different standpoints of Bohr and Heisenberg by studying their conceptions of the notion of the Observer.

1. PHYSICS AND THE OBSERVER

How did the observer enter Physics ? In The Order of Things /8/ the french nominalist structuralist poststructuralist archeologist Michel Foucault studies three epochs of the Western Human History he names Renaissance, The Age of Representation and The Age of Man (also used by G.Vico 1725 to label the third stage of the evolution of mankind). To the people of The Age of Representation, Man was a creature created by God, who could uncover through the sciences the secret meanings God had endowed the Nature. In our age, The Age of Man, Man is turned into the ground of knowledge and enters into the description (Representation) of the Nature as both an object and a subject[2]. Maybe then, the apparence of the observer in Physics was exactly what to be expected in this Age of Man. Note further that the few instances where, in classical physics, observers are spoken of these have been usually referred to as "demons"; Maxwell`s demon, Laplace`s demon, etc. Heisenberg discusses the observer in /9/. Heisenberg stresses the notion of the observer is necessary in the interpretation of quantum theory, and underlines that "the introduction of the observer must not be understood to imply that some kind of subjective features are to be brought into the description of nature". Yet he has to introduce what he names a "subjective element" to explain the reduction of the wave packets i.e. "the fact that the wave function representing the system changes discontinously when the observer takes cognizance of a result of a measurement." Close to the end of his article Heisenberg assures us that "the observer himself does not produce the transition" (the wave packet reduction). Either Heisenberg is careless or he is mantaining that ; first we have a result of a measurement then a cognizance of these results followed by the reduction of wave packets which in turn produces the results of the measurements (as we learn from the textbooks). Maybe, Heisenberg`s article is just an example of his "lighthearted words" to which C.F. von Weizsäcker referred to two years ago (/10/ p. 65), as also suggested by the following example taken from Heisenberg`s Chicago lectures /11/ where it is observed that one may compute the location (x) and the momentum (p) of an electron for past times such that $\Delta x \Delta p$ will be smaller than the usual limiting value. Heisenberg writes that "it is a matter of personal belief whether such a calculation concerning the past history of the electron can be ascribed any reality or not."[3] - Is the ontology of quantum mechanics just a question of personal belief ? Light hearted or not, Heisenberg insists on the importance of the concept of the observer in contrast with Bohr who was to introduce the concept of a detached observer /17/ which provoced criticism from Pauli[4] who on this point was more in line with Heisenberg. Bohr used the term "subjective" differently than Heisenberg who writes about the subjective element that "alters the mathematical representation discontinously" /9/. Bohr used the term subjective to emphasize that the description of experiments depends on the specific frames of reference ; a state of affairs he often compared with the theory of special relativity (cf. /15/,/18/).

A second dividing line among the Copenhagen People was related to the question of ontology. Following Bohr, Heisenberg stressed the important consequences of the irreducible quantum actions, whereas Pauli emphasized the acausal features of quantum mechanics. Bohr considered quantum mechanics a kind of a rational generalization of classical mechanics (cf. eg. /33/). Pauli went much further, revealing his ontological commitment, when he declared quantum mechanics pointed the way to a new conception of reality - a conception whose formation he considered as one of the most urgent tasks of our age (letter to Fierz 12.8.48, publ. in /13/). In this respect Pauli comes close to the mysticism implied in Tractatus by L. Wittgenstein, where the mystical concerns that which lies outside our world, outside the logical space where to we may only point. Bohr did not show any explicit[5] interest into ontological questiones. Instead he underlined the purposefullness of our descriptions of our experiences. However, such a description has to be effected within the bounds of a language. It is in this realm where Bohr located the center of gravity of the important philosophical issues. This motivates a comparision with Wittgenstein`s Tractatus and semiotics where language is accorded a central role.

2. BOHR AND WITTGENSTEIN

Tractatus Logico-Philosophicus /21/ was published as a book in 1922 - the same year James Joyces` Ulysses appeared and M. Schlick took over the chair after E. Mach in Vienna[6]. The world of Tractatus is created in seven main theses, the first one announcing the world is everything that is the case, i.e. the totality of facts - not things. When Bohr declares Physics is not the uncovering of an intrinsic substance or essence of the phenomena but an attempt at a purposefull ordering of our experiences, he is in a close agreement with Tractatus[7]. The world of Tractatus is a fragmented set of independent facts ; the only Necessity is logical necessity ; the only Order is the order imposed by Man, driven by a psychological need ; the belief in a causal nexus is superstition (/21/ lines 6.37;6.3631;5.1361). While the logical calculus of Tractatus cannot unite the world, Bohr loads the symbolic language of mathematics with such an unifying power. Tractatus closes with the seventh thesis announcing the Philosopher should rest : "Wovon man nicht spreche chen kann, darüber muss man schweigen" (Whereof one cannot speak, thereof one must be silent). Bohr and Wittgenstein took off from the same problem - the ambiguity of language. Wittgenstein set himself the task of constructing an ideal language, a logical grammar, that would complete the Begriffsschrift of Frege and the logical calculus of Russell (/21/ line 3.325). The young Bohr was enthusiastic about his discovery of a philosophical method of eliminating ambiguities of language through a construction analogous to the Riemann surfaces in complex function theory /23/. Quantum mechanics became an ingenious vehicle for Bohr to carry his philosophical ideas further. The problem of ambiguity was resolved by the concept of "complementarity" /20/, or contextuality. Thus, one is forbidden to speak of the spin (eg. s_x) of an electron in general - we should only attribute spin to an electron in the context of a certain experimental setup.

According to Tractatus, objects can only be mentioned in connection with some definite property. It is forbidden to say "x is an object" whereas an assertation like "x is a red object" could pass. In the Tractatus we read : "To underst-and a proposition means to know what is the case, if it is true" (4.024). Bohr, we might say, substituted "the experi-mental context" for Wittgenstein`s "what is the case". Wittgenstein attempted to save us from ambiguity by forbid-ding us to speak of the world as a whole[8]. His thinking in this direction may have been spurred by the "Russell`s para-dox", whereas Bohr got water on his wheel by the development of quantum mechanics.

3. SEMIOTICS AND THE ABSENT STRUCTURE[10]

To both Bohr and Wittgenstein language was fundamental philosophically.[9] Bohr is quoted to have said that "we are suspended in the language" (/23/,/10/ p. 55) and according to Wittgenstein ones language means the limit of ones world (line 5.6). Bohr viewed the language as a t o o l for a) an objective description of our experiences, and b) a purpo-sefull ordering of our experiences in a logically noncontra-dictory way. This matches a semiotic conception of language as a base for a`) communication, and b`) structuring the hu-man experiences forming a worldview (Weltanschaung). Indeed, Bohr considered an objective description to be equivalent to an unambigious communication /17/. Semiotics, as a branch of science, was envisaged by the french linguist F. de Saussure and the american thinker Ch. S. Peirce (cf. the contribution by P. V. Christiansen, this volume) among others around the turn of the century. F. de Saussure had a vision of a scien-ce he baptized "semiology" from the greec word `semeion` for `sign` /25/. This science would study the life of the signs in a social context. F. de Saussure counted semiology, the science yet to be, as a subdivision of general psychology, but here we will reckon semiotics as belonging to Philosophy proper. Indeed, the semiotic tradition originated with

Philosophy itself - it is just we are late to recognize it such. If Philosophy begang with the question : "What is the Being ?", then structuralism explicitely added : "How is the Being manifesting itself through the structures of language, nature, social order, or social disorder, etc ?" Thus, semiotics has to face ontology right from the beginning. If we map Bohr into this context we might say he forgot the Being - as have many structuralists too. Bohr wanted to ensure scientific objectivity through unambigious communication. The ordering of our experiences would be an attempt at arranging our accounts of the experiments into a structural unity.
- But what ensures we can do this, where do the structures come from ? Moreover, are not our objective accounts and designs of our experiments colored by our theories, our structural principles ? Bohr makes the following move: he grounds the objectivity in the language of classical physics, but takes the quantum language to be that which structures what we describe in classical language. Classical language as a structuring system restricted its speakers to a deterministic worldview with a fixed subject/object separation, according to Bohr[11]. Would we then not also expect that the quantum language restrict its speakers to a certain Weltanschaung as well ? But here, in a sense, Bohr reverses the argument, or the implication - the quantum language frees us from the rigid bounds of classical language. In the EPR discussion, between Einstein and Bohr on the completeness of quantum mechanics, Bohr is implying we have to live forever within the limits of the quantum language[12] - the following generations will be suspended in the principle-of-complementarity language.

There is a common apologetical strain in the thinking of Bohr, Pauli and Heisenberg. In the article /9/ cited above, Heisenberg dismisses a theory of D. Bohm becouse it is physically equivalent to the orthodox theory - "its exact repetition in a different language". Heisenberg denounces the "hopes" of Bohm the orthodox theory will be ever proved wrong. He continues, referring to Bohr, that such hopes are similar in structure to the sentence : "We may hope that it will later turn out that sometimes 2 x 2 = 5 , for this could be of great advantage for our finances". However, Heisenberg sees no cogent logical reasons why putting 2 x 2 = 5 would be of any help to us financially. Thus, the conclusion is that it would be pointless to try to transcend the orthodox thery. We are remembered of Wittgenstein writing : "The limits of my language means the limits of my world" (/21/ line 5.6). Pauli was also irritated by Bohm`s attempts in the "wrong direction" as well as by the "neurotic missunderstandings" by Einstein and by his "backward clinging" to field physics[13].

We might say, perhaps, that quantum mechanics was dear to Pauli becouse he thought the Being was speaking through it revealing insights that had been clogged by the rigid system of Newton. We may see Pauli in this respect as an hermeneutic who by psychoanalyzing quantum mechanics, and the history of physics, hoped to reach the deeper strata of reality. Pauli opposed Bohr who had, in a sense, substituted "objectivity" for "meaning", and he aligned himself with a semiotic tradition alongside Kepler and C. G. Jung. And here the views of Pauli and Bohm do intersect in a way. Indeed, in his book Wholeness and The Implicate Order /37/ Bohm outlines structuralist-Heideggerian thoughts. Like Heidegger Bohm attempts at outlining a new mode of language which would restore, or further the Harmony between Man and Being[14]. Bohm, Heidegger and Wittgenstein are united by their unique criticism. To them Philosophy is a mode of thinking - not the fighting with petty arguments, nor merely a method as the logical empiricists contended. To Heidegger the history of Western Philosophy is the sad story of the Logos being turned into an empty embellishment branded as Logic - Tractatus becomes the sunset of Logos where Symbolism takes its last breath awaiting its petrification into a mute formalism. In Tractatus Wittgenstein surveys the possibilities of thought. - Have we epigons f i x e d the limits of thought ? The semiotic tradition has always been in a tension between formalism and meaning - a tension also reflected in the thoughts of the Copenhagen People. In linguistics F. de Saussure /25/ compared language the play of chess whereas B. L. Whorf, an american linguist and a first rate fire-proof inspector, defined linguistics as "the quest of meaning" /42/. - Could we neutralize this tension between content and form by turning to Bohr`s Principle of Complementarity Diplomatical : Everyone is right from a complementary point of view ? Evidently Bohr did not explicitely treat the problem of formalism vs. meaning, or content vs. form, even if his central problem of the variable subject/object separation is not unrelated to these, suggesting Bohr had considered content and form to be relative (cf. /17/,/23/). Still these questions face everyone concerned with language and philosophy[15]. Notably, Bohr seems to take a Kantian position where ordinary language and the possibility of unambigious communication somehow becomes unproblematical. - But, can there be an unambigious communication ? For instance, the general Principle of Complementarity expressed in Ordinary Language is beyond unambigious communication - we could say it is outside the world paraphrasing Wittgenstein.

The above considerations directs us to the issue of linguistic relativity. - Thus, is the language of the Hopi people, which lacks the past-present-future categorization typical of the Indo-European languages, or the language of the Nootka people which lacks verb-noun classification, to be reckoned as an Ordinary Language upon which we may found Scientific Objectivity ? If not, then Scientific Objectivity would just be another name for ethnocentricity. - If we had base Physics on a Language with a different System of Categories, could we be sure we had arrive at the same Physics ?

In fact, Whorf`s principle of linguistic relativity holds that "all observers are not led by the same physical evidence to the same physical picture of the universe, unless their linguistic backgrounds are similar, or can in some way be calibrated" /44/. This thesis should make us sensitive to the questions above[16]. But it seems as if the Observer of quantum mechanics stands for an interpretant well versed in the orthodox theory guaranteeng any such disturbing questions do not arise.

3.1 Disputatio

* - Is language merely a t o o l for our communication and orientation in our daily life, as suggested by Bohr ? This is challenged by anthropological linguistics (Humboldt, Malinowski and Firth, Sapir and Whorf, etc). - How could we separate culture and language, or thought and language ? The Figura Ethymologica of Heidegger and Bohr contests such a separation. Moreover, supposing language is merely a tool for communication it is difficult to understand why it has to be so complicated (cf. /56/).

* - Is the Ordinary Language a Kantian categorial system beyond which we cannot reach ; are we suspended in the Language ; does my language mean the limit of my world ? - Is the Ordinary Language a biological evolutional a priori we must not question ? (Cf. note 17.)

* - How do propositions aquire meaning ? - By their possible experimental verifications ? Bohr suggests nothing beyond this kind of empirical semantics. This is the semantics elaborated in Tractatus (eg. line 4.2) - if we interpret it positivistically - with the consequence the Order of the Universe becomes meaningless. Here Bohr invokes the Symbolism of Mathematics, but it is difficult to understand how this can provide meaning if Mathematics is, as Bohr claims /17/, merely "a refinement of the common languge" (providing exact definitions). - Is the meaning a biological evolutional fact ; are we suspended in the meaning too ?

* Even if Bohr did not subscribe to the Russell`ian view of mathematics as a meaningless play with symbols, he did fail to emphasize that mathematics is essentially a continuing construction and definition of abstract structures. It is the structural conception that provides, what we may call, a s t r u c t u r a l s e m a n t i c s for mathematics and physics - the sense of a concept is defined by its structural relations to other concepts. Kleins Erlangen Program (1872) can be said to have inaugurated structural semantics within mathematics. Klein showed that the isolated geo-

-metries of Euclid, Riemann and Lobachevski-Bolyai-Gauss-Saccheri could be unified through the structures of symmetry groups. The sense of Geometry was seen to be a group structural relationship. (Cmp. note 18.)

* According to Bohr, the task of a theory is to eliminate all the apparent contradictions, and this is all we can demand within a new field of experience (cf. /33/). - Does the experiment preceed the theory ? - Did Bohr overlook the fact that theories to a great extent determine which experiments we will make - and especially which experiments we won't make ? A theory covers a limited part of reality, and if we stay within the discourse of the theory we may fail to recognize possible experiments that could contradict the theoretical predictions. We thus need competing theories, becouse theories can't be tested by empirical verifications alone. It is from this point of view we contest the apologetical strain of the Copenhagen People, and support the criticism expounded by P. Feyerabend /47/ and D. Bohm /46/ instead.

* The Principle of Complementarity saves the unambiguity of language functioning as a rule of consistency. - But did Bohr fail to recognize the possibility of deepening the semantics instead of merely regulating the syntax through a rule of consistency ? Examples of this second possibility is given by Bohm's hidden variable theory ; Jung's and Lacan's theories in the human sciences ; the extended standard theory of Chomsky in linguistics ; Marx' theory of History ; etc. The most promising deepening of the quantum semantics is given by the developing structural semantics of geometric quantization. It is realized that classical mechanics is characterized by the symmetry group of symplectic transformations (/31/,/51/), whereas quantum mechanics arises essentially through a restriction of this group of symmetry by an action of the unitary group U(1). (Cf. /32/,/50/,/53/.)

3.2 The Absent Structure

The heading of this section is taken from the title of the italian edition of a book by U. Eco : La Struttura Assente /27/. It is an expression for the thought that the structures are constructions; not things existing "out there" to be discovered. - Why may it be suppoesd these constructions reveal anything about reality, then - or, why are some structures relevant (employing the Bohm'ian idiosyncrasic usage of "relevant" /37/) ? -Why not becouse we are "suspended" in the reality, and not merely in a language ? There have been plenty of people trying to explicate this point in terms of biology, the theory of evolution, neurophysiology, the arch-

-etypes of Jung, etc. Yet, these are only branches of the stem of the ancient insight that our constructions are relevant to the extent the Logos is allowed to speak through us - the creators of the structures (cmp. also /41/). This is not a "positive statement" but an expression of a mode of thinking with important consequences for the question of the Foundations of Physics. This question becomes : "Is Physics exist ?"[19], or "Why may we suppose physical theories are, to some extent, true descriptions of reality, and not merely fancy hallunications ?". Experiments cannot prove theories right nor wrong - at most we test simple statements. Further, asserting the structures are absent we are directed to the question : "What does it mean to construct theories and abstract objects ?". Here we may make a link to Bohr. According to Bohr, science is founded on the concept of objectivity, i.e. unambigious communication. As such this formulation makes no suggestions what a scientific theory should be like. In fact, a theory does not just deal with unambigious communication, but with meaningful communication. Theories operate on the level of structural semantics, and our relevant constructions are constructions of relevant structures. It is this level Einstein referred to by his second of his two criterias for a theory : 1. External Confirmation, and 2. Internal Perfection (Naturalness) (/48/ p. 23).

Following Heidegger /38/ we formulate the general ontological question as followes : "Why are things instead of not to be ?". As an example of a physical ontological question we may ask : "Why are there electromagnetic fields ?". Indeed, this question has long occupied physicists trying to unite the forces of nature. Yet it is usually thought that Galilei laid the Foundations of Modern Science by banning the "metaphysical" why-question, only permitting the "positive" how-question. Such an attitude is no longer tenable when we realize that we are not reading the ready written Great Book of Nature, but that we are actually writing it.

4. NOTES

1*. P. Forman /1/ has piled up evidence proving that the non deterministic irrational nature of quantum theory became to be considered unavoidable by German physicists trying to adapt to the prevailing anti-scientific Zeitgeist of the Weimar Culture. This thesis is supported by the fact that the British physicists engulfed by a Zeitgeist intellectually less hostile and irrational, were less disposed to abandon causality as a basic principle /2/. A case /3/ in point is Dirac who considered quantum mechanics a mathematical generalization of classical physics. About Bohr`s philosophical ideas Dirac remarked /4/ he "did not understand them completely...my own line of thinking was really to put emphasis on thoughts which could be expressed in the forms of equations and much of Bohr`s thoughts were of a more general character and rather remote from mathematics" (quoted in note 77 /3/). Sal de Restivo presents in /5/ a critique of the preaching of "eastern mysticism" which claims to be the philosophical quintessence of the "new physics" and the Copenhagen Interpretation (cmp. authors like G. Zukav and F. Capra). For another aspect of this topic cf. /6/ and /7/.

2*. This Cartesian move is rejected by Foucault. Rajchman /12/ describes The Order of Things as "an argument for... the rejection of Man as a basic foundational principle that governs or grounds all phenomena" (p. 109).

3*. Heisenberg divides /9/ the world into two parts : The world of the Actual (the macroscopic world), and the world of the Possible (the microscopic world). The quote may reflect the view that the world of Possibilities (Tendencies, Potentialities, etc) is only comprehensible through Mathematics for which no ontology exists comparable with the "materialist ontology" of the macroscopic world. Heisenberg`s confusing discussion of the observer is partly due to the fact he connects the wave function with the knowledge of an observer (who makes discontinous alterations of a mathematical representation). Heisenberg`s observer is a kind of a metaphyical subject which ensures the world passes from the Possible to the Actual; - the "reduction of the wave packet".

4*. Bohr introduced the concept of the detached observer explicitely in /17/ 1954 to which Pauli replied in a letter 15.12.1955(?1954?) (publ. in /13/) that the measurement apparatus is a kind of an extension of the observer, whose unpredictable influences on the object under observation implies we have to abandon the idea of an isolated observer separated from the external physical events (cf. also /14/). Similarly, Heisenberg emphasizes /35/ that a subjective element is introduced into the description of atomic phenomena becouse an observer has constructed the measurement apparatus. In his lecture Bohr underlines that the very introduction of mathematical abstractions eliminates the subjective element. This may be compared with the position taken by R. Carnap in his Aufbau regarding structures vs. entities /58/.

5*. Bohr took the existence of atoms, electrons, etc for granted, yet he wanted to stress that these have no independent properties. Likewise for Wittgenstein, the existence of objects is as necessary for Philosophy as the names are for Language.

6*. In a cultural context Tractatus may be considered the most singular expression of the modernist mood, alongside with Ulysses and T. S. Eliot`s Wasteland, say (which also appeared in 1922). Cf. J. K. Braman /22/.

7*. "We encounter here in a new light the old conception, that when describing the nature it is not a question of uncovering the essential nature of the phenomena but only a question of detecting an order in the manifold of our experiences as far as it is possible" (transl. fr. the schw. ed. /19/, introd.). "Mechanics is an attempt to construct according to a single plan all t r u e propositions which we need for the description of the world" (/21/ line 6.343).

8*. Thus, asserting "x is an object" is to assert "x belongs to the set of all objects", making the forbidden reference to the whole. Cf. the introduction by B. Russell /21/.

9*. Following Bohr we had probably say that Language is the Foundation of Physics. Wittgenstein patterned the structure of the world on the structure of the language, allotting a fundamental role to the subject-predicate formalism of Frege-Russell-Leibniz-Aristotles-etc.

10*.Here we use "semiotics" and "structuralism" interchangeably, though "structuralism" has become considered a part of "semiotics" /27/. Naturally the ancient greecs already dealt with structures /26/ ; the Pythagorean school held that reality is mathematical ; Plato accorded great weights to mathematical structures (forms) in his cosmology ; Aristoteles` fundamental model of a structure was the growing organism ; the Stoics disscussed penetratingly the concept of the sign /28/. The gist of structuralism is revealed by the following thesis /26/ : "The relation is more important than the parts". The french line of sociological structuralism originated with A. Comte and was continued by E. Durkheim, C. Levi-Strauss and J. Piaget, among others. In Germany there was an influential structural thinker by the name of K. Marx. Linguistic structuralism (cf. /29/) was inaugurated by F.de Saussure in Europe, and by E. Sapir and L. Bloomfield in USA. They considered language a s y s t e m where words should not be studied in isolation - the object of study became the system of relations between words. The Copenhagen School of Linguistics, founded by L. Hjelmslev and V. Brøndal, developed the ideas of F de Saussure further. Hjelmslev, influenced by Carnap`s and Ch. Morris` semiotic theories, saw as the goal of linguistics to create a universal Language - a Metalanguage - which could serve science as a base for exact definitions. N. Chomsky /57/ revolutionized linguistics through his Transformation Theory where he set himself the task of finding the generating rules that produce all the correct expressions of a language. Mathematical structuralism may be said to have originated with F. Klein`s Erlangen Program (1872) (cf. /32/). In /59/ B. Russell contended to have given the very first exact definition of the concept of "structure" (based on the concept of "isomorphism"), which became the base for Carnap`s "structuralism" in /58/ (where it is claimed that all scientific propositions are propositions about structures). The structuralism of Russell and Carnap was hampered by it being modeled on logical structures, whereas the structures of interest in Science are the n a t u r a l structures - as emphasized by Einstein and Piaget. The members of the Bourbaki Group may be said also to have developed a mathematical structuralism. Perhaps the theory of fiber bundles /60/ and differential geometry /61/ are the studies where mathematical structuralism is most consciously developed (neglecting the theory of "categories"), becouse of their integrative scope. They are naturally linked to physical structuralism which originated with Einstein`s theory of Relativity and vigorously developed in the unification programmes. A notable french structuralist in the field of Physics is J-M Souriau ; cf. /31/ & /32/. Also Heisenberg was on the structuralist bandwagon, as noted in /26/, through his Heracleitean Platonic Cosmology ; cf. /35/. According to Heisenberg language is a system which cannot be exhausted by logical analysis becouse words have no exact definitions, which in turn ensures that language can "touch reality" and yield the Truth. Heisenberg`s structuralism is a dialectism where Contradictions are the creative principles of the world (cf. Ch X & XI /35/). There seems to be some Quinean strands in Heisenberg`s thoughts ; cf /62/. For a readable acount of the various aspects of structuralism consult J. Piaget`s book /30/.

11* Though it seems as if Bohr assumed that experience preceeds theory generally (cmp. "Here it cannot be the question of a physical explaination in the ordinary sense but of an elimination of every apparent contradiction, which is all we can demand within a new domain of experience" (transl. fr. the schwe. ed. /33/). Cf. also note 7). Yet we still need a language prior to experience in order to be able to describe the experience. Bohr equates this language with the Ordinary Language (of which Classical Mechanics is just a refinement, according to Bohr, within the conceptual frame of everyday life; cf. eg. /63/).

12*. "In fact, it is only the mutual exclusion of any two experimental procedures, permitting the unambigious definition of complementary physical quantities, which provide room for

for new physical laws, the coexistence of which might at first appear irreconcible with the basic principles of science" . (/34/ p. 700 ; cf. also /64/.) It is interesting that both Wittgenstein and Bohr think they "finally solved" (/21/ preface), or they have found the "complete resolution" (/15/) to the philosophical episthemological problems.Wittgenstein however adds : "And if I am not mistaken in this, then the value of this work secondly consists in the fact that it shows how little has been done when these problems have been solved".

13*. Cf. Pauli`s letter to Fierz dated 10.8.1954 (publ. in /13/). Pauli often uses psychologicalising arguments against the critics of the Copenhagen Interpretation. Eg. in a letter to Fierz, dated 6.1.1952 (ibid.), Pauli explains that the french communists and catholics detest complementarity becouse they are psychologically bound by certain "eschatological expectations". In /9/ Heisenberg accuses Bohm, using a Marxist jargong, of a superfluous "ideological superstructure". Furthermore, Heisenberg is bound to regard the Copenhagen Interpretation as "unavoidable" /9/ and as a "unique physical interpretation" in regard to which we no longer have any "freedom" /35/.

14*. Although Bohm makes no reference to Heidegger there are some striking parallels between them ; cmp. /37/,/38/,/39/ & /40/. When Bohm traces the crisis of our age to the "fragmentation" of our thought, Heidegger points to the "forgetting" of the Being as the cause. Both employs a Figura Ethymologica becouse "the study of origins of words may be regarded as a sort of archeology of our thought processes, in the sense that the traces of earlier forms of thought can be found by observations made in this field"(/37/ p. 54). It is suggested these earlier forms of thought were less fragmented, or that that were in an a better harmony with the Being. In particular both Bohm and Heidegger rejects the formalist conception of language, according to which language is nothing more than a tool for communication. In this respect they approach the linguistic tradition of Humboldt and Whorf who held that the structures of language reflect and affect the mentality of their speakers. The structuralist inclination of Bohm is apparent already from the title of his book. In this realm Bohm and Heisenberg might have spoken the same language.

15*. The problem of content vs. form (or expression) was pivotal to the Copenhagen School of Linguistics (vide note 10). It is of some interest to note that Brøndal shared H. Høffding with Bohr as one of his most influential teachers (/29/ §361). Høffding directed Brøndal`s interest to mathematics and physics which made him a fervent defender of the logical-philosophical approach to linguistics. J. A. Wheeler writes about Bohr that his "lifetime advocacy of complementarity owed much to the philosophy of his great teacher,

Harald Høffding" /24/. Bohr had in common with the Copenhagen School of Linguistics a formalistic conception of language (neglecting meaning) as a tool for communication whose most refined form was represented by mathematical symbolism.

16*. There may be a general misconception of the Whorfian standpoint, according to which people with different languages would see different things. However, Whorf is not contending they would see different things, but that they might see the same things as related in a different way. The point is not that a Hopi physicist would not understand "our physics", but that he might understand it in a different way, seeing relevance where we see irrelevance, and vice versa. Even if two theories have the same "translation" - into predicate logic, say - they may not be equivalent. They may have different meanings. We thus disagree with Howard /45/ who states that "the particular manner in which terms are defined in the formal development of a theory is somewhat arbitrary (eg. should "force" or "mass" be taken as primitive in a formal development of the Newtonian mechanics) and thus, do not shed much light on deeper philosophical problems. At best our choice of a way to formally develop a theory merely ratifies choices that have been made previously for philosophical reasons" (p. 362). Howard seems to have got the things upside down. Even Bohr emphasized the importance of a correct choice of definitions (cf. /15/). It is our philosophy which detemines our formal development which in turn explicates our philosophical position. If our formal development is bad, then so was our philosophical "choice". In order to be able to gauge whether a formal development is bad or good we need competing theories (not just experimental tests) as has been emphasized Feyerabend /47/. The theory which has a richer (deeper) semantics is then to be preferred : such a theory minimizes the accidences (contigencies) and maximizes the essences - it should answer why the world is as it is (not just state that it is what it is).

17*. "We know, that all understanding must in the end rest on the ordinary language, becouse only there we may be certain of touching the Reality ; and therefore we must remain sceptical in regard to every kind of scepticism toward the ordinary language and its essential concepts." (Transl. fr. /35/ p. 196.)

18*. Consider the vector-potential A_μ of classical electrodynamics. It is related to the electromagnetic fields through $F_{\mu\nu} = \partial_\nu A_\mu - \partial_\mu A_\nu$ which determine the motion of a charged particle of charge q, mass m and momentum p^μ according to $\dot{p}^\nu = (q/m) F_{\nu\mu} p^\mu$. These relations do not determine the meaning of the vector-potential, however. In fact, it was long cosidered as merely a technical device. Only through the U(1)-gauge theory by H. Weyl /52/ did the vector-potential aquire a structural fundamental meaning (as a connection in a U(1)-bundle), whereas the electromagnetic field was seen as a derived notion (curvature of the connection /54/). The theory of fiber bundles, and global mathematical theories in general, has a strong structuralistic strand which partly complements our following shema for structuralistic science;

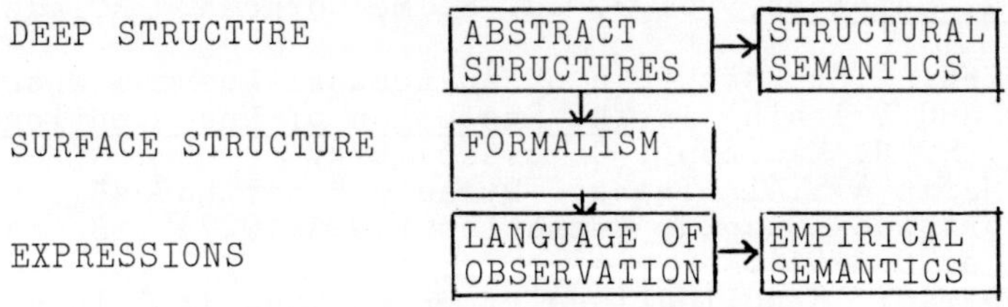

(for another use of "structural semantics" cf. /65/).

19*.This is an example of the characteristic structuralist question ; in /32/ Souriau asks "Does quantum mechanics exist ?"; Foucault asked "Does Man exist ?" ; the Whorfian question could be formulated "Does Language exist ?", cf. /44/ ; G. Simmel asked "How is society possible ?" (Soziologie 1908, quot. in /26/) ; etc. Structuralist science has thus regained the why-question. At the bottom looms the question "Does the Structure exist ?". A Bohrian move would be to say `we are suspended in the structure`, which is not very satisfactory. Nor are the structures residing in the Platonic Universe (as it has been traditionally thought of). We are left with the basic thesis of Piaget /30/ : Structures are constructions emerging in a continous process of formation. This we may take as the basic thesis of methodical structuralism. Here we are not far from the views of Einstein in /49/ where his `layer structure of scientific systems` strongly resembles the `hierarchy of structures` of Piaget. Moreover, nor Einstein nor Piaget were content with logical structures - what mattered to them were natural structures. Yet, if there is no general method of obtaining the natural structures, then we cannot agree with Piaget that structuralism is a method and not a Philosophy. Methodical structuralism is a mode of thinking - a philosophy - which urges us to f o r m the natural structures. By this we have come to the watershed between Einstein and Bohr. In the Bohr-Einstein discussion Bohr and Einstein viewed things from different standpoints. To Einstein a Physical Theory meant the ambition to comprehend the whole of Universe (eg. /48/p. 23). Naturally then, Einstein did`t accept quantum mechanics (or any other then known theory) as a complete theory - his was the vision of a unified field theory that would explain the Universe. Bohr on the other hand was concerned with the completeness of a theory with respect to a limited field of experience (cf. eg. /66/). Thus they had different conceptions of what a "theory" i s - no wonder they never arrived at a complete agreement on matters Philosophical. Simplifying, Einstein was a visionary, Bohr was an apologetic.

5. BIBLIOGRAPHY

/1/ P. Forman: `Weimar Culture,Causality and Quantum Theory, 1918-1927; Adaption by German Physicists and Mathematicans to a Hostile Cultural Environment` in Historical Studies in the Physical Sciences, vol 3, ed. R. Mc Cormach. Philadelphia 1971.
/2/ P. Forman: `The Reception of an Acausal Quantum Mechanics in Germany and Britain` in The Reception of Unconventional Science,ed. S. H. Mauskopf. Colorado 1978.
/3/ M. de Maria & T. la Trana: `Dirac`s "Unorthodox" contribution to orthodox Quantum Mechanics (1925-1927)`. Scientia 118 (1983) p. 595-611.
/4/ P. M. Dirac: `Recollections of an Exiting Era` in History of Twentieth Century Physics, ed. C. Wiener. Academic Press 1977.
/5/ S. de Restivo: The Social Relations of Physics, Mysticism and Mathematics. Reidel 1983. (Episteme vol. 10.)
/6/ L. Graham: `Reception of Quantum Mechanics in Soviet Union` in /24/.
/7/ S. Müller-Markus: `Niels Bohr in the Darkness and Light of Soviet Philosophy`. Inquiry 9 (1966).
/8/ M. Foucault: Les Mots et les Choses. Paris 1966. Engl. transl. The Order of Things. New York 1970.
/9/ W Heisenberg: `The Development of the Interpretation of of the Quantum Theory` in Niels Bohr and the Development of Physics, ed. W. Pauli. London 1955.
/10/ Discussion Sections. Symposium on the Foundations of Modern Physics, 16-20 June 1985, Joensuu, Finland. Report Series; Department of Physical Sciences, Uni. of Turku 1985.
/11/ W. Heisenberg: The Physical Principles of Quantum Theory. New York 1949. (Chicago lectures 1929)
/12/ J. Rajchman: Michel Foucault: The Freedom of Philosophy. New York 1985.
/13/ K. V. Laurikainen: Atomien Tuolla Puolen (Beyond the Atoms). Mänttä 1985.
/14/ W. Pauli: `Die Philosophische Bedeutung Der Idee der Komplementarität`. Experimentia VI/6 (1950).
/15/ N. Bohr: `Natural Philosophy and Human Cultures` (1938) included in /16/:
/16/ N. Bohr: Atomic Physics and the Human Knowledge. New York 1958. Schwed. ed. , Stockholm 1959.
/17/ N. Bohr: `The Unity of Knowledge` (1954) incl. in /16/.
/18/ N. Bohr: `The Quantum of Action and the Description of Nature` (1929) incl. in /19/.
/19/ N. Bohr: Atomic Theory and the Description of Nature. Cambridge 1934, 1961. Schwed. ed. Stockholm 1967.
/20/ N. Bohr: `The Quantum Postulate and the Recent Development of Atomic Theory` (1927) incl. in /19/. This paper is known as the "Como lecture".
/21/ L. Wittgenstein: Tractatus Logico-Philosophicus. London 1922. First published as `Logisch Philosophische Abhandlung` 1921 in Annalen der Naturphilosophie.

/22/ J. K. Braman: Wittgenstein`s Tractatus an The Modern Arts. New York 1985.
/23/ A. Petersen: `The Philosophy of Niels Bohr`, Bull.Atom. Scientist 19 (sept. 1963), 8-14. Repr. in /24/.
/24/ A. P. French & P. J. Kennedy (ed.): Niels Bohr. A Centennary Volume. Harvard University Press 1985.
/25/ F. de Saussure: Course de Linguistique Generale. Posth. publ.1916. Engl. transl. 1959.
/26/ T. Bottomore & P Nisbet: `Structuralism` in A History of Sociological Analysis, ed. Nisbet & Bottomore.London 1979.
/27/ U. Eco: La Struttura Assente. Milano 1968. A Theory of Semiotics. 1979.
/28/ U. Eco: Semiotics and The Philosophy of Language.London 1984
/29/ M. Ivič: Pravci u Linguistiki(1963). Trends in Linguistics. Mouton 1965.
/30/ J. Piaget: Le Structuralisme. Paris 1968. Structuralism London 1971.
/31/ J-M Souriau: Structures des Systemes Dynamiques. Dunod 1969
/32/ J-M Souriau: `Physics and Geometry` in Quantum, Space and Time - The Quest Continues, ed. A. O. Barut, A. van der Merwe & J-P Vigier. CUP 1984.
/33/ N. Bohr: `On Atoms and Human Knowledge` (1954) in /16/.
/34/ N. Bohr: `Can Quantum-Mechanical Description of Physical Reality Be Considered Complete ?`, Phys.Rev. 47 (1935), 696-702.
/35/ W. Heisenberg: Physik und Philosophie. Stuttgart 1978 ; Physics and Philosophy. New York 1958.
/36/ W. Heisenberg: `Reminiscences From 1926 and 1927`in /24/.
/37/ D. Bohm: Wholeness and The Implicate Order. London 1981.
/39/ M. Heidegger: Sein und Zeit (1927). Tübingen 1967. 11.
/40/ M. Heidegger: Die Technik ünd die Kehre. Pfullingen 1962.
/38/ G. Steiner: Heidegger. Fontana Paperback 1978.
/41/ C. F. von Weizsäcker: `Allgemeinheit ünd Wissheit` in Martin Heidegger zum Siebzigsten Geburtstag, ed. G. Neske 1959
/42/ B. L. Whorf: `A Linguistic Consideration of Thinking in Primitive Communities` (1936). First published in /43/.
/43/ J. B. Carroll (ed.): Language, Thought and Reality. Selected Writings of Benjamin Lee Whorf. MIT Press 1956.
/44/ B. L. Whorf: `Science and Linguistics`, Technol. Rev. 42 (1940) 6. Reprinted in /43/.
/45/ D. A. Howard: Complementarity and Ontology; Niels Bohr and The Problem of Scientific Realism..(Diss. Boston Univ. 1973). University Microfilms International 1983.
/46/ D. Bohm: `Bohr`s Views Concerning the Quantum Theory` in /24/.
/47/ P. Feyerabend: `How to be a Good Empiricist` in Philosophy of Science, the Delaware Seminar vol. 2, ed. B. Baumrin. New York 1963.

/48/ A. Einstein: `Autobiographical Notes` in Albert Einstein: Philosopher-Scientist, ed. P. A. Schilpp. The Library of Living Philosophers, Inc. 1959.
/49/ A. Einstein: `Physics and Reality`, The Journal of the Franklin Institute, 221 (1936) 3.
/50/ A. Heslot: `Quantum Mechanics as a Classical Theory`, Phys.Rev. D 31 (1985) 6, 1341-1348.
/51/ R. Abraham & J. E. Marsden: Foundations of Mechanics. 2. Mass. 1978.
/52/ H. Weyl: Gruppentheorie und Quantenmechanik (1928). Engl. ed. Dover 1950.
/53/ N. Woodhouse: Geometric Quantization. Oxford 1980.
/54/ E. Lubkin, Ann. Phys. 23 (1963) 233 ; T. T. Wu & C. N. Yang, Nucl. Phys. B107 (1976) 365 ; for a nontechnical but vivid acount of fibre bundles in physics cf. H. J. Bernstein & A. V. Phillips: `Fiber Bundles and Quantum Theory`, Sci. Am. 245(1981) 1.
/55/ B. L. Whorf: `Languages and Logic`, Technol.Rev. 43 (April 1941). Reprinted in /43/.
/56/ N. Chomsky: Problems of Knowledge and Freedom, The Russell Lectures. Pantheon Books 1971.
/57/ N. Chomsky: Syntactic Structures. Mouton 1957.
/58/ R. Carnap: Der Logische Aufbau der Welt. Berlin 1928.
/59/ B. Russell: Introduction to Mathematical Philosophy. London & New York 1919.
/60/ N. Steenrod: The Topology of Fiber Bundles. Princeton 1951.
/61/ S. Kobayashi & K. Nomizu: Foundations of Differential Geometry. Interscience Publ. Vol I 1963, vol II 1969.
/62/ W. V. O. Quine: `Two Dogmas of Empiricism` (1951), reprinted in Quine: From a Logical Point of View. Cambridge, Mass. 1953.
/63/ N. Bohr:`Biölogy and Atomic Physics` (1957) in /16/.
/64/ N. Bohr: `Physical Science and The Problem of Life` (1957) in /16/.
/65/ A. J. Greimas: Semantique Structurale. Paris 1966.
/66/ N. Bohr: `Atomteorin och Grundprinciperna för naturbeskrivningen` (1929) (Atomic Theories and The Principles for Description of Nature), in the schwed. ed. of /19/

** Acknowledgement

to the Niilo Helanders Stiftelse whose support for my project "The Future of History" gave an extra momentum to this contribution _.

CONTEXT AND NON-LOCALITY - A PEIRCEAN APPROACH

PEDER VOETMANN CHRISTIANSEN
Institute of Mathematics and Physics
Roskilde University Center
P.O. Box 260, DK 4000 Roskilde
Denmark

ABSTRACT

The aim of the Copenhagen interpretation of quantum mechanics is to provide a semantics for the symbols of the mathematical formalism. Niels Bohr's early philosophy which has a direct connection with his later ideas of complementarity is inspired both by danish existentialism and american pragmatism. The founder of pragmatism, C.S.Peirce, was, however, not known to Bohr, and Peirce's complete classification of non-linguistic signs has never been considered in connection with the Copenhagen-semantics, which is therefore in certain respects incomplete. In this paper it will be argued that Peirce's semiotic is a natural basis for quantum-semantics and that his "synechistic" concept of local realism can be tested experimentally. This opens a possibility of avoiding some of the extravagant ontological interpretations that have emerged in later years in consequence of the experimentally observed violations of Bell's inequalities.

The Copenhagen interpretation is often regarded as identical with the pure mathematical formalism and many textbooks nowadays neglect the epistemological questions that were so important before the formalism got its final shape in the late twenties. This neglect of philosophical questions may be partly due to Bohrs attitude in his later years where he emphasized that ordinary language is insufficient and even misleading and that it must be replaced by the mathematical formalism. This attitude seems to have been connected with scepticism towards philosophy in general, at least it has been interpreted in this way by philosophers of a somewhat positivistic inclination who tend to disregard the important role of philosophy as inspirator for new developments in physics[1)].

However, in his younger years Bohr was very engaged in philosophical speculations, and there is a strong historical evidence[2)] indicating that the philosophy of complementarity was evolved long before quantum mechanics and is influenced by several streams of philosophy as focesed by Bohrs philosophical teacher Harald Høffding. Max Jammer[3)] has pointed out the influence from the danish existentialist Søren Kierkegaard as important for an understanding of Bohr's ideas of the quantum jumps, and Witt-Hansen[4)] has drawn attention to the continuity-thinking of Leibniz, as mediated by Høffding, which seems closely connected with Bohr's correspondence-principle.

In this paper we shall look upon the influence from american pragmatism as exposed by William James (and Høffding) and the regrettable lack of influence from its originator, Charles S. Peirce, who was unknown to Bohr and has been neglected by and large in this century, until quite recently. The author's interest in these questions of history of philosophical influences on quantum mechanics is not so much connected to historical philosophy as to the conviction that the difficulties of interpretation that confront quantum mechanics nowadays, especially after the Aspect-experiments, are real and serious and concerning semantics, not the mathematical formalism.

Many weird ontologies, involving backwards causality, splitting universes, telepathic synchronicity, or just a plain old-fashioned pre-established non-local harmony are invoked in these years in order to explain the observed violations of Bell's inequalities. The author believes that such ongologies, although some of them are interesting and may be valuable in other contexts, have nothing to do with quantum mechanics, which is a simple and logical theory with the very limited scope of explaining elementary phenomena of the atomic world in a consistent way.

It will be argued that, according to Peirce, a local realism is not identical with the existence of local hidden variables and therefore not automatically leading to Bell's inequalities. The Aspect experiments are convincing proofs of the validity of the quantum formalism under the circumstances set in the laboratory, and, although some weak loopholes remain, they may be regarded as counterproofs of theories of local hidden variables. This does not mean, however, that the concept of local realism, or Einstein-separability, has been disproved. The "synechistic" sort of local realism, as proposed by Perice, has never been subjected to an experimental test, although it could easily be by a small alteration that would bring the experiments in closer relation to the thought experiment proposed by Einstein, Podolski and Rosen. Until now the experiments have just pointed to the need for a smiotic reconsideration of the quantum formalism but cannot be taken as evidence for the validity of extravagant ontologies.

HØFFDING, AND THE YEARS OF ECLIPTICA.

Niels Bohr began the study of physics at the Copenhagen University in the year 1903. At that time all students were initiated to the academic world through an introductory course in "propedeutic philosophy". The professor of philosophy, Harald Høffding, was an old friend of Bohr's father and had been a frequent guest in their home where lively discussions of science, philosophy and politics took place. Niels and his brother Harald were thus well prepared to take part in a continuation of the cross-cultural spirit of Copenhagen from "the golden age" in the 19th century. The fertile ground created by such giants as N.F.S.Grundtvig and S.Kierkegaard was still blossoming at the turn of the century with a new poetic realism and impressionism in the litterature and a cultural radicalism in philosophy and politics, opposed to the dictatoric conservative Estrup government and with Høffding and the brothers Brandes as leading figures.

In the autumn of 1904 Høffding made a journey to USA and England. After having returned full of new ideas he started a series of philosophical colloquia at the university. A group of 12 students attending these colloquia, among them Niels and Harald Bohr, formed a circle called Ecliptica with the purpose of further discussions of the philosophical topics that Høffding called their attention to. We have no detailed written reports of the discussions at these meetings, but we know that the atmosphere was stimulating, although heavily polluted with tobacco smoke. Not less than three internationally recognized "Copenhagen schools" have emerged from this small circle, viz. Niels Bohr's in physics, Edgar Rubin's in psychology, and Viggo Brøndal's (and Louis Hjelmslev's) structural linguistics.

One of the Copenhagen philosophers of the 19th century

that must have played a great role in the discussions is Søren Kierkegaard, a highly religious thinker who exhausted himself in violent attacks against the established church. Høffding was an expert in Kierkegaard's philosophy and considered his existentialistic writings on the choice and "the qualitative jump" to be of importance far outside theology, in the general logic of concepts and psychology. Kierkegaard was very critical against Hegel's dialectical philosophy, especially the thesis of quantity turning over into quality. In several books from around 1845, e.g. "Begrebet Angest" (The Concept of Dread) he analyzes the free choice as something that cannot be reduced or explained away by quantitative means. The choice is a qualitative jump confronting the subject with a vacuum that creates dizziness, nausea and dread. Kierkegaard employs a peculiar circular or "bootstrapping" logic in describing the jump, which is said to be a choice between possibilities that are presupposed or preset by the jump. Høffding's first series of colloquia in the spring of 1905 were discussions of the concept of free will and Kierkegaard's ideas must have had a central place. Niels Bohr has in several letters from 1909 reported how his own readings of Kierkegaard made a deep impression on him, and it is rather easy to see how many of his more cryptic remarks on complementarity bear a close resemblance to Kierkegaard's formulations.

Another important influence on Bohr's thinking in the early years of Ecliptica comes from America, viz. William James. During his stay in USA in the autumn 1904 Høffding had been the guest of William James and in his memoirs[5)] he describes what great inspiration the contact with James had been to him. In an interview given to Thomas Kuhn and Aage Petersen[1)] the day before his death Bohr tells that he read William James in his young years and that James' description of "the stream of consciousness" had made a lasting impression on him.

Besides being a pioneer of depth-psychology James is also considered the founder of the pragmatic school of philosophy. This school is founded on the so called pragmatic criterion of meaning which is closely related to Bohr's semantic thesis that the meaning of the quantum mechanical symbols is set by the experimental context. James himself, however, did not consider himself the inventor of pragmatism but pointed to his old friend Charles Sanders Peirce who had first formulated the pragmatic criterion of meaning in a paper "How to make our ideas clear" from 1878. There is no indication that Høffding or Bohr knew anything of Peirce's philosophy, but there is an indirect connection so far as James' influence is important and is inspired by a life-long acquaintance with Peirce.

Another indication of an indirect connection from Peirce to Bohr is found in the fact that Høffding after his stay in America went to England in the late autumn of 1904 where he stayed as a guest in the house of another of Peirce's close friends, the Lady Victoria Welby. A great part of Peirce's general theory of signs, semiotic, is evolved in letters to Lady Welby who herself had made original contributions to the theory of semantics. Peirce is known to have warned Lady Welby "perhaps you are in danger of falling into some error in consequence of limiting your studies so much to language"[6]. Unfortunately, this warning never went through to Bohr who remained "suspended in language"[7].

WHAT WENT WRONG IN COPENHAGEN?

In his discussions with Einstein at the Solvay meetings 1927-30 Bohr maintained the pragmatic attitude that the meaning of the symbols for position and momentum is set by the apparatus of measurement which itself is subject to the fluctuations described in Heisenberg's uncertainty relations. These fluctuations are thus inherent in the quantum mechanical semiosis, and we are not allowed to say that some "true" values are "disturbed" by the measurement process, because true values are created by the measurement and must not in general be assumed to exist by themselves in nature. This is all very much in accordance with Peirce's epistemology as described in an article from 1892[8)]. Peirce would also agree with Bohr in the view that physicists must allow God the freedom to play dice. This view, that chance is a genuine and irreducible factor in the physical universe was advocated as "tychism" in the above mentioned article.

Einstein tried at the Solvay meetings to demonstrate that quantum mechanics was inconsistent, i.e. that it contained an intrinsic logical flaw that would turn up in situations clearly within the scope of the theory. However, after 1930 he seems to have been convinced of its consistency within a limited field. His next attack, in 1935 with Podolski and Rosen[9)], is an attempt to demonstrate the incompleteness of the theory, i.e. that there are real situations where the theory cannot be applied or where it would lead to false predictions if one tried to apply it. In the light of Gödel's incompleteness theorem from 1931 the accusation of being incomplete would not seem so serious, because in mathematics this property is a necessary consequence of the consistency of the theory. Probably Einstein thought that the situation was analogous in physics, although he did not refer to Gödel's theorem. He had at that time just obtained a permanent position in Princeton and was joined by Gödel shortly afterwards.

In Copenhagen, however, the distinction between consistency and completeness was hardly noticed. L. Rosenfeld who took dictations from Bohr in his reply to Einstein, Podolski and Rosen (EPR) has in the memorial volume "Niels Bohr"[10)] given a vivid report on the efforts which leaves the impression that Bohr was trying to live up to his image as the great champion of the fights with Einstein rather than trying to understand in what respects the EPR paper presented a new viewpoint. Consider, for example the following quotation from Rosenfeld's report (my translation from danish):

> "Day after day, week after week the whole argumentation was patiently investigated by means of simpler and more transparent examples. Einstein's problem was reshaped and its solution formulated again with such precision and clarity that the weakness in the reasoning of the critics became evident and their whole argumentation, in spite of all its fake spirituality, was shattered to pieces. "They do it nicely," was Bohr's comment, "but what counts is to do it correctly"."

The myth about the "precision and clarity" of Bohr's reply to EPR[11)] does not stand for a closer scrutiny. Einstein never understood it and consequently the debate came to a stalemate. J.S.Bell has declared[12)] that Bohr's reply is "totally obscure" and similar but more vague declarations of uneasiness have been issued by Dirac and Feymann. Even Bohr himself had to admit when he reread the paper in 1949 that he "strongly felt the inadequacy of expression"[13)]. It is not the entire article that is obscure but a definite place that Bohr and Bell are referring to.

In the first part of the article Bohr pursues the line of argumentation that had proven so successful in the Solvay discussions:

> "<u>The finite interaction between the object and the measuring agencies</u> conditioned by the very existence of the quantum of action entails - because of the impossibility of controlling the reaction of the object on the measuring instruments if these are to serve their purpose - the necessity of a final renunciation of the classical ideal of causality and a radical revision of our attitude towards the problem of physical reality."

This is still a physical way of reasoning and it does not exclude a realistic attitude but just a semiotically unreflected sort of realism that would ascribe definite numerical values to quantities of the undisturbed reality. There is a strong resemblance between the Bohr-quotation above and the way Peirce argues against "the doctrine of necessity" in 1892[8)] (but of course Peirce does not mention the quantum of action):

> "Try to verify any law of nature and you will find that the more precise your observations, the more certain they will be to show irregular departures from the law. We are accustomed to ascribe these, and I do not say wrongly, to errors of observation; yet we cannot usually account for such errors in any antecedently probable way. Trace their causes back far enough, and you will be forced to admit they are always due to arbitrary determination, or chance."

In connection with this view on observations in general Peirce criticizes "the necessitarian position - that certain continuous quantities have exact values" and concludes that "any statement to the effect that a certain continuous quantity has an exact value, if well-founded at all, must be founded on something other than observation." In short: Bohr and Peirce agree that the analysis of the measurement process calls for a semiotic reconsideration of our notions of the undisturbed physical reality, but Bohr's argument is ontological, based on the existence of the quantum of action, whereas Peirce's argument is epistemological and based on our inability to account for errors of measurements "in any antecedently probable way". One may conjecture that Peirce, had he known about the quantum of action, would have supported Bohr's argument wholeheartedly.

However, in the later part of Bohr's reply to EPR there is a sudden turning away from the, still potentially realistic, attitude in the first quotation. This is the exact place where the article becomes abscure (both according to Bell and Bohr). In discussing the EPR thought experiment where a measurement on one particle can be said to affect the state of another particle Bohr writes:

> "Of course there is in a case like that just considered no question of a mechanical disturbance of the system under investigation during the last critical stage of the measurement procedure. But even at this stage there is essentially the question of <u>an influence on the very conditions which define the possible types of predictions regarding the future behavior of the system</u>."

This is almost certainly a place where Peirce would have objected, because Bohr tries to circumvent both realism (anti-nominalism) and synechism, i.e. the conception that signs exist outside the subjective mind and propagate through a continuum.

It is not clear what connection there could be between the "finite interaction" in the first Bohr-quotation and the "very conditions" in the second. When we make a measurement on particle 1 we have a free choice of measuring one or the other of two complementary properties of this particle, e.g. position or momentum. The choice between the two different ways of interacting with particle 1 according to Bohr prohibits the use of the EPR-term "the same reality" for particle 2, even though this particle is not affected by any mechanical disturbance, and this is because quantum mechanics forces us to regard the whole phenomenon of preparation and measurement as possessing an "individuality completely foreign to classical physics". The "very conditions" in the second quotation thus seems to be the formalism of quantum mechanics associated with the philosophy of complementarity, and Bohr is trying to persuade the reader to accept that quantum mechanics defines how the term "physical reality" may be correctly used. This is of course difficult to accept from a realistic standpoint: when the question is whether the quantum mechanical description of physical reality is complete, then the answer that quantum mechanics itself defines what "physical reality" is looks like a philosophical shortcircuit or cheating in the game of debate.

Apart from breaking with realism the introduction of the "very conditions" also breaks with the locality principle, or synechism. When we measure one or the other of the two complementary properties of particle 1 without disturbing particle 2 then it is true that quantum mechanics gives an unambiguous prediction for the future behavior of particle 2, viz. a wave function corresponding to a pure state of that particle. The "finite interaction" is in this case only involving particle 1, but it produces via the "very conditions" a change of state of particle 2. Bohr seems to forget that we still have the possibility of making an independent measurement on particle 2 and that this would amount to a test of the formalism that can be performed whether one accepts the influence via the "very conditions" or not. What about the "finite interaction" with particle 2 that would be introduced by such a second measurement? Can we be sure that it doesn't produce a conflict with the "very conditions" if the two measurements really are physically independent ? These questions are unanswered in Bohr's article.

According to the orthodox Copenhagen interpretation (after 1935) one is really not allowed to ask such questions. Rosenfeld, as the guardian of the orthodoxy, states[10] that "the refusal of Einstein's criticism does not add any new element to the concept of complementarity". The rest is silence in Copenhagen. Attempts to formulate a realistic concept of the quantum world have been met with cryptic formulations stressing both that we are suspended in language, that the language is inadequate, and that our problem is to use the language correctly, which in most cases mean to abstain from the use of words and stick to the mathematical formalism, or to quote Bohr directly.

The author believes that the orthodoxy after 1935 is due to the fact that nobody (including Bohr himself) ever understood what the "very conditions" are. Something went wrong in Copenhagen in the very moment they were introduced. Until then the Copenhagen interpretation had pursued the reasonable goal of creating a "minimal semantics" as v. Weizsäcker states it, but afterwards even that task became impossible, because the "very conditions" create the illusion that the mathematical formalism makes up reality instead of describing it. A reconstruction of the original goal is very much needed nowadays, because the experimental violation of Bell's inequalities has created confusion.

THE IRREDUCIBLE SIGN RELATION.

Following now the line of thought that the missing link in the minimal semantics of the Copenhagen interpretation is

1) a way to include the concept of non-linguistic signs of an undisturbed physical reality in the formalism, and

2) a description of conditions for their transformation to symbolic results of measurements,

we shall see how this may be done with the use of Peirce's general theory of signs, semiotic.

The first part of this program is in fact already achieved by Dirac with his introduction of the "kets" and the "bras", the concept of a state vector without reference to representation. With Peirce's terminology the ket is an example of a "degenerate" type of sign called an <u>index</u> and the representation is an <u>interpretant</u> which in combination with the index creates a non-degenerate sign, a <u>symbol</u> (the bra-ket).

The second part amounts to a proper theory of measurement, and this is not easily achieved, but one may hope that Peirce's idea of continuity and connectedness of sign relations, synechism, which were first expressed in a paper from 1892[14], may add some simple criteria of connectedness as a necessary condition for a measurement apparatus to constitute an interpretant in a quantum-semiotic sign relation. At least this consideration points to the relevance of the connectedness of Aspect's equipment, the use of coincidence counters, to the contextuality of the measured polarizations, i.e. the non-locality exhibited by the violation of Bell's inequalities.

The application of Peirce's ideas to quantum mechanics of course requires some reshaping of concepts and cannot be totally faithful to the original formulations. This is, however, not as bad as it may sound. Peirce, unlike Hegel, never created a completed system, and his aim was to develop scientific methods that could be applied by future researchers in an unending development. His own development of semiotic is a good example, it is easy to find contradicting statements in his writings, but his ideas are continually gaining in integrity and generality, and from about 1885 to his death in 1914 they seem free of contradictions. It is not necessary to read everything he wrote in order to use his ideas (which would be impossible, as he wrote about 80.000 pages), a few hints are sufficient in order to grasp the method, and from there one may proceed according to need of new development. The best starting point is the logic of relations and the triadic doctrine of categories. These concepts lead in a rigorous way to hierarchical systems of sign-classification that very soon become general enough to encompass all conceivable types of signs. A very useful classification based on a twofold application of the triadic categories (with some "selection rules") gives ten classes of signs[15]. This is already too much for our purpose; as long as we are mainly interested in the way the sign refers to the object a single trichotomy is sufficient.

Peirce's triadic doctrine of categories is based on the observation that a network of relations is more than one dimension requires at least triadic relations. On the other hand it can be shown that relations of order higher than three always can be reduced to triadic relations. A sign is defined as a genuine triadic relation involving

1. A primary sign or sign vehicle.
2. An object.
3. An interpretant.

Peirce's epistemology is based on the idea that everything we can speak and think about may be classified as signs. This then is reflected in his ontology as an objectively idealistic, but realistic (anti-nominalistic) conception of the non-transcendent part of the universe as consisting of a network of sign relations. The three factors of the sign relation become ontological categories or modes of existence. Peirce calls the three categories 1, 2 and 3 (firstness, secondness, thirdness) and they are defined in the following way:

1. Independent existence.
2. Existence as reaction to a first.
3. Existence as mediation between a first and a second.

The three ontological categories may then be used to classify each of the three factors of the sign relation. This principle would lead to 27 (3^3) classes, but a simple selection rule reduces the number of classes to 10. However, in physics we are mainly interested in describing the reference to the object, and therefore we need only apply the categories to the second place in the sign relation whereby we get three classes of signs:

1. Icons.
2. Indices.
3. Symbols.

Icons refer to the object ty "likeness", as pictures, indices by "pointing", and symbols by context and conventions. The first two classes are "degenerate" sign relations where the interpretant is absent or latent; icons may even exist in the absence of an object, like music and non-figurative paintings.

When we apply these classes in physics, e.g. for describing an atomic particle, it goes like this: The bare concept of a particle is an icon, its state vector is an index, and the state vector's coordinates in various representations are symbols. Symbols are created by the concrete physical setting of an interpretant, i.e. a measuring apparatus. In quantum-semiotic the sign relation must be considered an irreducible concept, which means that symbols don't exist in nature in the absence of a physical sign relation including a measuring apparatus. The concept of hidden variables and localizable causes within an elementary sign relation is therefore not consistent with quantum-semiotic.

In order to describe what constitutes an elementary sign relation we can apply Peirce's categories in connection with the well known classification of systems:

1. Active systems (output independent of input).
2. Passive/reactive systems (output determined by input).
3. Dissipative systems (output partly passive/reactive and partly active).

The quantum mechanical fluctuation-dissipation (FD) theorem[16)] can be regarded as an indication of semiotic consistency of the linear Hilbert-space formulation of quantum mechanics. We cannot expect this limited theory to be able to describe in detail what happens when the reactive quantum system under the guidance of active, classical fields is connected to a dissipative registration device, but the FD-theorem ensures that the deterministic description is no longer valid and chance enters the picture. This is where the "collapse" sets in as a crude expression for the fact that a sufficient, but irreducible sign relation has been established. The semiotic description based on the three categories of systems above is in accordance with the description of the measurement-process given by J. Rayski[17)] who, however, points out that the second (passive/reactive) and the third (dissipative) component of the analysis often are hopelessly intermingled. The author has earlier shown[18)] how the FD-theorem for the zero-point noise in an idealized linear physical sign relation can be used for calculating waiting-time distributions for a descrete event - "the qualitative jump".

The FD-theorem is only a shadow of what goes on in the measurement process. A proper theory of measurement requires an extension of the formalism into the realm of dissipative processes where pure states and hamiltonians cease to be sufficient and have to by replaced by more general concepts, e.g. density matrices and super-operators. Recent approaches like the points stated by H. Primas at this conference[19)] indicate that a fruitful development is going on.

For the purpose of creating a minimal semantics a full theory of measurement is probably not necessary. We can safely believe in advance that any proper theory may build on well known physical mechanisms and not on ad hoc concepts of a pre-established non-local harmony, splitting universes, faster-than-light signals, or backwards causation, that owe their existence to difficulties in the interpretation of ordinary quantum mechanics. Quantum semiotic may help to clarify the situation around the applicability of Bell's inequalities by pointing to the irreducibility of the sign relation as the reason why causes cannot be localized and described with hidden variables below this level. This does not mean that local realism has to be abandoned, only those versions that build on symbolic hidden variables. Peirce's idea of synechism implies that irreducible physical sign relations are constituted by continuous connections through space. In other words: the previous concept of <u>localizability of causes</u> must be replaced with a new kind of local realism based on <u>continuity of interaction</u>.

CONTEXTUALITY AND BELL'S INEQUALITIES.

The original derivation of Bell's inequalities[20] is based on the notion of local causality through hidden variables and therefore not directly applicable in quantum semiotic. However, as shown by E. Santos[21], the inequalities follow from classical logic calculus and just require that the measured properties of one particle can be regarded as independent of the context of measurement for another, remote particle. If we can choose between measuring two non-compatible properties a_1 and b_1 for particle 1 and have a similar choice between properties a_2 and b_2 for particle 2, and if the a priori probabilities of all four properties is ½ then the condition of mutual non-contextuality between the two particles leads to the following version of Bell's inequalities that involves only coincidence probabilities:

$$1 \geq p(a_1 \wedge a_2) + p(a_1 \wedge b_2) + p(b_1 \wedge a_2) - p(b_1 \wedge b_2) \geq 0$$

This is the type of predictions that have been tested and found violated in Aspect's experiments[22]. However, the derivation of the inequalities presupposes that the measured properties of the two particles exist independently and that the coincidences are counted retrospectively by comparison of the measured single particle properties. This is not the case in any of the experiments that have violated Bell's inequality. All registrations that enter the statistics take place in "central black boxes" like coincidence counters. The Aspect experiments have thus shown that a proper irreducible sign relation can be set up with 6 meter long wires connecting the single detectors to a central registration device. In order to falsify synechism an experiment without these connections would be needed.

REFERENCES.

1. Favrholdt, D., Danish Yearbook of Philosophy, 13, 206 (1976).
2. Faye, J., "The Bohr-Høffding Relationship Reconsidered", Preprint (1987), Institute of Mathematics and Physics, Roskilde University Center, P.O. Box 260, DK 4000, Roskilde.
3. Jammer, M., The Conceptual Development of Quantum Mechanics, McGraw-Hill, London (1966), pp 172-173.
4. Witt-Hansen, J., Danish Yearbook of Philosophy, 17, 31 (1980).
5. Høffding, H., Erindringer, Gyldendal, København (1928), pp. 202-214.
6. Lieb, I. C., Charles S. Peirce's Letters to Lady Welby. New Haven, Conn. (1953), p. 39.
7. Petersen, Aa., Quantum Physics and the Philosophical Tradition. M.I.T. Press (1968), p. 188.
8. Peirce, C.S., The Monist, II, 321 (1892), pp. 328-329.
9. Einstein, A., Podolsky, B., and Rosen, N., Phys. Rev., 47, 777 (1935).
10. Rosenfeld, L., in Niels Bohr, J.H.Schultz Forlag, København (1964), p. 124.
11. Bohr, N., Phys. Rev. 48, 696 (1935).
12. J.S.Bell in an interview in the film "Atomfysik og Virkelighed" by Lars Becker-Larsen, Statens Filmcentral, København (1985).
13. Bohr, N., Atomfysik og Menneskelig Erkendelse, J.H. Schultz, København (1957), p. 76.
14. Peirce, C.S., The Monist, II, 534 (1892).
This paper, "The Law of Mind" as well as ref. 8, "The Doctrine of Necessity Examined" can be found (slightly abbreviated) in the "Collected Papers", and in: Buchler, J., Philosophical Writings of Peirce, Dover,

N.Y. (1955).

15. Short, T.L., Semiotica, <u>60</u>, 103 (1986).
16. Callen, H.B., and Welton, T.A., Phys. Rev. <u>83</u>, 34 (1951).
17. Rayski, J., <u>Proc. 2nd Int. Symp. Foundations of Quantum Mechanics</u>, Tokyo (1986), pp. 282-286.
18. Christiansen, P.V., Text no. 93 (1985), IMFUFA, Roskilde University Center.
19. Primas, H., "Contextual Quantum Objects and Their Ontic Interpretation", abstract/paper delivered at this conference, Joensuu, august 1987.
20. Bell, J.S., Physics, <u>1</u>, 195 (1965).
21. Santos, E., Phys. Letters, A <u>115</u>, 363, (1986).
22. Aspect, A., and Grangier, P., <u>Proc. Int. Symp. Foundations of Quantum Mechanics</u>, Tokyo (1983), pp. 214-224.

MATHEMATICAL ASPECTS OF THE UNCERTAINTY RELATION CONNECTED WITH ITS PHYSICAL INTERPRETATION - REVIEW

Marian Grabowski
Institute of Physics
Nicholas Copernicus University
Toruń, Poland

ABSTRACT

We give the survey of mathematical aspects of the uncertainty relation which says that the more we know about the position of a particle the less we know about its momentum. The application of the Heisenberg inequality to explain the stability of an atom is also discussed.

Since the discovery of the commutation relation $QP-PQ=i\hbar I$ and the corresponding uncertainty relation $\Delta P \Delta Q \geqslant \hbar/2$ up to these days they have received much of attention in the literature. The mathematics lying behind these ideas, their physical meaning, philosophical implications have been extensively studied. The uncertainty relation and its interpretation was the important part of the Copenhagen interpretation anniversary of which we celebrate now.

Coming back to the first Heisenberg paper [20] we can find a twofold explanation how to understand the uncertainty relation. Firstly he stated that complementary physical quantities like momentum P and position Q can simultaneously be measured only with the accuracy allowed by the uncertainty relation. This interpretation led to discussing the question of a quantum measurement and joined measurements of two non-commuting observables. The literature devoted to those problems is enormous and an attempt to give full information is hopeless. The first subject I would like to put aside and I can only mention its very detailed analysis made by Lahti and Busch [8].

Secondly Heisenberg says

the more precise the position is determined the less we know about momentum and vice versa. (I)

The uncertainty relation describes this fact by means of the variances - the measure of localization of position and momentum. We would like to present within the historical perspective how different mathematical frameworks have been worked out to express this property. Of course the various mathematical formulations of the above fact are only one stream in the investigations of commutation relation for position and momentum. We omit all problems connected with the different formulation of canonical commutation relation (Heisenberg, Weyl, Schrödinger) [36] the famous von Neumann theorem [36] and all questions concerning the Weyl group. Those problems are still alive. For instance the sharp inequalities for L^p-norms of matrix coefficients of square integrable representations of the Weyl group has been given [28], [42]. It appeared they are connected with the entropy [48], [28] and the usual form of the Heisenberg relation [19]. In turn, Cartier [10] showed that the representation of the canonical commutation relation on a two dimensional torus is closely related to the theta-function.

As a matter of fact all results presented below are deduced from the properties of the Fourier transform. To specify these properties we must introduce a few mathematical notions.

Let $\mathcal{H} = L^2(\mathbb{R}^3)$ be the Hilbert space of our quantum system. For simplicity we assume n=3 ($\hbar$=1 throughtout). Almost all results are true for arbitrary finite n. We need here the Banach space $L^p(\mathbb{R}^3)$. It is the space of p-integrable functions with the norm $\|\psi\|_p = \left(\int|\psi(x)|^p dx\right)^{1/p}$ for $0 < p < \infty$ and when $p = \infty$ $\|\psi\|_\infty = \text{ess sup}\{|\psi(x)|\,;\, x \in \mathbb{R}^3\}$. Because ψ is a state of quantum system $\|\psi\|_2 = 1$. The Fourier transform of ψ is defined as

$$(F\psi)(p) = \hat{\psi}(p) = (2\pi)^{-3/2}\int e^{-ipx}\psi(x)dx. \qquad (1)$$

Equation (1) appears in the Heisenberg paper [20] but it is recognized not as a Fourier transform but as an implication from Dirac-Jordan theory. In the famous von Neumann's book (1) is presented as the Fourier integral as well as in the Stone's one [44].

Now we enumerate the properties of (1) from which we will obtain different formulations of the Heisenberg inequality.

Property A. In $L^2(\mathbb{R}^3)$ the position operator is defined as follows

$$(Q\Psi)(x) = x\Psi(x) \tag{2}$$

on a suitable domain $D(Q)$. Momentum P is defined as differentiation of Ψ on a domain $D(P)$. Both are self-adjoint on properly defined domains (it was firstly showed in [30], [44]). The Fourier transform F is a unitary mapping of $L^2(\mathbb{R}^3)$ (Plancherel theorem [34], [37]). Position and momentum are unitary equivalent. This equivalence being established by the Fourier transform

$$P = F^{-1}QF \tag{3}$$

is present in both von Neumann's and in Stone's books [30], [44].

Property B. From the definition (1) and the Plancherel theorem F is a following map

$$\begin{aligned} F &: L^1(\mathbb{R}^3) \rightarrow L^\infty(\mathbb{R}^3), \\ F &: L^2(\mathbb{R}^3) \rightarrow L^2(\mathbb{R}^3), \end{aligned} \tag{4}$$

where $\|F\Psi\|_2 = \|\Psi\|_2$ and $\|F\Psi\|_\infty \leqslant (2\pi)^{-3/2}\|\Psi\|_1$ respectively for $\Psi \in L^1$ or $\Psi \in L^2$. Using the Riesz-Thorin theorem [37], [38] F can be extended as a linear bounded map from $L^p(\mathbb{R}^3)$ to $L^q(\mathbb{R}^3)$, where $1/q+1/p=1$, $q \geqslant 2$ and the norm of this extention is the following [3]

$$C(q,p) = \left(\frac{2\pi}{q}\right)^{3/2q}\left(\frac{2\pi}{p}\right)^{-3/2p}.$$

Property C. Let $\varphi \in L^2(\mathbb{R}^1)$ and let it has a bounded support $\operatorname{supp}\varphi = \{x;\ \varphi(x)\neq 0\} \subset [-A,A]$, where $A>0$. We can define

$$\hat{\varphi}(z) = (2\pi)^{-3/2}\int e^{izx}\varphi(x)\,dx, \tag{5}$$

for $z\in\mathbb{C}$. Because $\varphi\in L^2(\mathbb{R}^1)$ and has a bounded support thus $\varphi\in L^1(\mathbb{R}^1)$ the integral (5) is absolutely convergent. It is easily seen that (5) is uniformly convergent on each compact set of $\mathbb{C}$. Thus $\hat{\varphi}(z)$ is an analytical and entire function. This property is frequently referred to as the Payley-Wiener theorem. As a matter of fact it is an easier part of this theorem [37], [32], [46]. This property is true for arbitrary n.

These three properties give different measures by means of which we can express the restrictions on the localization of positions and momenta.

I. Standard Approach and Its Extension.

After considerations of the uncertainty principle for Gaussian functions by Heisenberg in his famous paper [20]* which were rather heuristic, Kennard [24], Robertson [39], Schrödinger [41] and others gave the precise description

*Prof. H.Primas remarked that R.Wiener had given a mathematically precise formulation of the variance-type time/frequency inequality at a Göttingen mathematics seminar in 1924, with Heisenberg present. Author would like to thank Prof. H.Primas for this interesting historical remark.

closely connected with the mathematical language of quantum mechanics. For the exellent historical review see [22]. Using the Schwarz inequality for two arbitrary observables A,B we have

$$\Delta A^2 \Delta B^2 \geqslant \frac{1}{4}\left|(\Psi,[A,B]\Psi)\right|^2 + \left(\langle AB+BA\rangle/2 - \langle A\rangle\langle B\rangle\right)^2, \quad (6)$$

where $\Delta A^2 = \left(\Psi, (A-\langle A\rangle)^2\Psi\right), \langle A\rangle = (\Psi, A\Psi)$. In particular for A=Q, B=P we have Heisenberg relation in $L^2(\mathbb{R}^n)$

$$\Delta P^2 \Delta Q^2 \geqslant \frac{n^2}{4}. \quad (7)$$

Dispersion ΔA^2 is a measure of uncertainty of the outcome of the measurement of A. For example ΔQ^2 may be taken as a measure of the concentration of $|\Psi(x)|^2$ in the neighbourhood of its mean. A small value of ΔQ^2 corresponds to the high degree of concentration. In the standard description (7) to state (I) we use the variances of P and Q as measures of uncertainty of localization of position and momentum of the particle.

To obtain (7) we only need property A and the Schwarz inequality. For the sake of completeness we present below the well-known calculations. Because of (3) we have $ip\hat{\Psi}(p) = (\hat{\Psi'})(p)$ for smooth and rapidly decreasing functions Ψ. The set of such functions is dense in L^2. Ψ' stands for the derivative of Ψ. To simplify the notation we restrict ourselves to $L^2(\mathbb{R}^1)$ and assume $\langle Q\rangle = \langle P\rangle = 0$.

$$\Delta Q^2 \Delta P^2 = \int |x\Psi(x)|^2 dx \int |p\hat{\Psi}(p)|^2 dp = \int |x\Psi(x)|^2 dx \int |\hat{\Psi}'(p)|^2 dp =$$

$$= \int |x\Psi(x)|^2 dx \int |\Psi'(p)|^2 dp \geqslant \left(\int |x\Psi'(x)\Psi^*(x)|\, dx\right)^2 \geqslant$$

$$\geqslant \left[\int \frac{x}{2}\left(\Psi'(x)\Psi^*(x) + \Psi'^*(x)\Psi(x)\right) dx\right]^2 = \frac{1}{4}\int x\frac{d}{dx}|\Psi(x)|^2 dx = \frac{1}{4}. \tag{8}$$

In the fourth step we use Plancherel theorem, in the fivth - Schwarz inequality. Equality in (8) holds for Gaussian functions.

After reintroducing $\langle Q\rangle, \langle P\rangle$ the set of these functions is parametrized with two variables $\langle Q\rangle, \langle P\rangle$ and now such functions are called coherent states. They have very interesting properties and were extensively studied [2], [12] as well as their generalization for arbitrary Lie groups [12], [33].

Equation (7) can be used conclusively in the interpretation of mutual restriction on ΔQ and ΔP only if two quantities are finite. Even when Ψ is a suitably normalized characteristic function of the unit ball, the uncertainty of momentum P is infinite. Roughly speaking the uncertainty of momentum or position can be infinite if the wave function or its Fourier transform oscillates very rapidly and sharply. For the states in $L^2(\mathbb{R}^n)$ the finiteness of ΔQ and ΔP demands restrictions on their vanishing rates at infinity and the vanishing rates of their Fourier transforms. For instance if a state Ψ is of polynomial form at infinity, that is of the form $\Psi \sim |x|^{-\alpha}$ as $|x| \to \infty$ finiteness of ΔQ requires $\alpha > (n+2)/2$. Similar restrictions are imposed on the transform of ψ to assure the finiteness of ΔP. In turn, elements of $L^2(\mathbb{R}^n)$ demand only $\alpha > n/2$. Price, Cowling [11] defined a family of measures of uncertainty which generalizes the usual notion

$$\Delta_{t,\theta} Q = \| |x - \langle Q\rangle|^{\theta} \psi \|_t, \tag{9a}$$

$$\Delta_{t,\theta} P = \| |p - \langle P\rangle|^{\theta} \hat{\psi} \|_t. \tag{9b}$$

Equations (9) are measures of the spreading of distributions about $\langle Q\rangle$ and $\langle P\rangle$. For $t=2$, $\theta=1$ we have the usual variances. Now, if $\psi \sim |x|^{-\alpha}$, $\alpha > n/2$ we can always find t, θ such that $\Delta_{t,\theta} Q < \infty$, namely those t, θ satisfying $1 \leqslant t \leqslant \infty$, $\theta > 0$ and $\alpha > \theta + n/t$. In [11] it has been shown that there exist constants C_1, C_2 such that for all states in L^2

$$\Delta_{t,\theta} Q + \Delta_{s,\varphi} P \geqslant C_1(s,t,\theta,\varphi) \tag{10}$$

if $\theta > 1/t^{\#}$, $\varphi > 1/s^{\#}$, where $x^{\#} = 2x/(x-2)$ or equivalently

$$\Delta_{t,\theta} Q^{\alpha} \Delta_{s,\varphi} P^{1-\alpha} \geqslant C_2(\alpha, s, t, \varphi, \theta), \tag{11}$$

where $\alpha(\theta - 1/t^{\#}) = (1-\alpha)(\varphi - 1/s^{\#})$. (10) and (11) generalize the classical uncertainty principle ($t=s=2$, $\varphi=\theta=1$) and the Hirschman inequality [21] ($t=s=2$, $\varphi,\theta>0$). (10) and (11) conserve the interpretation (I). The class of states is enlarged but unfortunately it implies a complication of the shape of the uncertainty relation. Moreover, the constants C_1, C_2 are not the best.

II.Entropic Uncertainty Relation

Białynicki-Birula and Mycielski [6] and independently Beckner [3] used the property B to obtain the entropic uncertainty relation for position and momentum. $C(p,q)$ - the norm of the Fourier transform from L^p to L^q is the smallest number for which the Hausdorff-Young inequality holds for all $\Psi \in L^p$

$$\|\hat{\Psi}\|_q \leq C(p,q)\|\Psi\|_p, \tag{12}$$

where $1/p+1/q=1$, $q \geq 2$. (12) can be rewritten in the form

$$W(q)=C(p,q)\|\Psi\|_p-\|\hat{\Psi}\|_q \geq 0. \tag{13}$$

$W(2)=0$ by Plancherel theorem. Because $W'(q)\big|_{q=2} \geq 0$ we have

$$S(|\Psi|^2)+S(|\hat{\Psi}|^2) \geq n\ln\pi e, \tag{14}$$

where

$$S(|\Psi|^2)=-\int|\Psi(x)|^2\ln|\Psi(x)|^2dx \tag{15}$$

represents the entropy of position and $S(|\hat{\Psi}|^2)$ entropy of momentum.

(15) is a new measure of the localization of a particle. It describes the lack of our information about the position. When the information increases the entropy will become negative. The probability distribution is more peaked. (14) says that the position and momentum cannot simultaneously be peaked. We see that (14) gives again (I) but by

means of new notion - the entropy.

(14) has an interesting history. It has been conjectured by Hirschman in 1957 [21] and by Everett [14] in his famous many-world interpretation of quantum mechanics. Leipnik [26] had tried to give the proof of (14) but his considerations had gap. The proof of (14) could be possible only after evaluating the best constant in the Hausdorff-Young inequality - the proof is due to Beckner [3].

An interesting fact is that (14) implies the usual version of the Heisenberg inequality (7). See [3], [6].

The notion (15) and its different generalizations were investigated in [17], [18]. It is a very useful concept because it gives the uncertainty relation for the pair of canonically conjugated observables not necessarily represented by self-adjoint operators such as angle - angular momentum [6], phase - number of quanta, time - energy [16], direction in space - L_z, L^2 [4]. Using the idea from [13], [31] the new reinterpretation of (14) has been given [4], [5].

III.Complementarity of Position and Momentum.

For each Borel set on the real line we define the operator in $L^2(\mathbb{R}^1)$ (an extension to arbitrary n is again simple)

$$(E_Q(\Delta)\psi)(x) = \chi_\Delta(x)\psi(x), \qquad (16)$$

where $\chi_\Delta(x)$ is a characteristic function of Δ. $E_Q(\Delta)$ is a projection operator. Via the spectral theorem E_Q is the spectral measure of Q. The probability of finding the particle in the state Ψ within Δ is the following

$$\mu_\Psi^Q(\Delta) = (\Psi, E_Q(\Delta)\Psi) = \int_\Delta |\Psi(x)|^2 dx. \tag{17}$$

In a similar way we can define E_P corresponding to the momentum P or simply $E_P(\Delta) = F^{-1}E_Q(\Delta)F$. (16) is a direct measure of localization of a particle within Δ in the interpretation scheme of quantum mechanics. If Ψ has the support within Δ then $\mu_\Psi^Q(\Delta) = 1$. We will say that the particle is localized within Δ. From property C we see that we cannot simultaneously localize the position and momentum of the particle with certainty. If so we could find bounded Δ, Σ such that $\mu_\Psi^Q(\Delta) = \mu_\Psi^P(\Sigma) = 1$ for some Ψ. It would mean that both Ψ and $\hat{\Psi}$ could have bounded supports which is impossible according to property C.

For any bounded Δ, Σ if $\mu_\Psi^Q(\Delta) = 1$ for some $\Psi \in L^2(\mathbb{R}^1)$ then $\mu_\Psi^P(\Sigma) < 1$ and conversely. From property C of the Fourier transform we deduce that we cannot localize the particle at the same time both in the coordinate space and in the momentum space. This fact is true for all bounded Δ, Σ.

The above described property is connected with the complementarity [7]. Recently Busch and Lahti [9] gave a very detailed analysis of this property for P,Q for arbitrary Δ, Σ.

We would like to stress here one fact. (17) is the most fundamental notion in quantum mechanics whereas ΔA^2 and (15) are notions built upon the density probabilities. E_P and E_Q are good notions to express the complementarity in the sense of mutual exclusiveness of the experimental arrangements that permit the unambigous operational definitions of position and momentum. But when we look for the measure of uncertainty of localization then for instance the entropy is a better quantity. Let we have a uniform distribution $\varphi(x) = \chi_\Delta(x)/\sqrt{\Delta}$. Then $\mu^Q_\varphi(\Delta) = 1$. This notion does not reflect a better or worse localization (small or big Δ), whereas $S(|\varphi|^2) = \ln\Delta$ and the entropy is sensitive to a better or worse localization (the particle becomes more or less "classical").

Thus property C gives only a very coarse nevertheless interesting description of (I) - states the impossibility of exact simultaneous localizations of the position and momentum of a particle.

IV. Local Heisenberg Inequalities.

There exist a large class of inequalities between the usual dispersion of position or momentum and various functions of others. They say that when the momentum uncertainty is small, not only is the position uncertainty large, but for example the probability of being localized at any particular point is small. The name "local" comes from Faris [15], who first gave the review of those methods.

Although all results in this section are only exercises in mathematics of Fourier transform the simplest way to prove them are the commutator inequalities [15], [27].

Let me allow demonstrate one simple example coming from Lèvy - Leblond [27]. We start from $[P,F(Q)]=-iF'(Q)$ and owing to (6)

$$\Delta P^2 \Delta F^2 \geqslant \frac{1}{4}|\langle F'\rangle|^2. \tag{18}$$

As F we take the Heaviside function

$$F(x) = \begin{cases} 0, & x < x_0 \\ 1, & x > x_0 \end{cases}.$$

The averange values are the following

$$\langle F^2\rangle = \langle F\rangle = \int_{x_0}^{\infty} |\Psi(x)|^2 dx = 1 - P(x_0), \tag{19}$$

where Ψ is the state of quantum system and $P(x_0)$ is the probability of localization of particle on the half-axis $(-\infty, x_0)$.

$$\langle F'\rangle = \langle \delta(Q-x_0)\rangle = |\Psi(x_0)|^2. \tag{20}$$

This step can be made in a mathematically rigorous way [15]. Finally we have the inequality

$$|\Psi(x)|^2 \leqslant 2\Delta P \left\{P(x)\left[1 - P(x)\right]\right\}^{1/2}. \tag{21}$$

Because $P(1-P) \leqslant 1/4$ we have

$$|\Psi(x)|^2 \leqslant \Delta P. \tag{22}$$

If the value of Ψ at some point x is large the dispersion of momentum must be large too. Below we list a few inequalities of the type (21).

i/ If φ has a compact support

$$\operatorname{supp}\varphi \Delta P \geqslant \pi/2 \ \text{(Lèvy - Leblond)}. \tag{23}$$

ii/ Uffink and Hilgevoord defined the overall width in position at level α [47]

$$W_\alpha(Q) = \min\left\{ |x_2 - x_1| ; \int_{x_1}^{x_2} |\Psi(x)|^2 dx = \alpha \right\} \tag{24}$$

for which they proved the inequality

$$W_\alpha(Q)\Delta P \geqslant \alpha. \tag{25}$$

iii/ Faris proved

$$\int_{|x-a|\leqslant b} |\Psi(x)|^2 dx \leqslant c(b\Delta P)^2, \tag{26}$$

where $c = 4/(n-2)^2$.

Other local inequalities can be found in [15], [27]. They all agree with the interpretation (I).

There are two sorts of local uncertainty principles closely connected with the stability of an atom.

iv/ Let's take $\mathcal{H} = L^2(\mathbb{R}^n)$. We define

$$I_\mu(\Psi) = \int r^\mu |\Psi(x)|^2 dx, \qquad k(\Psi) = \int |\nabla\Psi|^2 dx, \tag{27}$$

$|\nabla\Psi|^2 = \sum_{i=1}^{n} \left| \frac{\partial\Psi}{\partial x_i} \right|^2$ for sufficiently smooth functions, μ is real, x is a point in $\mathbb{R}^n$ and r denotes its distance from the origin. V.Bargmann was the first who showed by arguments very similar to the proof of the usual uncertainty relation the following inequality [1]

$$k(\Psi)\, I_{2\mu+2}(\Psi) \geqslant \left(\tfrac{1}{2}(n+\mu)\, I_{\mu}(\Psi)\right)^2. \tag{28}$$

For $\mu=0$ it is a form of an uncertainty relation. The case $\mu=-2$ was discussed earlier by Kato [23]. For $\mu=-1$ we have

$$\langle P^2 \rangle^{1/2} \geqslant \frac{n-1}{2} \langle \frac{1}{r} \rangle. \tag{29}$$

This inequality says that the states cannot be too concentrated near $r=0$ without forcing a large kinetic energy. Because always $\langle r^{-1}\rangle^{-1} \leqslant \langle r \rangle \leqslant \langle r^2 \rangle^{1/2}$ (29) is qualitatively stronger than the Heisenberg principle.

v/ Sobolev [43] provided the inequality the meaning of which is in the spirit of iv/. This inequality is strongly dimension-dependent. We write it for n=3

$$\langle P^2 \rangle \geqslant K_s \|\Psi\|_6^2, \tag{30}$$

where $K_s = 3(\pi/2)^{4/3}$. The rigorous derivation of (30) can be found in [40] and for arbitrary n in [45]. Proof of (30) is much more complicated than that of (7). Equation (30) gives a lower bound for the kinetic energy in terms of some integral of Ψ which does not involve derivatives. It says that if we try to compress a wave function anywhere then the kinetic energy will increase. iii/ and v/ are useful to prove no-binding theorem [15].

V. Heisenberg Inequality and the Stability of Atoms.

The Heisenberg relation (7) was frequently used in textbooks of quantum mechanics to explain the stability of an atom. Because there exist many missunderstandings here we discuss it in details.

We ask why the atom does not collapse? Why is it stable? To answer this question we must explain why the ground state energy of an atom is finite. Why

$$(\psi, H\psi) \geqslant E_0 \tag{31}$$

for some $E_0 > -\infty$, where H is the Hamiltonian of atom?

As an example we take the hydrogenic atom

$$H = -\tfrac{1}{2}\Delta - \frac{Z}{r}. \tag{32}$$

We use units in which $\hbar = 1$, $m=1$, $|e|=1$. The usual explanation of (31) for the Hamiltonian (32) is based on the Heisenberg relation $\langle P^2\rangle \langle Q^2\rangle \geqslant 9/4$, where $\langle P^2\rangle$ is the kinetic energy. The arguments are as follows: if the electron tries to get within a distance R of the nucleus, the kinetic energy is at least as large as R^{-2}. Thus $(\psi, H\psi) \geqslant \frac{1}{2}R^{-2} - \frac{Z}{R}$ and the last expression has minimum $-Z^2/2$ for $R=1/Z$.

This argument is false [15], [29]. We cite here a counterexample coming from Lieb [29]. Let's consider ψ consisting of two parts $\psi = \psi_1 + \psi_2$. ψ_1 is a narrow packet of radius R centered at the origin with $\int |\psi_1|^2 = 1/2$. ψ_2 is a spherically symmetric packet and has support in a narrow shell of mean radius L and $\int|\Psi_2|^2=1/2$. If L is large then roughly $\int|x|^2|\Psi(x)|^2dx\sim L^2/2$ whereas $\int\frac{1}{r}|\Psi|^2dx\sim\frac{1}{2R}$. Thus from the Heisenberg relation we conclude only that $\langle P^2\rangle\geq 9/2L^2$ and hence that $(\Psi,H\Psi)\geq 9/4L^2-Z/2R$. Now we can make E_0 an arbitrary small negative number taking $R\to 0$. In other words we can find states with the averange $\langle\frac{1}{r}\rangle$ conceivably large without $\langle Q^2\rangle$ being small.

But an equation (29) or (30) is enough to overcome this difficulty. For example using (29) we have

$$(\Psi,H\Psi)\geq\frac{1}{2}\left\langle\frac{1}{r}\right\rangle^2-Z\left\langle\frac{1}{r}\right\rangle\geq-\frac{Z^2}{2}. \tag{33}$$

By Sobolev - like inequality for fermions Lieb established the stability of bulk matter [29]. It is an important result because in our example there exists a solution of Schrödinger equation which gives $E_0>-\infty$, whereas for n-particle system such result does not exist.

We see that the study of Heisenberg relation even in our days is quite interesting from the mathematical point of view as well as from the physical one.

References

[1] Bargmann, V., Helv. Acta Phys. 45 249 (1972).

[2] Bargmann, V., Butera, P., Girandello, L., Klauder, J.R., Rep. Math. Phys. 2 221 (1971).

[3] Beckner, W., Ann. Math. 102 159 (1975).

[4] Białynicki-Birula, I., Madajczyk, J.L., Phys. Lett. 108A 384 (1985).

[5] Białynicki-Birula, I., Phys. Lett. 103A 253 (1984).
[6] Białynicki-Birula, I., Mycielski, J., Commun. Math. Phys. 44 29 (1975).
[7] Busch, P., Lahti, P.J., Philosophy of Science 52 64 (1985).
[8] Busch, P., Lahti, P.J., Phys. Rev. D 29 1634 (1984).
[9] " , Report Series No R87, Dep. of Phys. Sciences, University of Turku, Finland.
[10] Cartier, P., Proc. Sympos. Pure Math. 9 361 (1966).
[11] Cowling, M.G., Price, J.F., SIAM J. Math. Anal. 15 151 (1984).
[12] Davies, E.B., Quantum theory of open systems, Academic Press, London 1976.
[13] Deutch, D., Phys. Rev. Lett. 50 631 (1983).
[14] Everett, H.III, Rev. Mod. Phys. 29 454 (1957).
[15] Faris, W.G., J. Math. Phys. 19 461 (1978).
[16] Grabowski, M., Phys. Lett. A (to appear).
[17] " Inter. J. Theor. Phys. 17 635 (1978).
[18] " Rep. Math. Phys. 14 377 (1978).
[19] " " 20 153 (1984).
[20] Heisenberg, W., Z.Physik 43 172 (1927).
[21] Hirschman, I.I.Jr, Amer. J. Math. 79 152 (1957).
[22] Jammer, M., The philosophy of quantum mechanics, N.Y. John Wiley 1974.
[23] Kato, T., Trans. Am. Math. Soc. 70 195 (1951).
[24] Kennard, E.H., Z. Phys. 44 326 (1927).

[25] Lahti, P.J., Inter. J. Theor. Phys. 19 789 (1980).

[26] Leipnik, R., Information and control 2 64 (1959).

[27] Lèvy-Leblond, J-M., Phys. Lett. 111 353 (1985).

[28] Lieb, E.H., Commun. Math. Phys. 62 35 (1978).

[29] " Rev. Mod. Phys. 48 553 (1976).

[30] v.Neumann, J., Mathematishe Grundlagen der Quantenmechanik, Berlin, Springer 1932.

[31] Partovi, M.H., Phys. Rev. Lett. 50 1883 (1983).

[32] Payley, R., Wiener, N., Fourier transforms in the complex domain, Amer. Math. Soc. 1934.

[33] Perelomov, A.M., Commun. Math. Phys. 26 222 (1972).

[34] Plancherel, M., Rend. Circ. Math. Palermo 30 289(1910)

[35] Price, J.F., J. Math. Phys. 24 1711 (1983).

[36] Putnam, C.R., Commutation properties of Hilbert space operators and related topics, Springer Verlag, Berlin 1967.

[37] Reed, M., Simon, B., Metods of modern mathematical physics, Vol.II, Fourier analysis, self-adjointness, Academic Press, N.Y. 1975.

[38] Riesz, M., Acta Math. 49 465 (1926).

[39] Robertson, H.P., Phys. Rev. 34 163 (1929).

[40] Rosen, G., SIAM J. Appl. Math. 21 30 (1971).

[41] Schrödinger, E., Sitzungber. Preuss. Akad. Wiss. 296 (1930).

[42] Scutaru, H., preprint FT-167-19 Central Institute of Physics, Bucharest P.O.B. 5206 Romania.

[43] Sobolev, S.L., Math. Sb. 46 471 (1938).

[44] Stone, M.H., Linear transformations in Hilbert space and their applications to analysis, Amer. Math. Soc. N.Y. 1932.

[45] Talenti, G., Ann. Mat. Pura Appl. IV 110 353 (1976).

[46] Titchmarsh, E., Introduction to the theory of Fourier integrals, Oxford, Clarendon Press 1937.

[47] Uffink, J.B.M., Hilgevoord J.Phys. Lett. 95A 474 (1983).

[48] Wehrl, A., Rep. Math. Phys. 16 353 (1979).

REMARKS ON S. KOCHEN'S INTERPRETATION OF QUANTUM MECHANICS*)

Th. Görnitz and C. F. v. Weizsäcker
Arbeitsgruppe Afheldt in der Max-Planck-Gesellschaft
Bahnhofplatz 4
D-8130 Starnberg

Simon Kochen has offered a new interpretation of the quantum mechanical formalism. In contradistinction to the "orthodox" interpretation he calls it the "Perspective Interpretation" (Kochen 1985). We consider it as an illuminating clarification of the mathematical structure of the theory, especially apt to describe the measuring process. We would, however, feel that it means not an alternative but a continuation to the Copenhagen Interpretation (Bohr and, to some extend, Heisenberg).

The word "orthodox" seems to be as misleading in physics as in religion, since it tends to palliate differences of opinions within orthodoxy. One of the present authors (C. F. v. W.) has tried to present a continuation of the Copenhagen Interpretation in his recent book "Aufbau der Physik" (1985; henceforth quoted as "Aufbau"). In spite of some differences in expression, the two continuations seem to converge in several interesting points. Kochen's presentation is more mathematically oriented, "Aufbau" on the other hand contains a more complete physical description. Both presentations also converge in a common open question.

---

*) Work supported in part by Deutsche Forschungsgemeinschaft

1) Mathematical Structure

Kochen's mathematical presentation of quantum mechanics is destined to describe the measuring process. He considers two interacting systems S' and S with associated Hilbert spaces H' and H. Lét the sets $\{\tau_i\}$ and $\{\sigma_i\}$ be orthonormal bases on H' and H respectively. Then the Hilbert space of the combined system S'+ S has an associated Hilbert space $H' \otimes H$ in which any pure state is represented by a vector

$$\Phi = \Sigma \ \alpha_{ij} \ \tau_i \otimes \sigma_j \ .$$

$H' \otimes H$ is the vector space erected on a basis which is the cartesian product of the two bases in H'and H. In this traditional description the two basic sets $\{\tau_i\}$ and $\{\sigma_j\}$ can be chosen independently.

If we want to describe a measuring process in which, say, S is considered as the object and S'as the instrument of measurement, then for every state of the composed system there exists a unique pair of bases, such that every σ_i of S corresponds uniquely to one τ_i in S', and finding τ_i to be the state of S' after the measurement we can conclude that the state of S is σ_i.

This purpose is achieved in two steps. First the tensor product is described as the space of all bounded linear maps Φ: H --> H'. This coordinate-free definition is expressed in the coordinates τ_i and σ_j by the set of all matrices a_{ij}. Secondly, any given map Φ is decomposed according to the Polar Decomposition Theorem for linear transformations which was first formulated by E. Schmidt (1906, 1907). The same decomposition was used by J. v. Neumann (1932), chapter VI. There is a unique decomposition

$$\Phi = U \ V,$$

where

$$V: H \text{ --> } H$$

is a positive Hermitean operator and

$$U: H \longrightarrow H'$$

is a (partial) isometry with domain the image of V. Φ will in general not be diagonalizable, but V is. U then defines the corresponding bases $\{\varnothing_i\}$ and $\{æ_i\}$ in H and H' such that, with respect to them, Φ is diagonal:

$$\Phi = \Sigma \quad \mu_i \ \varnothing_i \otimes æ_i$$

Thus a given pure state of the composed system S+S' always defines an unique pair of corresponding bases in S and S' which permit considering a state $æ_i$ in S' as a measurement of the corresponding state $\varnothing_i$ in S.

Kochen further shows that in a measurement process $\Phi(t)$ can be continuously decomposed into $\varnothing_i(t)$ and $æ_i(t)$ such that after the end of the interaction, when the two systems are far from each other, $\Sigma |\mu_i|^2 P(æ_i)$ can be considered as a diagonalized density matrix of the measuring instrument. An observation of the state of the instrument will then give $æ_i$ with the classical probability $|\mu_i|^2$, thus automatically fulfilling Bohr's condition that the result of the measurement ought to be describable in classical terms.

2. An Analogous Mathematical Description of Measurement

In Aufbau (chapter II.2d, pp. 531 - 534) the same problem is treated in a somewhat different manner. We attempt to describe the measurement of an observable L belonging to the object X_1 by an instrument X_2 . We assume that L has a simple spectrum and that the measurement takes place by an interaction around some time t_0 . Then the Hamiltonian of the composed system will be

$$H = H_0 + H_I$$

with

$$H_0 = H_1 \otimes id_2 + id_1 \otimes H_2$$

as the Hamiltonian of the two free motions, and H_I the interaction which is large only in a time span $\triangle t$ around t_0. For simplicity, we assume that $H_0 \ll H_I$ during $\triangle t$. id_i is the identity operator in the Hilbert space of X_i.

Let $p(\lambda)$ be the probability of finding the eigenvalue λ of L in the measurement. Then there must exist a property of X_2 which corresponds to λ; we call it μ_λ. The probability of finding μ_λ if we observe X_2 by a second instrument X_3 ought then to be $p(\mu_\lambda) = p(\lambda)$.

In the general case there will be a subspace of the Hilbert space of X_2 corresponding to μ_λ instead of only a single vector. Let the coset space of this space with respect to this equivalence be de-noted by $H(X_2)$, only this space is of interest for the measuring process. The coset space is formed by a projection and therefore it is a subspace of the original Hilbert space. The correspondence between λ and μ_λ is then one-to-one and is specific for the given measuring process and the instrument.

Let the state $|x_b\rangle$ of the composed object before t_0 be a product state:

$$|x_b\rangle = |x_{b1}\rangle|x_{b2}\rangle \in H(X_1) \otimes H(X_2) \qquad (1)$$

The eigenvectors $|\lambda_i\rangle$ of L define a basis in the Hilbert space $H(X_1)$:

$$L|\lambda_i\rangle = \lambda_i\ |\lambda_i\rangle$$

therefore $|x_{b1}\rangle$ can be developed with respect to this basis:

$$|x_{b1}\rangle = \Sigma\ a_i\ |\lambda_i\rangle$$

and

$$p(\lambda_i) = |a_i|^2 .$$

To the states $|\lambda_i\rangle$ there correspond unique states $|\mu_i\rangle$ which also constitute a basis, in $H(X_2)$ of course.

The action of the operator L on $|x_{b1}\rangle$ can be represented by

$$L\ |x_{b1}\rangle = \Sigma\ a_i\ \lambda_i\ |\lambda_i\rangle$$

If we want to regard L as a "Hamiltonian" then the $p(\lambda)$

remain unchanged by the induced time evolution:

$$\exp(iLt)\ |x_{b1}\rangle = \Sigma\ a_i \exp(i\lambda_i t)\ |\lambda_i\rangle$$

and

$$|a_i| = |a_i \exp(i\lambda_i t)|$$

The initial state (1) is given already in its polar decomposition. On the other hand $|x_b\rangle$ has to change

$$|x_b\rangle \dashrightarrow |x(t)\rangle$$

such that after $\triangle t$ it becomes $|x_f\rangle$:

$$|x_f\rangle = \Sigma\ a_i\ |\lambda_i\rangle\ |\mu_i\rangle$$

Now we see which form the measuring interaction should have. If $|x_{b1}\rangle$ is already an eigenstate $|\lambda_i\rangle$ of L, then it should not be changed during the measuring process and $|x_{b2}\rangle$ must be twisted into $|\mu_i\rangle$. Therefore for each i there has to exist a unitary operator V_i in $H(X_2)$ which performs this operation. Since this must happen for each i, V_i depends on the choice of $|x_{b2}\rangle$.

We can connect this fact with the description of measurement given by Primas (1981, 1987). If we treat the environment of the system $X = X_1 + X_2$ as classical, then the dependence of V_i on $|x_{b2}\rangle$ means a non-linear equation for the development of X. The non-linearity might be reduced to the strictly linear Schrödinger equation by including the environment into the quantum description - strictly speaking including the universe. In the present paper, however, we make no explicit interpretational use of this remark.

Let

$$L = \Sigma\ \lambda_i\ P_i$$

be the spectral decomposition of L. Then for a fixed initial state of the instrument and for any state of the object the time development for the measuring procedure can be given by the unitary operator

$$U(t_0+\delta t) = U(\delta t) = \Sigma_i \exp(i\lambda_i\delta t)P_i \otimes V_i(\delta t)$$

with

$$0 \leq \delta t \leq \triangle t.$$

For each operator V_i there is a spectral decomposition

$$V_i(\delta t) = \Sigma_k \exp(i\sigma_{ik}\delta t)\pi_{ik}$$

with projection operators π_{ik} from $H(X_2)$. So we get

$$U(\delta t) = \Sigma_i \Sigma_k \exp[i(\lambda_i+\sigma_k)\delta t] P_i \otimes \pi_{ik}$$

The Hamiltonian H_I of the interaction is therefore

$$H_I = \Sigma_i \Sigma_k (\lambda_i+\sigma_{ik}) P_i \otimes \pi_{ik}$$

This operator can be seen as an abstract description of an operator which would be called a local operator if P and π were projection operators on the position of a particle. Precisely if two particles interact only when they are at the same place their interaction will provide us with a position measurement

$$H(x_0) = \int dx\, dy \cdots \delta(x_0 - x)\delta(x - y)$$

Now we propose the working hypothesis that Kochen's and our description will turn out to be identical in principle.

The polar decomposition of Ø is unique. Then there must exist a Hamiltonian for the measuring process which provides such a decomposition. Our Hamiltonian provides a decomposition which corresponds to the physical idea of the measurement. Hence it ought to be at least an example for the class of Hamiltonians he needs. However, we shall not carry out this analysis here in detail.

3. Interpretation of the Measuring Process

The Copenhagen Interpretation may be condensed into the phrase:

> ***"If we accept that quantum theory describes human knowledge there is no problem. If we do not accept this there is no solution."***

"Knowledge" here means knowledge of somebody about something, not fantasy or a "subjective state of mind".

This refers especially to the state vector. It expresses the probabilities of all possible results in all possible measurements at the object. The key concept here is probability (Aufbau ch.3). We define probability as the prediction of a relative frequency. This definition refers explicitly to the structure of time (chapter 2) as expressed in the difference of the "now", the factual past and the possible future. This structure is a presupposition of any possible meaning of the word "experience", e.g. defining experience as having learned from the past for the future. This "modes of time" are presupposed by any empirical science. On the other hand they cannot be deduced from physics in which time is treated as a real variable, not even the Second law of thermodynamics (chapter 4). This should not be surprising. All fundamental laws of physics are assumed to be universally valid, i.e. valid at any time; hence they cannot possibly define a special moment of time as "now".

The so-called "collapse of the wave function" is no problem at all in this view and - in our view - permits no description at all if we discard probability as meaning knowledge. Any new observation will change our probabilities. This is so in classical physics already and refers to any description of knowledge on an objective world in time. As an example, let us cast two good dice, one after the other. Before the first cast, the probability to find "twelve" as the sum of their eyes will be 1/36. After the first cast it will be 1/6, if there was "six" on one die, and zero if another number had appeared. The difference in quantum theory is only that the assumption of the result being objectively predetermined does not seem permissible any more (see however Aufbau, chapter 13).

This situation is not changed by Kochen's analysis. $\Sigma|\mu_i|^2P(\ae_i)$ is a diagonalized density matrix after the measuring interaction. This means only that $|\mu_i|^2$ can now be treated according to the classical calculus of probability. The density matrix of the instrument must be changed when it becomes known which eigenvalue i (μ_λ in our language) of the measuring observable is present in the instrument. This newly acquired knowledge changes ("collapses") the probability, i.e. here the classically understandable wave function of the instrument. There is no change as compared to the Copenhagen Interpretation.

In our temporal description, Bohr's statement that observation must be described in classical terms permits a simple interpretation. An observation states a fact. A fact, when stated, belongs to the past and is irreversible. The irreversibility of the numerical result is approximately warranted by the immense complexity of any macroscopic instrument. The difficulty which many authors have found in the quantum theory of measurement seem to be derived from the fact that it is impossible to describe an irreversible process <u>as irreversible</u> by a unitary transformation in Hilbert space. But this is not requested anyhow. An irreversible process <u>as irreversible</u> means loss of knowledge by the observer (Aufbau chapter 11,2cβ, pp. 523 - 526). Statistical thermodynamics shows that the reverse process is not impossible but, given the knowledge we posses after the rough classical observation, equipped with an extremely small probability; <u>we</u> have no reason to expect its happening.

<u>4. The Perspective Interpretation</u>

While thus not shaking the Copenhagen description of human knowledge in time, Kochen's Perspective Interpretation opens up a most interesting vista of unexploited philosophical possibilities. They are connected with the idea of a quantum theory describing the knower himself (Aufbau chapter 11,2e; pp. 535 -538, chs. 12 - 14).

Bohr's description of observations, or of "phenomena", to use his later term (Bohr 1935) (Aufbau chapter 11, 2cΓ, pp.526 - 531) rests precisely on a sober distinction between the knower and the known. For the sake of methodological clarity he precisely excludes any quantum description of the observer himself, let alone the observer's "mind" or "consciousness". Consciousness is just <u>not</u> introduced into the class of the objects of physics by Bohr; physics, only, is considered by him consistently as something <u>we</u> know, i.e. within the frame of what we can possibly know, not beyond that frame. "What is a chair for a divine spirit? We don't know. But for us it is a solid body on which we can sit." (This is not a literal quotation from Bohr, but, hopefully, expressing his epistemology.)

But evolutionary biology has made it increasingly probable that humankind is part of nature like anything else and might well admit a description by physics. Then one would have to go beyond the limits which Bohr so carefully respected. One would have to ponder first on whether a phenomenon like consciousness might permit a description within the frame of quantum theory. We suppose this to be possible. We reconstruct "abstract quantum theory" as a theory on probability predictions referring to <u>any</u> empirically decidable alternative (Aufbau chapter 8). If I can empirically (e.g. by introspection) distinguish between two different states of my own mind, this is an empirically decidable alternative and might admit the application of abstract quantum theory. Quantum theory is not a special theory on atoms, on particles and waves, ect. Quantum theory is not a microscopic theory; only classical theory is a macroscopic theory. But quantum theory may tentatively assumed to be universal.

The Perspective Interpretation is a step in this direction. As Kochen points out, in his picture there is a symmetry between the two systems S and S'. Either one may be described as "observing" the other. Any object in the world may be considered as being "perceived" by the rest of the world, and vice versa. The symmetry is not quite complete. Mathematically, U is a <u>partial</u> isometry, mapping part of one Hilbert space on the null vector of the other one. Physically, the irreversibility needed for defining fixed facts of the past rests on the complexity of the instrument, while the object needs not to be complex, thereby not necessarily possessing the ability to memorize results.

The open question in Kochen's approach is just connected with the act of perception, of becoming aware. If a system S' can be aware of its own state, then it can conclude what was (or is) the state of the S it observed. But to become aware that the state, say $æ_i$, means that it observes itself and thereby changes its density matrix $\Sigma |\mu_i|^2 P(æ_i)$ into one single state $æ_i$ which was only predictable with probability $|\mu_i|^2$. The inner dynamics of consciousness cannot simply be replaced by assuming that consciousness knows itself. This was clearly before Bohr's mind. The difficult meaning of self-knowledge was perhaps his main example for complementarity. But the way towards an analysis of these problems within quantum theory is open.

The difference between Kochen's and the orthodox interpretation begins only here. J. v. Neumann (1932, chapter VI) distinguishes three systems: I the object, II the instrument, III the observer. The classical description of measurement is for him the observation of I by II+III, the quantum description the observation of I+II by III. The "collapse" of the wave function happens between I and II in the first description, between II and III in the second one. Kochen differs from both descriptions. Page 160 - 161 he says: "We now take our major step in the new interpretation. In place of an official human observer, we assume that each system acts as a witness to the state of the other. When the polar decomposition assigns the mixed states

$$\Sigma \lambda_i^2 P_i \quad \text{and} \quad \Sigma \lambda_i^2 P_i'$$

to the two interacting Systems S and S', then S actually has exactly one of the properties P_i as witnessed by S', and S' has the corresponding property P_i' as witnessed by S. In other words, if

$$\Phi = \Sigma \lambda_i \, \emptyset_i \otimes \Psi_i \,,$$

the S and S' are in one of the corresponding states Ψ_i and $\emptyset_i$." Page 165 he considers two different decompositions of the total system into an observed system S and two "observing systems" S_1 and S_2, and the merely description transition of one of these decompositions into the other. This transition will in general change the state of S. He concludes: "Thus the collapse of the wave function is not a real physical effect in this interpretation but is simply the result of a change in perspective from one witnessing to another."

This description of the "collapse" does not, however, make the real difference explicit. This difference lies in the statement that "S actually has exactly one of their properties P_i as witnessed by S'". This is an ontological assumption which explicitly differs from Bohr's and v. Neumann's epistemological caution of not making arbitrary assertions which cannot possibly be verified or falsified by experiment that would permit a description by the only available theory, which is quantum theory. From Kochen's paper it is not evident whether he assumes that S actually jumps to and fro between its possible states, or whether its state remains fixed and only continually changing with time like a classical magnitude.

We believe that a satisfactory description would at first of all need some abstract but intelligible model of a possible quantum description of the process of self-awareness of an observer; without that we remain in the field of mere undiscussed and perhaps empty assertions.

Acknowledgement

One of us (Th. G.) is grateful to the Deutsche Forschungsgemeinschaft for financial support.

References

Bohr, N. (1935): Phys. Rev. 48, 696 - 702

Kochen, S. (1985): A new interpretation of quantum physics, in: Symp. on the foundations of modern physics; P. Lahti and P. Mittelstaedt (eds.) World Scientific

Neumann, J. v. (1932) Die mathematischen Grundlagen der Quantentheorie, Springer Verl. Berlin

Primas, H. (1981): Chemistry, Quantum Mechanics and Reductionism, Springer Verl. Berlin

Primas, H. (1987): Contextual Quantum Objects and their Ontic Interpretation, to app. in Symp. on the foundations of modern physics; P. Lahti and P. Mittelstaedt (eds.) World Scientific

Schmidt, E. (1906): Entwicklung willkürlicher Funktionen nach Systemen vorgeschriebener, Math. Annal. 63, 433

Schmidt, E. (1907): Auflösung der allg. Integralgleichung, Mathematische Annalen 64, 161

Weizsäcker, C. F. v. (1985): Aufbau der Physik, Hanser, München

A QUANTITATIVE CRITERION OF THE APPLICABILITY OF THE CLASSICAL DESCRIPTION WITHIN THE QUANTUM THEORY

L.A.Khalfin, B.S.Tsirelson

Steklov Mathematical Institute, Leningrad D-11, USSR

1. INTRODUCTION

The Copenhagen interpretation stipulates for a joint application of the two description modes, the classical and the quantum ones, to the physical world, which apparently is "connected" in the sense that it cannot be naturally (from the nature of things) divided into two realms, woll-separated with some gap, one being covered with the quantum description, the other - with the classical one. The Copenhagen interpretation specifies an interface between the application domains (not "applicability domains"!) of the two description modes; being sharp, this interface is conventional and movable. It can be shifted within the limits of a "neutral domain", where both description modes are applicable and conform; this is nothing but the correspondence principle for the quantum and classical theories. The "neutral domain" is supposed to be roomy enough to contain all physical phenomena immediately perceivable by human observers and also all those which can be successfully described by classical physics and treated as a part of the classically described macroscopic environment.

The physical contents of the correspondence principle is obviously connected with the roominess of the "neutral area". Is it really so vast as indicated above? That was declared rather than proved. Two main approaches are known to the problem of the quantum-classical correspondence: "the quasi--classical limit for $h \to 0$" and the Everett's approach.

The first approach seems thorough, being based on Hamiltonian mechanics and on the short-wave asymptotics for the Schrödinger equation. However, Hamiltonian dynamics can turn (rather quickly, during the so-called Ehrenfest time) a microscopically localized wave function into a macroscopically delocalized one. Indeed, fast divergence of trajectories is ordinary for non-well-integrable systems of classical mechanics.

The second (Everett's) approach obtains the correspondence principle from a certain scenario of the quantum correlation propagation. However, the mentioned scenario is over--simplified. The correlation propagation is described by highly abstract models for interactions; their correspondence to more realistic models remains problematic, as well as quantitative criteria.

We propose a new approach to the problem of the quantum-classical correspondence, combining features of the two mentioned approaches and by virtue of that free from their drawbacks pointed above. Our approach uses Hamiltonian mechanics and the short-wave asymptotics for the Schrödinger equation, and also the quantum correlation propagation. The former saves us from over-simplifying the interactions description (our description is realistic in some cases, but, of course, not in all cases), the latter - from the delocalization of wave functions.

We have arrived at this approach when developing investigations of Bell-type inequalities for quasi-classical systems, which can be described by localized wave functions moving along the classical trajectories; in this case the quantum theory provides small corrections to classical predictions. It was natural to suppose that there are quantum corrections to the Bell inequalities too. Just so was this problem treated in our previous paper[1]. It was shown there that the maximum value of the possible violation of the Bell-CHSH inequality for a quasi-classical system depends essentially on such factors as the accuracy of measurements and the thermal fluctuations intensity; these factors introduce discrepancies. It was shown that in the absence of discrepancies the Bell-CHSH inequality may be violated for quasi-classical systems to the same extent as for spin systems. In the opposite case, when the quantum uncertainties are small in comparison with the discrepancies, we have obtained an upper bound vor violations of the Bell-CHSH inequality showing that these violations are small. However, no lower bound was given. Further investigations presented here led us to an unexpected result. It turns out that, if the discrepancies exceed the quantum uncertainties, then a classical description with hidden variables is possible and therefore the Bell inequalities hold exactly, without any corrections. (All hidden variables are assumed local throughout this paper).

Our approach inherits from investigations of Bell inequalities the idea that the informal notion of the possibility of a classical description may be formalized as the possibility of a description with hidden variables. And hidden variables are treated here somewhat non-traditionally; instead of a continuous-in-time description, we obtain a "dotted-in-time" one. The continuous evolution of the quantum state of a quasi-classical system, weakly interacting with a heat bath, is accompanied (in the sense explained throughout this paper) by a discrete-in-time evolution of the classical state. The time pitch is nothing but the characteristic time for the process of establishing quantum correlations between the system and the bath. And it is determined by the thermal fluctuations intensity. This time can be estimated easily by means of the method presented here and involving fluctuation-dissipation relations. Several examples are given. For macroscopical systems belonging to the above-mentioned "neutral domain" the time pitch turns out to be small, and the sequence of classical states turns out to approximate the classical trajectory.

The mathematical examination of the presented approach is due to B.S.Tsirelson. The general outlines were formed in the process of authors' collaboration.

2. QUANTUM MEASUREMENTS OF LIMITED ACCURACY AND BELL INEQUALITIES

Apparently there is no accepted rigorous definition of the quantum measurement accuracy. We accept that a measurement of a coordinate with an accuracy Δq is equivalent (of course, in the outcomes' distribution, not in affecting the object) to the exact coordinate measurement followed by adding $\zeta\Delta q$ to the outcome; here ζ is a normally distributed random variable with zero mean and unity variance. A measurement of a momentum with an accuracy Δp is understood similarly. We want to realize, how the possibility of the Bell inequality violation is related to $\Delta q\Delta p$. This problem was considered in our paper[1]; here we obtain essentially more complete results by virtue of use of the "measurement compatibility" notion. Measurements of limited accuracy are, of course, not ideal ones, therefore it is better to avail of the general treatment of a quantum measurement (see for example[2]) as a transformation of quantum states into classical ones:

$$M : L_1(H) \rightarrowtail L_1(X);$$

H is the Hilbert space, describing the quantum system; $L_1(H)$ denotes the Banach space of all Hermitian trace-class operators in H; X is a measurable space describing possible measurement outcomes; $L_1(X)$ denotes the Banach space of all measures on X having densities with respect to some prescribed "background" measure (or, what is the same, the space of all functions integrable with respect to this "background" measure). The map M, describing a measurement, must be linear, positive and normalization preserving; that is, if $W \in L_1(H)$, $f \in L_1(X)$ and $f=M(W)$, then $Tr(W)=\int f$ and $W \geq 0 \Rightarrow f \geq 0$. We set aside some possible mathematical refinements related to the infinite dimension, because our conclusions are in fact insensitive to them. The conjugate transformation $M^*: L_\infty(X) \rightarrowtail L_\infty(H)$ maps the commutative algebra $L_\infty(X)$ of classical observables into the non-commutative C^*-algebra $L_\infty(H)$ of quantum observables. For the measurement outcome the probability of hitting a subset Y of X equals $\int_Y \rho$, where $\rho = M(W)$, W is the density matrix; this probability can be written as $\langle M^*(\mathbb{1}_Y)\rangle_W = \mathrm{Tr}(W \cdot M^*(\mathbb{1}_Y))$, where $\mathbb{1}_Y$ is the indicator of the event Y (the function that equals 1 on Y and 0 out of Y). The possibility of a violation of the Bell-CHSH inequality (one of the family of Bell inequalities) for quantum objects means in this terms that there exist two Hilbert spaces H_1, H_2, four measurements $M_{nm}: L_1(H_n) \rightarrowtail L_1(X_{nm})$ $(1 \le n \le 2, 1 \le m \le 2)$, four classical observables $f_{nm} \in L_\infty(X_{nm})$ taking values ±1 only, and a density matrix on the tensor product $W \in L_1(H_1 \otimes H_2)$ such that

$$| \langle A_1 \otimes B_1 + A_1 \otimes B_2 + A_2 \otimes B_1 - A_2 \otimes B_2 \rangle_W | > 2,$$

where $A_k = M^*_{1k}(f_{1k}) \in L_\infty(H_1)$, $B_l = M^*_{2l}(f_{2l}) \in L_\infty(H_2)$. The existence of such W is equivalent to the relation

$$\| A_1 \otimes B_1 + A_1 \otimes B_2 + A_2 \otimes B_1 - A_2 \otimes B_2 \| > 2,$$

which can be written as

$$\| \Sigma_{kl}\, \tau_{kl} M^*_{1k}(f_{1k}) \otimes M^*_{2l}(f_{2l}) \| > 2;$$

here and in what follows $\tau_{11} = \tau_{12} = \tau_{21} = +1$, $\tau_{22} = -1$. In the case of arbitrary measurements M_{nm} the above norm can be equal to $2\sqrt{2}$, but cannot exceed this boundary (see [1]).

2.1. Definition. Two measuremets $M_k: L_1(H) \rightarrowtail L_1(X_k)$, k=1,2, are *compatible*, if there is a commutative diagram

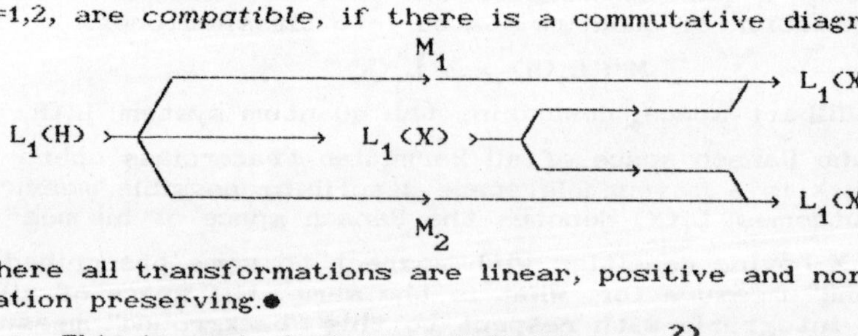

where all transformations are linear, positive and normalization preserving.●

This is equivalent to a definition in [2] (see Ch. 2, § 6). The Bell inequalities can result from the measurements compatibility.

2.2. Theorem. Let H_1, H_2 be Hilbert spaces and $M_k: L_1(H_1) \rightarrowtail L_1(X_k)$ are compatible measurements (k=1,2). Then for every $f_k \in L_\infty(X_k)$ such that $\|f_k\| \le 1$ and every $B_l \in L_\infty(H_2)$ such that $\|B_l\| \le 1$ $(l=1,2)$

$$\|\Sigma_{kl}\ \tau_{kl} M_k^*(f_k) \otimes B_l\| \le 2.$$

Proof. By virtue of the compatibility, the two measurements M_l can be replaced by one measurement $M : L_1(H_1) \rightarrowtail L_1(X)$, two functions f_k being transferred to X (see the diagram in 2.1). It is sufficient to prove that $\| \Sigma \tau_{kl} M^*(f_k) \otimes B_l \| \le 2$ for $f_k \in L_\infty(X)$, $\|f_k\| \le 1$. For any $x \in X$ $\| \Sigma \tau_{kl} f_k(x) B_l \| =$ $= \| f_1(x)B_1 + f_1(x)B_2 + f_2(x)B_1 - f_2(x)B_2 \| \le |f_1(x)+f_2(x)| \cdot \|B_1\| +$ $+ |f_1(x)-f_2(x)| \cdot \|B_2\| \le 2$, hence $\| \Sigma \tau_{kl} f_k \otimes B_l \| \le 2$. It only remains to prove that the transformation $M^* \otimes \mathbb{1} : L_\infty(X) \otimes L_\infty(H_2) \rightarrowtail$ $\rightarrowtail L_\infty(H_1) \otimes L_\infty(H_2) = L_\infty(H_1 \otimes H_2)$ is positive (then it cannot increase the norm, because it preserves the identity operator). And this follows from the Stinespring's[3] Theorem 4, according to which the positive transformation M^* with a commutative domain is completely positive.●

The proved theorem shows that $\|\Sigma \tau_{kl} M_{1k}^*(f_{1k}) \otimes M_{2l}^*(f_{2l})\| \le 2$, that is, the Bell-CHSH inequality holds provided at least one of two measurement pairs $\{M_{11}, M_{12}\}$, $\{M_{21}, M_{22}\}$ satisfies the compatibility condition. Comparing the above proof with the classical proof of the Bell inequality we see that a point x of X is nothing but a hidden variable.

Armed with the mathematical tools presented above, we return to measurements of limited accuracy. The measurement of a coordinate with an accuracy Δq is the transformation $L_1(H) \rightarrowtail L_1(-\infty,+\infty)$ which transforms W into ρ according to the equation

$$\rho(q) = \frac{1}{\sqrt{2\pi} \cdot \Delta q} \left\langle \exp\left[-\frac{1}{2}\left(\frac{Q-q}{\Delta q}\right)^2\right] \right\rangle_W ;$$

here Q is the coordinate operator. And the same holds for a momentum. It is known[2] that these measurements are compatible provided $\Delta q \Delta p \ge \hbar/2$. We see that the violation of Bell inequalities in a spinless coordinate-momentum Gedankenexperiment, discussed in [1], Sect. 3, requires the measurement accuracy high enough. In this case <u>inadequate accuracy cannot be compensated with any reiterations of the experiment</u>.

3. OPERATIONS DESTROYING QUANTUM CORRELATIONS

The quantum theory asserts that Bell inequalities can be violated in an experiment with two particles or, more generally, two quantum objects, jointly prepared with one source and then separately measured with two instruments. And it is of principle that the quantum object transmission from the source to the instrument cannot be replaced with any classically describable signal transmission whatsoever (since otherwise the classical proof of Bell inequalities would have been applicable). On the other hand, in quasi-classical situations quantum states can be approximated by classical ones. Can this reason be used to obtain Bell inequalities?

Any quantum object transmission can be approximated to a certain extent by a classical signal transmission according to the following plan. At first, one subjects the quantum object to a quantum measurement, getting some aggregate of numbers (or, more generally, some classical state). Then one prepares a quantum state (of the same or another but identical object), using the above numbers as classical parameters affecting the prepared state. As it were, the classical system absorbs and re-emits the quantum object. An operation which transforms one quantum state into another according to such plan will be called here a classically factorizable operation. The rigorous definition follows (H, X, $L_1(H)$, $L_1(X)$ are as in Sect. 2).

3.1. Definition. A linear transformation $O : L_1(H) \rightarrowtail L_1(H)$ is called a *classically factorizable operation*, if there is a commutative diagram

$$L_1(H) \xrightarrow{\quad O \quad} L_1(H), \qquad L_1(H) \xrightarrow{O_1} L_1(X) \xrightarrow{O_2} L_1(H)$$

where transformations O_1, O_2 are linear, positive and normalization preserving. ●

The class of all classically factorizable operations over H will be denoted by $\mathrm{Opf}(H)$. It is a part of the class $\mathrm{Op}(H)$ of all operations over H (see [4] for a definition); this follows easily from Stinespring's[3] Theorems 3, 4, showing that any positive transformation is completely positive provided at least one of the two algebras (the domain and the range) is commutative. It is clear that O_1 describes a measurement (or absorbtion), O_2 - a preparation (or re-emitting).

If $O \in \mathrm{Opf}(H)$, then measurements of the form M_1O, M_2O are compatible for arbitrary measurements $M_k : L_1(H) \rightarrowtail L_1(X_k)$; this is evident by virtue of the diagram

$$L_1(H) \rightarrowtail L_1(X) \rightarrowtail L_1(H) \rightarrowtail \begin{cases} L_1(X_1) \\ L_1(X_2) \end{cases}$$

(In fact, this property of an operation is equivalent to its classical factorizability.) This statement coupled with Theorem 2.2 leads to the following result.

3.2. Theorem. Let H_1, H_2 be Hilbert spaces, and $O \in Opf(H_1)$. Then for any Hermitian operators A_1, A_2 in H_1 and B_1, B_2 in H_2, having norm ≤ 1,

$$\| O^*(A_1) \otimes B_1 + O^*(A_1) \otimes B_2 + O^*(A_2) \otimes B_1 - O^*(A_2) \otimes B_2 \| \leq 2. \bullet$$

So <u>any violation of Bell inequalities becomes impossible if at least one of the two quantum objects is subjected to a classically factorizable operation</u>. And it is not at all necessary that the object be actually absorbed and re-emitted by a classical system. We will see that in many cases an equivalent effect is caused by thermal fluctuations in a medium between the source and the instrument.

Theorem 3.2 was formulated essentially as a property of the transformation $O^* \otimes 1$ in $L_\infty(H_1 \otimes H_2)$; it transforms observables of the form $\sum \tau_{kl} A_k \otimes B_l$ into observables with the norm ≤ 2 (τ_{kl} was introduced in Sect. 2). It is natural to ask, what is remarkable in the transformation $O \otimes 1$ in $L_1(H_1 \otimes H_2)$. It turns out that it transforms any density matrix into a mixture of non-correlated (that is, having the form $W_1 \otimes W_2$) density matrices. And this means destroying quantum correlations between two correlated objects when at most one of them is subjected to a classically factorizable operation. Only classical correlations remain, which are describable by means of a classical parameter affecting both objects. Of course, this rules out the possibility of any violation of the Bell inequalities.

The class Op(H) of all operations is vast, as well as its subclass Opf(H). In the rest of this paper we confine ourself to "Gaussian" operations defined below. A special part played by such operations will be argued in Sect. 4. The irreducible representation of the finite-dimensional canonical commutation relations is supposed given in the Hilbert space H. The Weyl correspondence is well-known[5)] between operators in H and functions on the phase space (which is finite-dimensional symplectic linear space); the function corresponding to an operator is called its Weyl symbol. The Weyl symbol of a density matrix is known as its Wigner distribution too. A density matrix is called[2)] Gaussian if its Wigner distribution is equal to the density of a Gaussian measure. And the last obeys a known condition[2)] connected with the Heisenberg uncertainty relation. It is convenient to formulate this condition in terms of two bilinear forms on the space L of all linear real-valued functions on the phase space, namely, the Poisson bracket $\{f, g\}$ considered as a skew real-valued bilinear form on L, and the covariation form $Cov_\gamma(f,g) = \int fg d\gamma - (\int f d\gamma)\cdot(\int g d\gamma)$ or rather the corresponding norm $\|f\|_\gamma = \sqrt{Cov_\gamma(f,f)}$. So a Gaussian measure γ on the phase space is a Wigner distribution if and only if

$$h\cdot\{f, g\} \le 2\|f\|_\gamma \|g\|_\gamma \tag{1}$$

for all f,g∈L.

3.3. Definition. An operation O∈Op(H) is called *Gaussian* if there is a Gaussian measure γ on the phase space such that for any $W\in L_1(H)$ the Wigner distributions w for W and w_1 for $W_1=O(W)$ are connected through the convolution relation $w_1 = \gamma * w$, that is, $w_1(x) = \int w(x+y)\gamma(dy)$. ●

It is easy to see that one and only one Gaussian operation O_γ exists for any Gaussian measure γ on the phase space. See also [1], Sect. 3.

3.4. Theorem. A Gaussian operation O_γ is classically factorizable provided

$$h\cdot\{f, g\} \le \|f\|_\gamma \|g\|_\gamma \tag{2}$$

for all f,g∈L.

Proof. Consider the Gaussian measure γ_0 such that $\gamma_0 * \gamma_0 = \gamma$. Then $\|f\|_{\gamma_0} = 2^{-1/2}\|f\|_\gamma$ for all f∈L. Accordingly to the criterion formulated above γ_0 is the Wigner distribution for some Gaussian density matrix W_0. We define the transformation O_1: $L_1(H) \rightarrowtail L_1(X)$, where X is the phase space, through the relation $Ws(O_1(W)) = \gamma_0 * Ws(W)$ for any $W\in L_1(H)$; here Ws denotes the Weyl symbol. Then $O_1(W)(x) = Tr(W_0(x)\cdot W)$, where $W_0(x)$ is the Gaussian density matrix whose Wigner distribution is γ_0 translated to x. Hence O_1 is positive. Further, we define the transformation O_2: $L_1(X) \rightarrowtail L_1(H)$ through the relation $Ws(O_2(f)) = \gamma_0 * f$ for any $f\in L_1(X)$. Then $O_2(f) = \int W_0(x) f(x) dx$, therefore O_2 is positive too. And $Ws(O_2(O_1(W))) = \gamma_0 * \gamma_0 * Ws(W) = \gamma * Ws(W)$, hence $O_2 O_1 = O_\gamma$. ●

The proved theorem yields a sufficient condition for a Gaussian operation to be classically factorizable; we do not know whether this condition is necessary. In the case of one degree of freedom, that is, two-dimensional phase space (q,p), any Gaussian operation is unitary equivalent (through a linear canonical transformation) to one of operations $O_{\Delta q,\Delta p}$ which are O_γ for γ such that $\|q\|_\gamma=\Delta q$, $\|p\|_\gamma=\Delta p$ and $Cov_\gamma(q,p)=0$. And such operations are mutually unitary equivalent for any fixed value of $\Delta q\Delta p$. They are classically factorizable provided $\Delta q\Delta p \geq h$ due to Theorem 3.4. We do not know whether the critical value of the parameter $\Delta q\Delta p/h$ separating the presence and the absence of the classical factorizability is equal to 1. But we know that such critical value exists in (0,1]. Its positiveness can be obtained from properties of $\sup \|\sum \tau_{kl} O^*_{\Delta q,\Delta p}(A_k)\otimes B_l\|$ considered as a function of $\Delta q\Delta p/h$, the supremum being taken over all A_k, B_l having norm ≤ 1. Our information about this function is scanty; but we can prove that it is decreasing, continuous on the right, constant and equal to 2 on $[1, +\infty)$; at the origin however it is equal to $2\sqrt{2}$ and continuous.

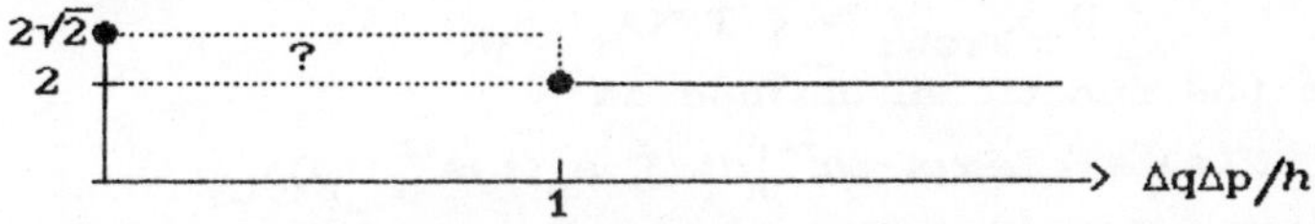

This behavior looks like a phase transition, the more so that the argument of this function turns out to be depending on the temperature in situations considered further.

4. QUASI-CLASSICAL SYSTEMS AS SPECIAL OPEN SYSTEMS

One is speaking about a classical Hamiltonian mechanical system when one neglects all deviations in the behavior of a physical system from the classical dynamics. When the deviations are to be taken into account as small corrections, then it is said about a quasi-classical system. Usually one bears in mind, however, only the deviations resulting from the quantum nature of the system. But other deviations prevail quite often, those occuring through the fact that the considered system is not well isolated, and first of all through the thermal fluctuations. In this section quasi-classical systems are treated as weakly non-isolated quantum systems of a special form. The first approximation to their dynamics is naturally classical Hamiltonian mechanics. The second approximation is a Gaussian operation introduced in Sect. 3 rather than conservative quantum mechanics or classical diffusion.

4.1. Elimination of Tensor Product from Chronological Descrition of Interaction

An interaction of two quantum systems is described by a unitary operator U on the tensor product $H_1 \otimes H_2$. Assuming a non-correlated initial state $W_1 \otimes W_2$ with a prescribed W_2, the quantum state change for the first system is described with the operation $O \in Op(H)$ defined by the equality $\langle M \rangle_{O(W_1)} = \langle U^* M U \rangle_{W_1 \otimes W_2}$ for all $M \in L_\infty(H_1)$. (We do not subject the second system to any measurements; accordingly O is non-selective, that is, normalization preserving). Further, $U = T\exp(-ih^{-1}\int dt H_{int}(t))$, where $H_{int}(t)$ is the interaction Hamiltonian, T is the chronological ordering, and the interaction representation is used. We accept the following form of interaction:

$$H_{int}(t) = \Sigma_{\nu} A^{(1)}_{\nu}(t)\otimes A^{(2)}_{\nu}(t). \qquad (3)$$

This is not too restrictive because operators of this form can approximate any operator in $H_1\otimes H_2$. And this form is convenient to formulate needed conditions for an interaction in terms of $A^{(1)}$, $A^{(2)}$.

A way to eliminate tensor products is suggested by the following formula:

$$\langle U \rangle_{W_1\otimes W_2} = \langle T\mathcal{F}(A_1) \rangle_{W_1}, \qquad (4)$$

where $\mathcal{F}$ is the functional defined as

$$\mathcal{F}(a) = \langle T\exp(-ih^{-1}\int dt\cdot\Sigma\, a_{\nu}(t)A^{(2)}_{\nu}(t))\rangle_{W_2}$$

for any path $a = \{a_{\nu}(t)\}$. From here on indices ν and sums over ν will be omitted (but meant), and $A_k(t)$ will be written rather than $A^{(k)}(t)$. Let us prove (4).

$$\mathcal{F}(a) = \left\langle \sum_{n=0}^{\infty} \left[-\frac{i}{h}\right]^n \int\!\!\dots\!\!\int_{t_n>\dots>t_1} dt_n\dots dt_1 a(t_n)\dots a(t_1)A_2(t_n)\dots A_2(t_1) \right\rangle_{W_2} =$$

$$= \sum_{n=0}^{\infty} \int\!\!\dots\!\!\int_{t_n>\dots>t_1} dt_n\dots dt_1 f_n(t_n,\dots,t_1)a(t_n)\dots a(t_1),$$

where $f_n(t_n,\dots,t_1) = \left[-\frac{i}{h}\right]^n \langle A_2(t_n)\dots A_2(t_1)\rangle_{W_2}$; so,

$$\langle T\mathcal{F}(A_1)\rangle_{W_1} = \left\langle \sum_{n=0}^{\infty} \int\!\!\dots\!\!\int_{t_n>\dots>t_1} dt_n\dots dt_1 f_n(t_n,\dots,t_1)A_1(t_n)\dots A_1(t_1) \right\rangle_{W_1} =$$

$$= \sum_{n=0}^{\infty} \left[-\frac{i}{h}\right]^n \int\!\!\dots\!\!\int_{t_n>\dots>t_1} dt_n\dots dt_1 \langle A_2(t_n)\dots A_2(t_1)\rangle_{W_2} \langle A_1(t_n)\dots A_1(t_1)\rangle_{W_1} =$$

$$= \left\langle \sum_{n=0}^{\infty} \left[-\frac{i}{h}\right]^n \int\!\!\dots\!\!\int_{t_n>\dots>t_1} dt_n\dots dt_1 A_1(t_n)\dots A_1(t_1)\otimes A_2(t_n)\dots A_2(t_1) \right\rangle_{W_1\otimes W_2} =$$

$$= \left\langle T\exp\left[-\frac{i}{h}\int dt A_1(t)\otimes A_2(t)\right] \right\rangle_{W_1\otimes W_2} = \langle U \rangle_{W_1\otimes W_2},$$

and (4) is proved. The following formula can be proved similarly:

$$\langle U^* M U \rangle_{W_1\otimes W_2} = \langle \hat{T}(M\mathcal{F}(A_1,A_1)) \rangle_{W_1}, \qquad (5)$$

where $\mathcal{F}$ is the two-place (with two path arguments) functional defined as $\mathcal{F}(a,b) =$

$$= \langle (T\exp(-\frac{i}{h}\int dt b(t)A_2(t)))^* \cdot (T\exp(-\frac{i}{h}\int dt a(t)A_2(t))) \rangle_{W_2}, \qquad (6)$$

and $\hat{T}$ denotes the ordering according to the following rules: (a) operators A(t) substituted in $\mathcal{F}$ in place of the first argument are ordered chronologically; (b) the inverse order for the second argument; (c) the operators mentioned in (a) acting before M (that is, are written on the right of M) and those mentioned in (b) acting after M (on the left of M). Note that the right-hand side of (6) can be considered as the scalar product of two vectors, one depending on a, the other - on b (in fact, any state defines a scalar product on the algebra of observables), hence there is an expansion $\mathcal{F}(a,b) = \sum \mathcal{F}_k(a)\overline{\mathcal{F}_k(b)}$ with some one-place functionals $\mathcal{F}_1, \mathcal{F}_2, \ldots$ Then (5) gives

$$\langle U^* M U \rangle_{W_1 \otimes W_2} = \Sigma_k \langle (T\mathcal{F}_k(A_1))^* \cdot M \cdot (T\mathcal{F}_k(A_1)) \rangle_{W_1}; \qquad (7)$$

unlike to (5), only the usual chronological ordering is used in (7). The formula (7) can be considered as a specific version of the formula (3.7) in [4]. Of course, the functionals $\mathcal{F}_k$ are not uniquely defined by $\mathcal{F}$, but their arbitrariness does not affect the right-hand side of (7). At last, the summation over the discrete auxiliary index k can be replaced with the integration over an auxiliary variable running over arbitrary space with a positive measure.

4.2. Gaussian Approximation for Environment Influence

Now we will render concrete the functional $\mathcal{F}$, defined by (6), for some class of situations. Namely, we suppose in what follows that the first system is a macroscopic mechanical system (for example, a moving rigid body) and the second system is a many-body system in the sense of statistical physics (for example, a gas around the rigid body). The influence of the second system on the first one manifests in fluctuations of the motion of the first system. Usually such fluctuations are normally (Gaussian) distributed random variables. The well-known central limit theorem of the probability theory suggests a reason for this special kind of distribution, namely, the integral effect can be considered as the sum of many (almost) independent contributions. The mathematical machinery is simple; the Fourier transform for the probability measure is the product of many functions:

$$f(\lambda) = \prod f_k(\lambda) = exp \sum ln\, f_k(\lambda) \approx exp \sum (ia_k\lambda - \tfrac{1}{2}b_k\lambda^2),$$

where $a_k = -i\frac{d}{d\lambda}\Big|_{\lambda=0} ln\, f_k(\lambda)$, $b_k = -\frac{d^2}{d\lambda^2}\Big|_{\lambda=0} ln\, f_k(\lambda)$. And it is important that $ln\, f(\lambda)$ admits a quadratic approximation in much broader domain than $f(\lambda)$ itself.

Similarly we argue that $ln\, \mathcal{F}(a,b)$ admits a quadratic approximation in much broader domain than $\mathcal{F}(a,b)$ itself. Instead of the classical notion of independent random summands we use the quantum notion of independent systems described by the tensor product of Hilbert spaces, W_2 being factorized into the tensor product of density matrices and $A_2(t)$ being expanded into the sum of subsystem observables. Then $\mathcal{F}(a,b)$ becomes factorized into the product of functionals corresponding to subsystems, and the quadratic approximation is obtained. In addition, it can be shown that $\mathcal{F}(a,b)$ is the Fourier transform of "quantum distribution for fluctuations", in a sense.

It is more simple to find coefficients for the quadratic approximation for $\ln \mathcal{F}(a,b)$; in fact, this is an usual calculation up to the second order in $a(t)$, $b(t)$ considered as small parameters in (6), giving

$$\ln \mathcal{F}(a,b) \approx -\frac{i}{h}\int dt f_1(t)(a(t)-b(t)) - \\ - \frac{1}{2h^2}\iint dt_1 dt_2 f(t_1,t_2)(a(t_1)-b(t_1))(a(t_2)-b(t_2)) + \\ + \frac{i}{2h}\iint_{t_2>t_1} dt_1 dt_2 g(t_1,t_2)(a(t_1)-b(t_1))(a(t_2)+b(t_2)), \tag{8}$$

where

$$f_1(t) = \langle A_2(t) \rangle_{W_2},$$

$$f(t_1,t_2) = \langle A_2(t_1)\circ A_2(t_2) \rangle_{W_2} - \langle A_2(t_1) \rangle_{W_2}\cdot\langle A_2(t_2) \rangle_{W_2},$$

$$g(t_1,t_2) = \langle \frac{i}{h}[A_2(t_1), A_2(t_2)] \rangle_{W_2}. \tag{9}$$

Of course, $A\circ B$ denotes $(AB+BA)/2$.

We will see that (8) implies that fluctuations form a Gaussian random process. It is very likely that the applicability domain for (8) coincides with the class of situations where fluctuations are (almost) Gaussian, regardless of how convincing the above reasoning with many independent subsystems seems. From here on we suppose that (8) holds.

4.3. What We Understand by Macroscopic Mechanical System

The canonical quantization of a macroscopic Hamiltonian mechanical system leads to a quantum system which reconstructs the original classical system as a result of well-known passage to the limit for $h \to 0$. A wave packet contracts into a point which moves along a classical trajectory. The passage to the limit needs some conditions, for example, that all used observables result from quantization of such functions on the phase space which are smooth and do not depend on h. However, we want apply such arguments to the world where h is a finite constant rather than infinitesimal. The small parameter is not h itself but its value in some "macroscopical" system of units suitable for a given situation. This idea can be formalized as follows. Let us introduce constants $\bar{S}$, $\bar{q}$, $\bar{p}$, $\bar{E}$, $\bar{t}$ having dimensions of the action, length, momentum, energy and time respectively, obeying

$$\bar{q}\cdot\bar{p} = \bar{E}\cdot\bar{t} = \bar{S} \gg h.$$

It was said in Sect. 4.2 that the first system is supposed to be a macroscopic mechanical system. Now we can define what we understand by this. Prior to the canonical quantization we have a classical Hamiltonian system, we restrict ourselves to a bounded domain Ω of its phase space, and we assume the following condition:

$$\left|\frac{\partial^{k+l} h(q,p)}{\partial q^{k} \partial p^{l}}\right| \le \frac{\bar{E}}{\bar{q}^{|k|} \cdot \bar{p}^{|l|}} \tag{10}$$

for all $(q,p)\in\Omega$, $|k|+|l| = 1,2,3$; here $h(q,p)$ is the Hamilton function and k, l are multiindices, that is, $k=(k_1,\dots,k_n)$, n is the number of degrees of freedom, q^{k} denotes $q_1^{k_1}\dots q_n^{k_n}$, $|k|$ denotes $k_1+\dots+k_n$, and the same for l. The inequality (10) is nothing but the C^3-smoothness with the constant 1 (in Ω) of the Hamilton function in the choosen system of units. Three of derivatives are essential in this paper; perhaps, in other cases a different number of derivatives will be necessary. First-order derivatives must satisfy (10) "without a wide margin", that is, the following relation must be violated:

$$\left|\frac{\partial^{k+l} h(q,p)}{\partial q^{k} \partial p^{l}}\right| \ll \frac{\bar{E}}{\bar{q}^{|k|} \cdot \bar{p}^{|l|}} \quad \text{for all } (q,p)\in\Omega,\ |k|+|l| = 1;$$

otherwise the Hamilton function would have been almost constant. It is possible to generalize the condition (10) to the case of a symplectic manifold as the phase space, but we do not dwell on this matter. Note that (10) implies that the considered system have no characteristic time less than $\bar{t}$; so it cannot perform many oscillations during $\bar{t}$, and trajectories cannot diverge very much during $\bar{t}$.

So the macroscopicity condition is formulated for an isolated classical system. The corresponding isolated quantum system is defined uniquely by the canonical quantization using the Weyl correspondence (mentioned in Sect. 3 above), accordingly we have no need to impose more conditions on quantum dynamics, but we require certain localization of a quantum state, because our technique is local. Later we will argue that such localization results from environment influence, but at first we accept the following condition: each considered quantum state is localized in some domain of the phase space, whose size is

$$\ll \bar{q}\cdot(h/\bar{S})^{1/3} \quad \text{and} \quad \ll \bar{p}\cdot(h/\bar{S})^{1/3} \tag{11}$$

for the coordinate and momentum dimensions respectively. Of course, our conclusions can be extended to arbitrary mixtures of such states. The localization of a state in a domain means fast decreasing of its Weyl symbol (= Wigner distribution) outside of this domain. The bounds stated by (11) arise from our technique used further. Note that they permit a quantum state to occupy many phase cells, namely

$$\sim \frac{\bar{q}\cdot(h/\bar{S})^{1/3}\cdot\bar{p}\cdot(h/\bar{S})^{1/3}}{h} = (\bar{S}/h)^{1/3} \gg 1$$

for each degree of freedom. So the domain is macroscopically small but microscopically large.

Further, we require some conditions on environment influence; it must conform with the macroscopic nature of the system itself. We impose a condition on the operators $A_1(t)$ introduced in Sect. 4.1. However, the interaction representation was used in 4.1, hence the Hamiltonian motion of the first system itself was included into explicit time dependence of $A_1(t)$, whereas here we take $A_1(t)$ in Schrödinger representation, hence this explicit time dependence does not include the Hamiltonian motion (and in conservative cases $A_1(t)$ does not depend on t). We accept the following smoothness condition:

$$\left| \frac{\partial^{k+l} a(q,p,t)}{\partial q^k \, \partial p^l} \right| \le \frac{\bar{a}}{\bar{q}^{|k|} \cdot \bar{p}^{|l|}} \qquad (12)$$

for all $(q,p)\in\Omega$, $|k|+|l| = 1,2$; here $a(q,p,t)$ is the Weyl symbol for $A_1(t)$, $\bar{a}$ is some constant such that first-order derivatives satisfy (12) "without a wide margin"; cf. with (10) above.

However, the operators $A_1(t)$ describe the interaction only together with the operators $A_2(t)$, and we need a condition on $A_2(t)$ or rather on the functional $\mathcal{F}$ defined by (6) to ensure that the environment influence is not too large, namely, that it deflects the first system from a classical trajectory within the bounds stated by (11). We will see in the next section that deflecting properties depend on the Fourier transform of $\mathcal{F}$ and that the formulation below is relevant. The functional $\mathcal{F}$ can be approximated, as well as any functional at all, by some linear combinations of functionals $\mathcal{F}_{c,d}$ of the form

$$\mathcal{F}_{c,d}(a,b) = exp\left(-\frac{i}{h}\int dt a(t)c(t) + \frac{i}{h}\int dt b(t)d(t) \right). \qquad (13)$$

We assume that our functional $\mathcal{F}$ admits such approximation with the following restriction on pathes c,d:

$$\left| \int dt \bar{a} c(t) \right| \ll \bar{S}\cdot(h/\bar{S})^{1/3} \quad \text{and the same for d;} \qquad (14)$$

here the integral is taken over arbitrary time interval of duration $\le\bar{t}$.

4.4. Canonical Commutation Relations near Classical Trajectory

In this section we will show that locally (near a given classical trajectory) the operators $A_1(t)$ can be approximated with some operators satisfying the canonical commutation relations (CCR) in the sense that their commutators are some "c-numbers", that is, operators of the form $\lambda\cdot\mathbb{1}$. Here $A_1(t) = A_1^{int}(t)$ are operators in the interaction representation; the corresponding operators in the Schrödinger representation will be denoted by $A_1^{Sch}(t)$, so that

$$A_1^{int}(t) = exp\left(\frac{i}{h}tH_1\right)\cdot A_1^{Sch}(t)\cdot exp\left(-\frac{i}{h}tH_1\right),$$

where H_1 is the Hamiltonian for the first system. (It was noted that $A_1^{Sch}(t)$ did not depend on t in conservative cases). The approximation will ensure the possibility to replace the operators under $\mathcal{F}$ in formula (5).

We use the following statement (apparently belonging to the folk-lore) showing that a quadratic Hamiltonian generates a linear transform of the phase space, quantum dynamics being reduced to a transfer of Weyl symbols by this transform, that is, along classical trajectories.

4.4.1. Statement. Let a real-valued function $h(q,p,t)$ be a polinomial of degree ≤ 2 in variables $q = (q_1,\dots,q_n)$ and $p = (p_1,\dots,p_n)$ (however its dependence on t remains arbitrary). For each t we consider the operator $H(t)$ whose Weyl symbol is $h(q,p,t)$. Let $U(t) = T\exp\left(-\frac{i}{\hbar}\int_0^t ds H(s)\right)$, and further $W(t) = U(t)WU^*(t)$, $A(t) = U^*(t)AU(t)$ for any density matrix W and any observable A. Then the Weyl symbols $w(q,p,t)$ for $W(t)$, $a(q,p,t)$ for $A(t)$ satisfy

$$w(q(t),p(t),t) = w(q(0),p(0),0),$$
$$a(q(0),p(0),t) = a(q(t),p(t),0)$$

for any classical trajectory $q(t)$, $p(t)$ defined by $\dot{q} = \frac{\partial h}{\partial p}$, $\dot{p} = -\frac{\partial h}{\partial q}$. ●

One important <u>corollary</u>: if $a(q,p,0)$ is a linear function, then $a(q,p,t)$ is linear for each t.

Let $q_{cl}(t)$, $p_{cl}(t)$ be a classical trajectory remaining in the domain Ω during the considered time interval. We have for any fixed t the following expansion for the Hamilton function h (which is the Weyl symbol for H_1):

$$h(q,p) = h(q_{cl}(t), p_{cl}(t)) + dh(q-q_{cl}(t), p-p_{cl}(t)) +$$
$$+ \frac{1}{2}d^2h(q-q_{cl}(t), p-p_{cl}(t)) + \bar{E}\cdot\mathcal{O}\left(\left|\frac{q-q_{cl}(t)}{\bar{q}}\right|^3, \left|\frac{p-p_{cl}(t)}{\bar{p}}\right|^3\right);$$

the first differential dh is a linear form, the second one d^2h is a quadratic form; of course, the differentials are taken at the point $q_{cl}(t)$, $p_{cl}(t)$. Rejecting the remainder term, we obtain a polynomial $\tilde{h}(q,p,t)$ of degree ≤ 2 in q,p; it is the Weyl symbol for some operator $\tilde{H}_1(t)$. Similarly,

$$a(q,p,t) = a(q_{cl}(t), p_{cl}(t), t) + da(q-q_{cl}(t), p-p_{cl}(t)) +$$
$$+ \bar{a}\cdot\mathcal{O}\left(\left|\frac{q-q_{cl}(t)}{\bar{q}}\right|^2, \left|\frac{p-p_{cl}(t)}{\bar{p}}\right|^2\right),$$

where $a(q,p,t)$ is the Weyl symbol for $A_1^{Sch}(t)$; rejecting the remainder term, we obtain a polynomial $\tilde{a}(q,p,t)$ of degree ≤ 1

in q, p and the corresponding operator $\tilde{A}_1^{Sch}(t)$. We define approximating operators for $A_1^{int}(t)$ as

$$\tilde{A}_1^{int}(t) = \tilde{U}^*(t)\tilde{A}_1^{Sch}(t)\tilde{U}(t), \text{ where } \tilde{U}(t) = Texp\left[-\frac{i}{\hbar}\int_0^t ds\tilde{H}_1(s)\right].$$

Statement 4.4.1 or rather its corollary shows that the operators $\tilde{A}_1^{int}(t)$ are linear combinations of the Q,P operators and hence satisfy CCR. It should be proved that the replacement of A_1 with $\tilde{A}_1^{int}$ under $\mathcal{F}$ in formula (5) results in a small error provided W_1 is localized near the considered trajectory, that is, the point $q_{cl}(0)$, $p_{cl}(0)$ belongs to the domain appeared in (11). Taking (13) into account we see that it is sufficient to prove that the replacement of A_1^{int} with $\tilde{A}_1^{int}$ in

$$\left[Texp\left[-\frac{i}{\hbar}\int dsA_1^{int}(s)c(s)\right]\right]\cdot W_1\cdot\left[Texp\left[-\frac{i}{\hbar}\int dsA_1^{int}(s)d(s)\right]\right]^*$$

results in a change which is small, that is, of norm $\ll 1$ in $L_1(H)$, provided c,d satisfy (14). In Schrödinger representation $A_1^{Sch}(s)$ is replaced with $\tilde{A}_1^{Sch}(s)$ and at the same time H_1 with $\tilde{H}_1(s)$ in

$$\left[Texp\left[-\frac{i}{\hbar}\int ds(H_1+A_1^{Sch}(s)c(s))\right]\right]W_1\left[Texp\left[-\frac{i}{\hbar}\int ds(H_1+A_1^{Sch}(s)d(s))\right]\right]^*.$$

We introduce an "interpolating" operator, depending on an auxiliary parameter t, by taking $\tilde{A}_1^{Sch}(s)$ and $\tilde{H}_1(s)$ for s<t but $A_1^{Sch}(s)$ and H_1 for s>t. It is enough to estimate the derivative of the "interpolating" operator with respect to t. This derivative is the sum of two similar terms ("the right" and "the left"); we consider one of them. Taking into account that the norm of an operator in $L_1(H)$ remains unchanged if the operator is multiplied (from the right or from the left) by any unitary operator, we see that it is enough to estimate the following:

$$\|(H_1-\tilde{H}_1(t)+(A_1^{Sch}(t)-\tilde{A}_1^{Sch}(t))c(t))\left[Texp\left[-\frac{i}{\hbar}\int_0^t ds(\tilde{H}_1(s)+\tilde{A}_1^{Sch}(s)c(s))\right]\right]\cdot$$
$$\cdot W_1\cdot\left[Texp\left[-\frac{i}{\hbar}\int_0^t ds(\tilde{H}_1(s)+\tilde{A}_1^{Sch}(s)d(s))\right]\right]^*\|_{L_1(H)}.$$

We use again the possibility of multiplying by any unitary operator to make a little trick, namely, to replace d(s) to c(s); in fact we reject the whole unitary factor containing d(s) and instead we write the similar factor with c(s). Now we have the product of two operators. The first has the Weyl symbol

$$\bar{E}\mathcal{O}\left(\left|\frac{q-q_{cl}(t)}{\bar{q}}\right|^3,\left|\frac{p-p_{cl}(t)}{\bar{p}}\right|^3\right)+\mathcal{O}\left(\left|\frac{q-q_{cl}(t)}{\bar{q}}\right|^2,\left|\frac{p-p_{cl}(t)}{\bar{p}}\right|^2\right)\bar{a}c(t).$$

The second results from the evolution of W under the "perturbated" Hamiltonian $\tilde{H}_1(s)+\tilde{A}_1^{Sch}(s)c(s)$, which is again quadratic; hence the statement 4.4.1 is applicable, and we see that this second operator is localized in the domain

$$\left|\frac{q-q_{cl}(t)}{\bar{q}}\right| \ll (h/\bar{S})^{1/3}, \qquad \left|\frac{p-p_{cl}(t)}{\bar{p}}\right| \ll (h/\bar{S})^{1/3}.$$

These estimations together with the smoothness of the Weyl symbol for the first operator show that the norm of the product is $\ll \bar{E}\cdot(h/\bar{S}) + (h/\bar{S})^{2/3}\cdot\bar{a}c(t)$. Using (14) and integrating over time we obtain that the norm of the difference is $\ll (\bar{t}\bar{E}h/\bar{S}+\bar{S}h/\bar{S})/h$, that is, $\ll 1$.

4.5. Gaussian Approximation for Operation

We had seen in 4.4 that the operators $A_1(t)$ in the interaction Hamiltonian $H_{int}(t) = A_1(t)\otimes A_2(t)$ can be assumed locally satisfying the canonical commutation relations. However $A_2(t)$ can be assumed CCR too. Indeed, only the functional $\mathcal{F}$ resulting from $A_2(t)$ according to (6) is essential, and this functional is Gaussian, that is, the exponential of a quadratic polynomial, according to (8). It is easy to see that any CCR operators result in a Gaussian functional provided the state (W_2) is Gaussian. And conversely, we can construct CCR-operators and a Gaussian state for a given Gaussian functional $\mathcal{F}$ provided that the latter satisfies the condition (1) or rather its infinite-dimensional version. But this condition always holds for a functional resulting from some observables and state according to (8,9).

Thus we accept that $A_1(t)$ and $A_2(t)$ are both CCR (and in addition that W_2 is Gaussian). Then the bilinear Hamiltonian $H_{int}(t) = A_1(t)\otimes A_2(t)$ can be considered as a special case of the quadratic Hamiltonian and the statement 4.4.1 is applicable. But it was formulated for finite dimension only, whereas operators $A_2(t)$ can generate infinite dimensional CCR; however, this is in fact of no importance here. The mentioned statement reduces quantum dynamics to classical. A classical trajectory for the couple of systems is the pair of functions $X_1(t)$, $X_2(t)$ valued in the phase space of the first and the second system respectively. The operator $A_1(t)$ has its Weyl symbol $a_1(x,t)$ which is a linear function of x running over the phase space of the first system, and the same for $A_2(t)$, $a_2(x,t)$. The classical equations of motion under the bilinear Hamiltonian $a_1(x,t)\cdot a_2(x,t)$ can be written (in the interaction representation) in the form

$$\frac{d}{dt}b_1(X_1(t)) = \{a_1(t),\ b_1\}\cdot a_2(X_2(t),t),$$

$$\frac{d}{dt}b_2(X_2(t)) = \{a_2(t),\ b_2\}\cdot a_1(X_1(t),t);$$

here b_1 is an arbitrary linear function on the phase space of the first system, b_2- of the second one; the Poisson brackets for two linear functions result in a constant function, therefore the argument x is omitted inside these brackets. It follows from the second equation

$$a_2(X_2(t),t) = a_2(X_2(0),t) + \int_0^t ds\frac{d}{ds}a_2(X_2(s),t) =$$

$$= a_2(X_2(0),t) + \int_0^t ds\{a_2(s),\ a_2(t)\}\cdot a_1(X_1(s),s);$$

substitution in the first equation leads to

$$\frac{d}{dt}b_1(X_1(t)) = \{h_{eff}(x,t),\ b_1\}, \quad \text{where}$$

$$h_{eff}(x,t) = a_1(x,t)\cdot\left[a_2(X_2(0),t)+\int_0^t ds\{a_2(s),\ a_2(t)\}\cdot a_1(X_1(s),s)\right].$$

We see that the influence of the second system on the first one can be described (on the classical level for the time being) by means of an "effective Hamiltonian" $h_{eff}(x,t)$ for the first system. Of course, it is not a true Hamiltonian, because it depends not only on x and t, but also on the history $X_1(s)$, $s\le t$, and on the initial state of the second system $X_2(0)$. But this quasi-Hamiltonian description provides a convenient transition from the interaction representation used above to the Schrödinger one; one must only add $h_{eff}(x,t)$ to the Hamiltonian of the first system.

It was shown in Sect. 4.2 that the three functions f_1,f,g are sufficient to describe the influence of the second system. Let us express h_{eff} in terms of these functions. We have

$$g(t_1,t_2) = \langle\frac{i}{\hbar}[\ A_2(t_1),A_2(t_2)]\ \rangle_{W2} = \{a_2(t_1),\ a_2(t_2)\};$$

$$f_1(t) = \langle A_2(t)\rangle_{W2} = \int dx a_2(x,t)w_2(x),$$

$$f(t_1,t_2)+f_1(t_1)f_1(t_2)=\langle A_2(t_1)\circ A_2(t_2)\rangle_{W2}=\int dx a_2(x,t_1)a_2(x,t_2)w_2(x),$$

where $w_2(x)dx$ is the Wigner distribution for W_2, which is a Gaussian measure. Accordingly, the function $\zeta(t) = a_2(X_2(0),t)$ can be considered as a Gaussian random process with the mean $f_1(t)$ and the covariance $f(t_1,t_2)$, if $X_2(0)$ is considered as a random point with the distribution $w_2(x)dx$. We obtain

$$h_{eff}(x,t) = a_1(x,t)\cdot\left[\zeta(t)+\int_0^t ds g(s,t)a_1(X_1(s),s)\right].$$

And the summation over the auxiliary index ν, which was omitted but meant from Sect. 4.1 up to here, can be now eliminated together with the functions $a_1(x,t)$. Namely, we put $\zeta(x,t) = \sum_\nu \zeta_\nu(t)a_{1\nu}(x,t)$ and the same for f_1,

$f(x_1,t_1;\ x_2,t_2) = \sum_{\mu\nu} f_{\mu\nu}(t_1,t_2)a_{1\mu}(x_1,t_1)a_{1\nu}(x_2,t_2)$
and the same for g; then

$$h_{eff}(x,t) = \zeta(x,t)+\int_0^t ds g(X_1(s),s;\ x,t). \tag{15}$$

It is clear that $\zeta(x,t)$ is a Gaussian random process with the mean f_1 and the covariance f. By the way, these new functions f_1, f, g can be expressed by means of operators $A_2(x,t) = \sum_\nu a_{1\nu}(x,t)A_{2\nu}(t)$. In fact from the very beginning we could have treated the interaction in terms of $A_2(x,t)$ rather than of $A_1(t)\otimes A_2(t)$.

Quantum dynamics can now be reconstructed due to the statement 4.4.1. Namely,

$$\langle M \rangle_{O(W_1)} = \langle U^*MU \rangle_{W_1\otimes W_2} = \mathbb{E}\int dx_0 m(X_1(t,x_0))w_1(x_0)$$

for any initial state W_1 of the first system and any observable M for the first system, which is measured at a moment t; here m is the Weyl symbol for M, w_1 - for W_1; $X_1(t,x_0)$ denotes the classical trajectory starting from an initial point x_0 and governed by the Hamiltonian including h_{eff};it depends on a path of the random process ζ, and $\mathbb{E}$ denotes the averaging on ζ.

Often we will describe dynamics in terms which seem to be classical (for example, the force of friction); however we will mean quantum dynamics corresponding to classical as stated here.

Now we abandon the second system itself (which was called also the environment) because its influence on the first system is taken into account by means of h_{eff}.

4.6. What We Understand by Quasi-Classical Approximation

Traditionally one treats the quasi-classical approximation as short-wave asymptotics for the Schrodinger equation. We propose a wider comprehension in the spirit of Wigner's[6)] ideas, including some environment influence taken in the approximation given in 4.5 above. Besides the Hamiltonian, three functions are used:

$$f_1(x,t),\quad f(x_1,t_1;\ x_2,t_2),\quad g(x_1,t_1;\ x_2,t_2),$$

x,x_1,x_2 running over the phase space. These functions provide a reduced description of an environment influence. An "evolution operation" describes quantum dynamics rather than a unitary evolution operator. The presence of two time arguments in f, g reflects possible memory effects in the environment, through which the quantum evolution is generally non-markovian. The markovian property and a semi-group of operations arises if f, g are localized near $t_1=t_2$; this case is of special interest. The opposite case, when a characteristic time for the environment is much greater than that for the system, falls outside of this paper; apparently it is connected with concepts of dynamical chaos (classical and quantum). Our approach can be used to investigate the long-time behavior of a system only in case when stochastic effects prevail over quantum ones for moderate time (and all the more for large time) and besides an environment gives no long-time memory effects. But this is the case for a wide class of macroscopic systems.

Due to the decay of memory effects we can consider classical dynamics for $t\in(-\infty,+\infty)$ rather than for $t\in(0,+\infty)$.

The functions f_1, f, g are assumed small and smooth on a macroscopic scale. This is why the evolution operation is locally Gaussian. Naturally, the corresponding coefficients are smoothly varying on the macroscopic scale. By the way, the functions f, g satisfy some inequality: the skew bilinear form defined by the kernel g is bounded by the positive symmetric bilinear form defined by the kernel f. This shows that a non--Hamiltonian (in particular, dissipative) evolution is indispensably accompanied by a noise. But we do not dwell on this matter, because we are mostly interested in such cases when the noise is much greater than its fundamental lower bound (due to non-zero temperature; see below).

4.7. Using Fluctuation-Dissipation Relations

From here on we suppose that the second system (the environment) is a heat bath. At the same time we suppose that the interaction does not depend explicitly on t. So A_1^{Sch}, A_2^{Sch} do not depend on t, and $[H_2,W_2]=0$ (H_2 is the Hamiltonian for the second system), hence

$$f_1(x,t) = f_1(x), \quad f(x_1,t_1;\, x_2,t_2) = f(x_1,x_2,t_1-t_2),$$
$$g(x_1,t_1;\, x_2,t_2) = g(x_1,x_2,t_1-t_2).$$

We suppose that W_2 is the Gibbs state for a positive temperature T or, more generally (to comprehend the case of continuous spectrum of H_2), the Kubo-Martin-Schwinger (KMS) state. We do not touch upon fundamental problems of statistical physics about the origin of such state, and we leave out of account the known non-Gibbsian terms. We accept the following fluctuation-dissipation relation connecting the Fourier transforms (with respect to time) for f and g:

$$\hat{g}(x_1,x_2,\omega) = -i\hat{f}(x_1,x_2,\omega)\; th\, \frac{h\omega}{2kT}. \tag{16}$$

In terms of f,g themselves this is a convolution relation:

$$g(x_1,x_2,t) = \int_{-\infty}^{+\infty} ds \left(\frac{h}{kT}\; sh\, \frac{\pi kT}{h}(t-s)\right)^{-1} f(x_1,x_2,s). \tag{17}$$

Note that the time kernel is exponentially small outside the domain $t-s=\mathcal{O}(h/kT)$. We suppose that $h/kT=\mathcal{O}(\bar{t})$ and that both functions $f(x_1,x_2,\tau)$, $g(x_1,x_2,\tau)$ are negligible outside the domain $\tau=\mathcal{O}(\bar{t})$. This is suited for continuous spectrum of H_2; for models with a discrete spectrum one can use decreasing in a domain which is finite but large enough.

It follows from (17) that $\int_{-\infty}^{+\infty}d\tau g(x_1,x_2,\tau)=0$. Using $g(x_1,x_2,\tau)=-g(x_2,x_1,-\tau)$ we have $\int_0^{+\infty}d\tau g(x_1,x_2,\tau) = \int_0^{+\infty}d\tau g(x_2,x_1,\tau)$ and hence

$$\frac{\partial}{\partial x^k}\bigg|_{x=x_0}\int_0^{+\infty}d\tau g(x,x_0,\tau) = \frac{1}{2}\,\frac{\partial}{\partial x^k}\bigg|_{x=x_0}\int_0^{+\infty}d\tau g(x,x,\tau);$$

this will be used below.

An environment influence results in a trend and fluctuations. The trend is determined by the differential of the "effective Hamiltonian" (15) averaged on ζ:

$$\frac{\partial}{\partial x^k}\bigg|_{x=X(t)}\mathbb{E}h_{eff}(x,t) = \frac{\partial}{\partial x^k}\bigg|_{x=X(t)}\left[f_1(x,t)+\int_{-\infty}^{t}ds g(X(s),s;\, x,t)\right] =$$

$$= \frac{\partial}{\partial x^k}\bigg|_{x=X(t)}f_1(x) - \frac{\partial}{\partial x^k}\bigg|_{x=X(t)}\int_0^{+\infty}d\tau g(x,X(t-\tau),\tau) =$$

$$= \frac{\partial}{\partial x^k}\bigg|_{x=X(t)}f_1(x) - \frac{\partial}{\partial x^k}\bigg|_{x=X(t)}\int_0^{+\infty}d\tau g(x,X(t),\tau) -$$

$$- \frac{\partial}{\partial x^k}\bigg|_{x=X(t)}\int_0^{+\infty}d\tau(g(x,X(t-\tau),\tau)-g(x,X(t),\tau)) =$$

$$= \frac{\partial}{\partial x^k}\bigg|_{x=X(t)}\left[f_1(x) - \frac{1}{2}\int_0^{+\infty}d\tau g(x,x.\tau)\right] -$$

$$- \int_0^{+\infty}d\tau\frac{\partial}{\partial x^k}\bigg|_{x=X(t)}\int_0^{\tau}d\theta\frac{d}{d\theta}g(x,X(t-\theta),\tau); \qquad (18)$$

the first term is a total derivative without any path dependence; hence it brings in a Hamiltonian correction to the motion. A non-Hamiltonian correction is introduced by the second term. It can be written as

$$\iint\limits_{0<\theta<\tau<\infty} d\tau d\theta \dot{X}^l(t-\theta)\frac{\partial^2}{\partial x_1^k\partial x_2^l}\bigg|_{x_1=X(t),x_2=X(t-\theta)} g(x_1,x_2,\tau),$$

and it is clear that the knowledge of this correction for all pathes $X(s)$ is equivalent to the knowledge of the function $\frac{\partial^2}{\partial x_1^k\partial x_2^l}g(x_1,x_2,\tau)$. Due to the fluctuation-dissipation relation this is equivalent to the knowledge of the function $\frac{\partial^2}{\partial x_1^k\partial x_2^l}f(x_1,x_2,\tau) = Cov(\zeta_{,k}(x_1,\tau),\ \zeta_{,l}(x_2,0))$, that is, of the covariance function for fluctuation forces. The non-Hamiltonian correction can be expressed through the fluctuation covariances as follows:

$$- Cov \left[\int_{-\infty}^{+\infty} d\tau \zeta_{,k}(X(t),t+\tau)\frac{1}{\pi}ln\left|cth\frac{\pi kT}{2h}\tau\right|, \int_{-\infty}^{t} ds\dot{X}^{l}(s)\zeta_{,l}(X(s),s)\right].$$

But this is too general for our purposes.

4.8. Markovian Approximation

It was supposed in 4.7 that the relaxation time for the heat bath is $O(t)$. Starting from here we suppose that it is $\ll t$. Correspondingly, the evolution becomes local in time, that is, markovian, described by differential equations rather than by integro-differential ones. A force of friction arises often which is proportional to the velocity. Sometimes it is proportional to the third derivative (for example, the radiative friction). It may occur a linear combination of different derivatives. Non-linear friction is also possible (for example, one proportional to the cube of velocity), but we do not examine such cases here.

So we linearize the non-Hamiltonian correction (18):

$$- \left.\frac{\partial}{\partial x^{k}}\right|_{x=X(t)}\int_{0}^{+\infty} d\tau (g(x,X(t-\tau),\tau)-g(x,X(t),\tau)) \approx$$

$$\approx - \left.\frac{\partial}{\partial x^{k}}\right|_{x=X(t)}\int_{0}^{+\infty} d\tau (X^{l}(t-\tau)-X^{l}(t))\left.\frac{\partial}{\partial y^{l}}\right|_{y=X(t)} g(x,y,\tau) =$$

$$=-\int_{0}^{+\infty} d\tau \left.\frac{\partial^{2}}{\partial x_{1}^{k}\partial x_{2}^{l}}\right|_{x_{1}=X(t),x_{2}=X(t)} g(x_{1},x_{2},\tau)\cdot\sum_{n}\frac{(-1)^{n}}{n!}\tau^{n}\left.\frac{d^{n}}{ds^{n}}\right|_{s=t} X^{l}(s)=$$

$$= -\sum_{n} C_{kl}^{n}(X(t))\left.\frac{d^{n}}{ds^{n}}\right|_{s=t} X^{l}(s),$$

where $C_{kl}^{n}(x)=\frac{(-1)^{n}}{n!}\int_{0}^{+\infty} d\tau\cdot\tau^{n}\cdot\left.\frac{\partial^{2}}{\partial x_{1}^{k}\partial x_{2}^{l}}\right|_{x_{1}=x,x_{2}=x} g(x_{1},x_{2},\tau);$

the sum over n is regarded as an asymptotic expansion rather than as a convergent series. On the other hand, (16) gives for $\omega\ll kT/h$ $\quad \hat{g}(x,y,\omega) \approx -\frac{h}{2kT}i\omega\hat{f}(x,y,\omega)$, hence

$$\int_{-\infty}^{+\infty} d\tau\phi(\tau)g(x,y,\tau) \approx \frac{h}{2kT}\int_{-\infty}^{+\infty} d\tau\phi'(\tau)f(x,y,\tau)$$

for any function ϕ whose Fourier transform is negligible outside the domain $\omega=O(kT/h)$. However

$$\int_{-\infty}^{+\infty} d\tau\phi(\tau)g(x,y,\tau) = \int_{-\infty}^{+\infty} d\tau g(x,y,\tau)\sum_{n}\frac{1}{n!}\phi^{(n)}(0)\tau^{n} =$$

$$= \sum_{n}\frac{1}{n!}\phi^{(n)}(0)\cdot\left(\int_{0}^{+\infty} d\tau g(x,y,\tau)\cdot\tau^{n} + \int_{0}^{+\infty} d\tau g(x,y,-\tau)\cdot(-\tau)^{n}\right) =$$

$$= \sum_{n}\frac{1}{n!}\phi^{(n)}(0)\cdot\left(\int_{0}^{+\infty} d\tau g(x,y,\tau)\cdot\tau^{n} - (-1)^{n}\int_{0}^{+\infty} d\tau g(y,x,\tau)\cdot\tau^{n}\right),$$

hence

$$\left.\frac{\partial^{2}}{\partial x_{1}^{k}\partial x_{2}^{l}}\right|_{x_{1}=x,x_{2}=x}\int_{-\infty}^{+\infty} d\tau\phi(\tau)g(x,y,\tau)=\sum_{n}\left((-1)^{n}C_{kl}^{n}(x)-C_{lk}^{n}(x)\right)\phi^{(n)}(0),$$

so we obtain

$$\int_{-\infty}^{+\infty} d\tau\phi'(\tau)Cov(\zeta_{,k}(x,\tau),\ \zeta_{,l}(x,0)) \approx \frac{2kT}{h}\sum\left[(-1)^n C^n_{kl}(x)-C^n_{lk}(x)\right]\phi^{(n)}(0),$$

that is,

$$Cov(\zeta_{,k}(x,\tau),\ \zeta_{,l}(x,0)) \approx -\frac{2kT}{h}\sum_n\left[C^{n+1}_{kl}(x)+(-1)^n C^{n+1}_{lk}(x)\right]\delta^{(n)}(t).$$

We see that the same coefficients $C^n_{kl}(x)$ determine both the non-Hamiltonian correction to dynamics and the correlation of fluctuations. The last is found only for coincided points x; however it can be shown that one may put

$$Cov(\zeta_{,k}(x(t_1),t_1),\ \zeta_{,l}(x(t_2),t_2)) \approx$$

$$\approx -\frac{2kT}{h}\sum_n\left[C^{n+1}_{kl}\left(x\left(\frac{t_1+t_2}{2}\right)\right)+(-1)^n C^{n+1}_{lk}\left(x\left(\frac{t_1+t_2}{2}\right)\right)\right]\ \delta^{(n)}(t_2-t_1)$$

with an error of the same order as the error introduced above by the linearization.

Let us note some restrictions for $C^n_{kl}(x)$ arising from the positive definiteness of a covariance function. Namely, the following matrix must be positive semi-definite for all ω, both positive and negative:

$$\frac{h}{2kT}\left.\frac{\partial^2}{\partial x_1^k \partial x_2^l}\right|_{x_1=x,x_2=x} \hat{f}(x_1,x_2,\omega) \approx -\sum_n (i\omega)^n\left[C^{n+1}_{kl}(x)+(-1)^n C^{n+1}_{lk}(x)\right] =$$

$$= -(C^1_{kl}+C^1_{lk})-i\omega(C^2_{kl}-C^2_{lk})+\omega^2(C^3_{kl}+C^3_{lk})-\dots .$$

It is clear that the first non-vanishing term here must be a term with an even power of ω, and this leading term must contain a positive semi-definite matrix.

The most simple and most widespread case is that of force of friction which is proportional to the velocity: $F_{fric}=-K_1\dot{q}$. Such case can be described by means of C^n_{kl} with n=1, and we arrive at the corresponding fluctuation forces F_{fluc} with the following covariation:

$$Cov(F_{fluc}(s),\ F_{fluc}(t)) = 2kTK_1\delta(s-t). \tag{19}$$

The next case is "radiative-type" force of friction which is proportional to the third derivative: $F_{fric}=K_3 q^{(3)}$; here C^3_{kl} is used giving

$$Cov(F_{fluc}(s),\ F_{fluc}(t)) = -2kTK_3\delta''(s-t). \tag{20}$$

The same fluctuations conform to more general friction $F_{fric}=K_2q^{(2)}+K_3q^{(3)}+K_4q^{(4)}$, because the symmetric part of the matrix C^n_{kl} with an even n makes no contribution to the low-frequency part of fluctuation spectrum. Note also that $K_3q^{(3)}$ is in fact an approximation for some expression which is non-local in time. This approximation is valid for trajectories smooth enough, and invalid for "self-accelerating trajectories" containing the factor $exp(mt/K_3)$.

5. TIME OF CLASSICAL FACTORIZATION

5.1. Definition and General Relations

It was shown in section 4 that the operation describing the evolution of a quasi-classical system is locally Gaussian, and a method was given to find its parameters. Therefore results of section 3 can be applied to decide whether this operation is classically factorizable. We define the *time of classical factorization* τ_f for a quasi-classical system as such time that the operation describing the evolution during a time t is classically factorizable for $t>\tau_f$ but it is not for $t<\tau_f$. This is treated locally, that is, near a classical trajectory. We regret to note that the theorem 3.4 yields only a sufficient condition for classical factorizability and we do not know whether it is necessary. Through this fact true values for τ_f can disagree with the values obtained below. However the ratio of this values is bounded with some constant of mathematical nature, so our calculations yields acceptable estimations.

Let us calculate the time of classical factorization for translational degrees of freedom for a body of mass m whose interaction with an environment results in a small force of friction $F_{fric}=-K_1\dot{q}$, other forces being absent. The friction coefficient K_1 interests us so far as it enables to find fluctuation forces. We neglect a damping effect of the friction because in considered cases τ_f is much less than the time $\tau_1=m/K_1$ characterizing the damping effect. Considering one degree of freedom, assuming zero initial coordinate and velocity and applying (19) we obtain the following (co-)variances:

$$D(p(t)) = 2kTK_1t; \qquad D(q(t)) = 2kTK_1t^3/3m^2;$$

$$Cov(q(t),\, p(t)) = 2kTK_1t^2/2m.$$

Calculating the criterion (2) we find that $\tau_f = \left(\sqrt{3}\frac{h}{kT}\cdot\frac{m}{K_1}\right)^{1/2}$. So

$$\tau_f^2 \sim \tau_{therm}\tau_1, \tag{21}$$

where $\tau_{therm}=h/kT$ is a characteristic period of thermal fluctuations, $\tau_1=m/K_1$.

The second case $F_{fric}=K_3q^{(3)}$ may seem completely similar but in fact it is not. In the first case the fluctuation force as a function of time behaves as the white noise, hence the momentum behaves as the Wiener process having continuous pathes, therefore the instant values of the momentum are well-defined. However in the second case the momentum behaves as the white noise and the coordinate - as the Wiener process (like the situation for the Brownian motion), hence the instant values of the momentum have too large variances. But this momentum averages to a small value during a long time. One can speak about intensive exchange of momentum between the body and the surroundings media where a momentum passed to the body returns progressively to the media. So it should be given an averaging time τ_{aver}, specifying indirectly thickness of a boundary layer of the media which is implicitly included in the considered quasi-classical system. Then we obtain

$$\tau_f \sim \tau_{aver}\left(\frac{\tau_{therm}}{\tau_3}\right)^2, \tag{22}$$

where $\tau_{therm}=h/kT$ as above, $\tau_3=K_3/m$.

Two effects were considered above. However their combination needs a special consideration. Indeed, if one effect results in large fluctuations of the momentum and small fluctuations of the coordinate, but the opposite for the other effect, then their combination leads to a time of classical factorization which is much less than that for each effect separately. We obtain

$$\tau_f \sim min\left(\tau_1^{1/2}\tau_{therm}^{1/2},\ \tau_1^{1/2}\tau_3^{-1/2}\tau_{therm},\ \tau_1^{1/3}\tau_3^{-1/3}\tau_{aver}^{1/3}\tau_{therm}^{2/3},\ \tau_3^{-2}\tau_{aver}\tau_{therm}^2\right). \tag{23}$$

The above formulas are valid for an almost free body. For a body in a non-linear potential specifying by a time τ_{mech} (it may be a period of oscillation or a time of divergence of trajectories) the following two terms must be included under the minimum sign in (23):

$$\tau_1^{1/6}\tau_3^{-1/6}\tau^{1/3}_{therm}\tau_{mech}^{2/3},\qquad \tau_3^{-1/2}\tau_{therm}^{1/2}\tau_{mech}. \tag{24}$$

However if the minimum is $\gg \bar{t}$ then we known that $\tau_f \gg \bar{t}$ and nothing more.

5.2. Several Examples

Let us consider a macroscopic body interacting only with a thermal electromagnetic radiation and with a rarefied gas. Its translation motion is accompanied by the following force of friction:

$$F_{fric} \sim -\left(\frac{v}{c}\cdot P_{rad} + \frac{v}{v_0}\cdot P_{gas}\right)\cdot S,$$

where v, c, v_0 are the velocities of the body, the light and the gas molecules respectively, P_{rad} (resp. P_{gas}) is the specific pressure of the radiation (resp. the gas), S is the sectional area of the body. Substituting

$P_{rad}\sim(kT)^4(hc)^{-3}$, $P_{gas}\sim\rho v_0^2$, $v_0\sim(kT/m_0)^{1/2}$ and $S\sim l^2$ we obtain

$F_{fric}=-K_1 v$ with $K_1 \sim h^{-3}c^{-4}k^4l^2T^4 + k^{1/2}m_0^{-1/2}\rho l^2T^{1/2}$;

here T is the temperature of the radiation and the gas (assuming equal), m_0 is the mass of the gas molecules, ρ is the specific density of the gas and l is the size of the body. Hence $\tau_1 = m/K_1 \sim min\,(h^3 c^4 k^{-4} ml^{-2} T^{-4},\ k^{-1/2} m_0^{1/2} \rho^{-1} ml^{-2} T^{-1/2})$. We take for example $m \sim 1$ g, $l \sim 1$ cm, $T \sim 1$ K, $\rho \sim 10^{-26}$ kg/m^3, $m_0 \sim 10^{-27}$ kg; then $\tau_1 \sim min(10^{24}, 10^{25}) \sim 10^{24}$ sec, $\tau_{therm} \sim 10^{-11}$ sec and (21) gives $\tau_f \sim 10^6$ sec.

Let the body have an electric charge Q and consider the radiative friction

$$F_{fric} = (6\pi\varepsilon_0)^{-1} c^{-3} Q^2 q^{(3)};$$

here $q^{(3)}$ denotes the third derivative of the body coordinate; the coefficient is adapted to the SI system of units. We have $K_3 \sim \varepsilon_0^{-1} c^{-3} Q^2$ and $\tau_3 = K_3/m \sim \varepsilon_0^{-1} c^{-3} Q^2 m^{-1}$. We take for example $Q \sim 10^{-9}$ C, then $\tau_3 \sim 10^{-31}$ sec and (22) gives $\tau_f \sim 10^{40} \tau_{aver}$. We see that the classical factorization caused by the radiative friction is practically impossible. The combined account of $\tau_1 \sim 10^{24}$ sec and $\tau_3 \sim 10^{-31}$ sec through (23) shows that τ_3 is negligible even for $\tau_{aver} \sim 10^{-12}$ sec.

Now let the body be in the Coulomb field of another body with a charge Q_2 at a distance r. Calculating the second derivative of the potential with respect to r we find $\tau_{mech} \sim \varepsilon_0^{1/2} m^{1/2} r^{3/2} Q^{-1/2} Q_2^{-1/2}$. We take for example $r \sim l$, $Q_2 \sim Q$, then $\tau_{mech} \sim 1$ sec. The six terms form τ_f (see (23), (24)), and the leading term is $\tau_1^{1/6} \tau_3^{-1/6} \tau_{therm}^{1/3} \tau_{mech}^{2/3} \sim 10^5$ sec, which is however not too different from the first term $\tau_1^{1/2} \tau_{therm}^{1/2} \sim 10^6$ sec. Since all the terms are $\gg \tau_{mech}$, one concludes that $\tau_f \gg \tau_{mech} \sim 1$ sec.

But the motion in a potential causes a varying deformation and hence a dissipation provided that the potential is not quadratic. Let the body be a dielectric crystal (say, quartz or sapphire). The dissipation in the body can be described by the relation $Q = (\omega\tau_{dis})^{-1}$, where Q is the quality of mechanical oscillations of the frequency ω and τ_{dis} is a parameter characterizing the material. Sapphire at $T \sim 1$ K has $\tau_{dis} \sim 10^{-15}$ sec; see [6], Chap. 2. For the case of the body in the Coulomb potential the dissipation through deformations can be described by a force of friction proportional to the velocity with $\tau_1 \sim E\tau_{dis}^{-1} r^2 l^{-1} m^{-1} \tau_{mech}^4$, where E is the modulus of elasticity. Taking $E \sim 10^{11}$ N/m^2 we obtain $\tau_1 \sim 10^{24}$ sec.

At last we consider the dissipation caused by the heat exchange between the body and its surrounding when the body is subjected to varying deformations. The thermodynamic cycle is almost adiabatic if $\varkappa c^{-1}\rho^{-1}l^{-2}\tau_{mech} \ll 1$ and almost isothermal in the opposite case; here $\varkappa$ is the heat conductivity for the material of the body, c is its specific heat and ρ is its specific density. For sapphire at $T \sim 1$ K we have $\varkappa c^{-1}\rho^{-1} \sim 10^4$ m^2/sec, this is why we consider only the almost isothermal case. The dissipation can be described again by a force of friction proportional to the velocity with $\tau_1 \sim \varkappa\alpha^{-2}T^{-1}r^2l^{-3}m^{-1}\tau_{mech}^4$, where α is the temperature coefficient of expansion. For sapphire at $T \sim 1$ K we have $\varkappa\alpha^{-2}T^{-1} \sim 10^{22}$ m·kg·sec^{-3} and again $\tau_1 \sim 10^{25}$ sec.

The mechanisms of dissipation have been considered which are the most difficult ones for being eliminated, and we see that some extraordinary but surely attainable isolation of mechanical degrees of freedom makes it possible to retain quantum correlations in macroscopic motion during a macroscopically considerable time which may be enough to execute rather a complicated movement. At the same time we see that even under such extraordinary circumstances a macroscopic delocalization of the wave function is unattainable. On the other hand, consider a macroscopic system under usual circumstances, say, the pointer of a galvanometer, which moves at $T \sim 300$ K with a friction such that $\tau_1 \sim 3$ sec. Then (21) gives $\tau_f \sim \tau_1^{1/2}\tau_{therm}^{1/2} \sim 10^{-6}$ sec. Of course, making use of (21) in such a way is questionable because the interaction is not weak. Nevertheless it is very likely that <u>the time necessary for a human observer to read an instrument is in any case much greater than the time of classical factorization</u> for the corresponding degrees of freedom of the pointer.

6. DISCUSSION AND CONCLUSION

6.1. About Accuracy of Mechanical Description

There is the judgement that any talk about the coordinate of a macroscopic body with the accuracy, say, 10^{-20} m, is a mathematical abstraction obviously devoid of any physical sense regardless of which mechanics are used, classical or quantum. To prove it they think enough to remind that a macroscopic body is nothing more than a bounded state of many particles described actually by a field theory, and the body is non-ideal in many respects, each of which is enough to introduce a fuzziness much greater than 10^{-20} m.

However a macroscopic body is a collective phenomenon

and it is naturally much less fuzzy than its small parts. The accuracy intrinsic to coordinates and momenta of a macroscopic body is the accuracy up to which the corresponding mechanical equations hold. And it is a fundamental limit for the accuracy of our measurements. If this limit is not attained, it becomes a challenge for the measurements technology rather than for the theory, unless some fundamental barrier is revealed.

For the extraordinary but attainable situation considered in 5.2 the mechanical equations hold during $t \sim 1$ sec with the following accuracy:

$$(D(p(t)))^{1/2} \sim \frac{t}{\tau_f}\sqrt{\frac{\hbar m}{t}} \sim 10^{-24} \text{ kg}\cdot\text{m/sec} \quad \text{for the momenta,}$$

$$(D(q(t)))^{1/2} \sim \frac{t}{\tau_f}\sqrt{\frac{\hbar t}{m}} \sim 10^{-21} \text{ m} \quad \text{for the coordinates.}$$

Of course, it is a matter of quantum mechanical equations, since $D(p(t))\cdot D(q(t)) \ll \hbar^2$. (This does not contradict to the uncertainty principle, because these parameters are related to an operation rather than a state.) We try to suggest some two-valued measurement with such accuracy. Unfortunately it destroys the quantum state and cannot be used to prepare a state localized as much. Thus, we place a body with a positive electric charge near the unstable stationary point between two fixed bodies with negative charges. Choosing the parameters as in 5.2 we have $\tau_{mech} \sim 1$ sec. Several tens of seconds later an obvious deviation will occur toward one of the two attractors. So we can measure the following two-valued observable:

$$sign\left(Q + \frac{\tau_{mech}}{m}\cdot P + \zeta\right),$$

where ζ is a random variable with $(D(\zeta))^{1/2} \sim 10^{-21}$ m.

6.2. About Bell Inequalities for Macroscopic Bodies

Is it possible to observe a violation of a Bell inequality in some macroscopic experiment? This intriguing problem was discussed in our paper [1] and will be discussed in what follows but remains still unsolved. The formal solution given in [1] leaves the three open questions: (a) how to prepare the needed initial state; (b) how to eliminate thermal fluctuations; (c) how to execute the needed measurements. The question (a) remains open; (b) is answered in Sect. 5.2; (c) will be now discussed. According to [1] and Sect. 2 here we need a way to measure (alternatively, of course) $sign(P+\zeta)$ or $sign(Q+\eta)$, where ζ,η are normally distributed random variables such that $D(\zeta)\cdot D(\eta) \ll \hbar^2$. Due to the unitary equivalence the pointed operators can be replaced with $sign(Q+aP+\zeta)$ and $sign(Q+bP+\eta)$ with arbitrary but essentially different a,b (the initial state being replaced accordingly). And such measurements were proposed in Sect. 6.1; τ_{mech} being the classical parameter like the famous orientation of a spin-measuring device.

Remark. Maybe, it is the preparation which is crucial here. Maybe, only the states with positive Wigner distributions (cf. with [8]) can be prepared for macroscopic systems during an acceptable time. This seems an interesting topic for mathematical examination.

6.3. About Hidden Variables

A "dotted-in-time" description with hidden variables was promised in the introduction and will be developed here. Consider a quasi-classical system with a finite time of classical factorization τ_f. One can partition the time axis into intervals of the duration τ_f (or more); then an aggregate of hidden variables (or, what is the same, a classical state) per interval arises as follows.

The quantum state change during one interval is described by some operation O of the form $O = O^{qc}O^{cq}$, where O^{cq} transforms quantum states into classical ones, and O^{qc} does the reverse. The evolution during a time $\gg\tau_f$ can be described by the iteration of O:

$$...OOO... = ...(O^{qc}O^{cq})(O^{qc}O^{cq})(O^{qc}O^{cq})...;$$

rearranging the brackets we obtain

$$...(O^{cq}O^{qc})(O^{cq}O^{qc})(O^{cq}O^{qc})...,$$

non-coupled boundary terms being omitted. However, $O^{cq}O^{qc}$ is nothing else but a Markov transition function; correspondingly, its iteration describes a Markov chain

$$... \rightarrowtail \lambda_i \rightarrowtail \lambda_{i+1} \rightarrowtail ...$$

which accompanies the quantum evolution; each λ_i is an aggregate of hidden variables. If τ_f is small then the Markov chain approximates a continuous-time Markov process. This is the case for immediately perceivable motion (say, the pointer of a galvanometer). The opposite case arises for extraordinary circumstances considered in Sect. 5.2. Executing quantum measurements of short duration in partitioning instants we obtain a Markov chain of the form

$$... \rightarrowtail \lambda_i \rightarrowtail (\lambda_i,x_i) \rightarrowtail \lambda_{i+1} \rightarrowtail (\lambda_{i+1},x_{i+1}) \rightarrowtail ...$$

where λ_i is the aggregate of hidden variables for the time interval preceding the measurement which gives the outcome x_i. Omitting one half of terms, one obtains again $... \rightarrowtail \lambda_i \rightarrowtail \lambda_{i+1} \rightarrowtail ...$ but the transition probabilities are changed by the measurements. Omitting the other half of terms one obtains a two-component Markov chain $... \rightarrowtail (\lambda_i,x_i) \rightarrowtail (\lambda_{i+1},x_{i+1}) \rightarrowtail ...$ with the second component observable but the first one not. Note that the random sequence $\ldots x_i, x_{i+1}, \ldots$ is not Markov in general; this is connected with the non-ideal nature of our measurements. Of course, in many cases one and the same interaction process forms the basis of both the classical factorization and the measurement. In this case one can take $\lambda_i = x_i$, and the hidden variables become open.

6.4. About Quasi-Classical Limit

Instead of the well-known quasi-classical limit for $h \to 0$ we propose a new limit: for $h \to 0$, $\lambda \to 0$ and $\lambda^2/h \to \infty$; here λ is the coupling constant of the interaction with the environment. The classical factorization leads to a Markov chain in the phase space. The condition $\lambda^2/h \to \infty$ ensures that $\tau_f \to 0$, hence the time discreteness vanishes in the limit. And the stochastics vanish too; the Markov pathes tend to classical trajectories.

This (h,λ)-limit is more adequate than the traditional h-limit for describing situations occuring usually for actual macroscopic systems, like a billiard-ball rolling on a billiard--cloth, or a planet rotating around the sun, or the pointer of a galvanometer moving along the indicator.

It is possible to calculate for given finite h and λ the corresponding Δq, Δp, Δt indicating the limits for the classical description.

In addition, the long-time behavior for finite λ is quite different from that for $\lambda=0$.

Table. Modes of the behavior of a macroscopic mechanical system.

isolation from environment \ accuracy of measurements	usual	extraordinarily high
usual	classical mechanics with dissipation and small fluctuations	hidden variables
extraordinarily high	classical mechanics	quantum mechanics

6.5. Information Propagation as Cause of Classical Factorization

Let us return to the operators $A_2(x,t)$ mentioned in Sect. 4.5 and put

$$U(x) = T\exp\left(-\frac{i}{h}\int dt A_2(x,t)\right)$$

for an arbitrary point x of the phase space of a quasi-classical system taken in the Heisenberg representation. Due to (6),

(8), (9) we obtain

$$h^{-2}D(\zeta(x_2)-\zeta(x_1)) \approx -2\ ln\ |\ \langle U^*(x_2)U(x_1)\rangle_{W_2}\ |.$$

Define the Riemann metric on the phase space by taking the above right-hand side as the square of the distance between infinitesimally close points x_1, x_2. Then the criterion (2) of classical factorizability can be written as

$$h\cdot\{f, g\} \leq |df|\cdot|dg|$$

for any smooth functions f,g on the phase space; here $|df|$ denotes the norm of the covariant vector df on the Riemann space. For the case of one degree of freedom this means that the phase volume of the ellipse, which is the unit circle for this Riemann metric, does not exceed πh.

In the case of a pure state W_2 we have

$$\langle U^*(x_2)U(x_1)\rangle_{W_2} = \langle\psi(x_1), \psi(x_2)\rangle,$$

where $\psi(x)=U(x)\psi$, ψ is the state vector for W_2. The state of the second system after the interaction is close to $\psi(x)$ provided the state of the first system is localized near x, namely, in a neighborhood where $\psi(x)$ is almost constant. However, this condition is impracticable in the case of classical factorization. We realize that our Riemann metric characterizes the interaction as "information propagation"; the second system reacts to the first one through the transition $\psi \longrightarrow \psi(x)$, neighboring points being almost indistinguishable. Of course, we take some liberty with the notion of information; we should rather to talk about quantum correlation setting capable to lead to an indirect quantum measurement...

The classical factorizability means nothing else but the inevitability of such information propagation. This leads to a new point of view on actions which are to be taken in order to prevent the classical factorization (see Sect. 5.2). The high vacuum is needed to prevent the information propagation to the surrounding gas. (But one may say equally right that it is needed to reduce the fluctuations.) The cooled sapphire is needed to prevent the propagation of the information about mechanical movements of the body through its deformations into the phonon gas inside the body or generally into internal degrees of freedom. And so on. The following idea arises: maybe it is better to do the experiment in a superconducting chamber to prevent the information propagation through the electromagnetic radiation? And this is really the case. This is a way to reduce the radiative friction and the friction on the thermal electromagnetic radiation.

We see that the "information propagation" point of view is heuristic. However, its inadequate use leads to discouraging conclusions. It is senseless to drown (to overrun) the radiation of the body with some noise. Speaking about information in this context we are perfectly indifferent towards the availability of this "information". Perhaps this kind of information should be named "hidden information". Note that it is undestroyable (cf. with the classical law of the conservation of the phase volume).

Remark. Our approach, being local, allows us to estimate $\langle U^*(x_2)U(x_1)\rangle_{W2}$ only for near points x_1, x_2. We assume that it is small all the more if these points are distant. We have no reason for doubt, however we would like to have a proof.

6.6. Several Remarks

The essence of the Everett's approach is an irreversible propagation of quantum correlations as the cause of the classical behavior. But as to "many worlds", this is rather Kolmogorov's approach. Indeed, the Kolmogorov's axiomatics of the probability theory (≈ 1930) treats any random variable as a function of a universal auxiliary variable ω. Enlarged to the whole physical world this leads immediately to one "possible world" per each value of ω. And Everett warned from the very beginning: worlds co-exist as possibilities rather than as physical systems; see [9], Note ‡ on page 459-460.

Our technique has a common part with that of Stapp [10], namely the Gaussian transform of quantum states into classical ones (see O_1 in the proof of Theorem 3.4 above). It enables to overcome the known continuum problem in the quantum measurement theory. On the other hand, we are not inclined to share Stapp's determination to postulate an exceptional ontological status for the light or for any physical system whatsoever. But we could have entitled this paper like the Stapp's one: "Heat as Foundation of Being". Indeed, just the thermal fluctuations are the main cause of the classical factorization; the latter would occur much slower at zero temperature. And the nature of the heat bath is of less importance. A case is possible where the electromagnetic radiation to the infinity is negligible and the classical factorization results mainly from the phonon radiation into the body.

Maybe there are such cases where the classical behavior results from some mechanism other than proposed here.

We are not inclined to insist on a completely sharp classical description. If we consider this description as secondary, derived from the quantum one, then we should not be surprised at some its fuzziness. However, we are glad that it satisfies the Bell inequalities exactly. Further, why the sole classical realm can be derived? We have no reason to contend its non-uniqueness, but we admit this possibility. If it is really sole according to Deutsch's [11] trend, then it is interesting, but not vital. Indeed, nobody believes it vital for a physical theory to predict the sole planet inhabited by people in the world. The exceptional status of some realm with respect to us people is rather natural.

Maybe a reader is displeased that we use "second-grade" notions of thermodynamic nature for foundations of the "first--grade" quantum theory. However, if we admit the possibility that the God is kindly disposed to dice, why not to heat also?..

Of course, we do not pretend for having solved here any problem of statistical physics. On the contrary, we need solutions for some of them for the foundation of our approach.

REFERENCES

1) Khalfin, L.A. and Tsirelson, B.S., "Quantum and quasi-classical analogs of Bell inequalities", in: Symposium on the foundations of modern physics, World Scientific, 441-460 (1985).

2) Holevo, A.S., "Probabilistic and statistical aspects of quantum theory" (in Russian), Moscov, "Nauka", 1980. English transl.: North-Holland, 1982.

3) Stinespring, W.F., "Positive functions on C^*-algebras", Proc. Amer. Math. Soc. 6, 2, 211-216 (1955).

4) Kraus, K., "Operations and effects in the Hilbert space formulation of quantum theory", Lecture Notes in Physics, Springer, 29, 206-229 (1974).

5) Berezin, F.A. and Shubin, M.A., "The Schrodinger equation" (in Russian), Moscow Univ. Press, 1983.

6) Wigner, E.P., "Some problems of our natural sciences", Intern. J. Theor. Phys. 25, 5, 467-476 (1986).

7) Braginskii, V.B., Mitrofanov, V.P. and Panov, V.I., "Systems with small dissipation" (in Russian), Moscov, "Nauka", 1981.

8) Bell, J.S., "EPR correlations and EPW distributions", preprint CERN, Geneva, January 1986.

9) Everett, H., ""Relative state" formulation of quantum mechanics", Rev. Modern Phys. 29, 3, 454-462 (1957).

10) Stapp, H.P., "Light as foundation of being", preprint LBL-19144, Berkeley, February 1985.

11) Deutsch, D., "Quantum theory as a universal physical theory", Intern. J. Theor. Phys. 24, 1, 1-41 (1985).

UNSHARP JOINT MEASUREMENT OF COMPLEMENTARY OBSERVABLES IN A PHOTON SPLIT BEAM EXPERIMENT

P. MITTELSTAEDT, A. PRIEUR and R. SCHIEDER
Physikalische Institute der Universität zu Köln
D-5000 Köln 41, Federal Republic of Germany

ABSTRACT

In a quantum mechanical two slit experiment one can observe a single photon simultaneously as particle (measuring the path) and as wave (measuring the interference pattern) if the path and the interference pattern are measured in the sense of unsharp observables. These theoretical predictions are confirmed experimentally by a photon split-beam experiment using a modified Mach-Zehnder interferometer.

1. INTRODUCTION

The notion of complementarity which Bohr proposed in his Como lecture can be illustrated very well by means of a beam split experiment. This experiment was suggested by Einstein during his discussion with Bohr on epistemological problems of quantum theory.[1] It shows that a classical description of the photons' behaviour in the apparatus leads to contradictions. The situation is the same as in the well-known double-slit experiment: If the validity of the uncertainty relation is presupposed, one cannot circumvent the conclusion that the measuring devices for the measurement of a particle (the path) and for that of a wave (the interference pattern) are mutually exclusive.

On the other hand, the same uncertainty relation defines the frame within which the principle of complementarity can be relaxed. From the practical point of view, this means that an unsharp measurement of the path does not destroy the interference pattern completely. Bohr emphasized this in his Como lecture in 1927[2], but it was first analysed in quantitative way by Wootters and Zurek in 1979[3]. Investigating the double-slit experiment they evaluated the visibility of the interference pattern for a given undeterminancy (in an objective as well as in a subjective way) of the photons' path.

In the following we apply the ideas of Wootters and Zurek to a photon split beam experiment which is realizable and which can be calculated exactly. In order to simultaneously get information about the path and the interference pattern (which are represented by the observables P(B) and P(A), respectively, as defined below), we perform a sharp (ideal first kind) measurement of a third observable P(C) which commutes neither with P(B) nor with P(A). As a result, both P(B) and P(A) are only incompletely objectified. They are, however, not completely unknown: Using information theory (see below) the expectation values of P(A) and P(B) in an eigenstate $|C\rangle$ of P(C), which is known after the measurement, determine the information about P(A) and P(B). The amount of information for one observable is of course less than the amount we would get by sharply measuring this observable, i.e. when it is completely objectified.

Only in this sense - namely that of an incomplete objectification - is the sharp measurement of P(C) an unsharp measurement of P(A) and P(B). These ideas can also be incorporated into the general theory of unsharp observables, which has been elaborated by P. Busch.[4] It is obvious that <u>unsharp</u> measurements of this kind must not be confused with <u>inaccurate</u> ones, since the unsharpness expresses the objective undeterminacy of the path and of the interference pattern in the given state, but not the observer's ignorance about these properties.

2. QUANTUM MECHANICAL CONSIDERATIONS

The photon split-beam experiment which is discussed here can be completely described within the framework of a two-dimensional Hilbert space H_2. The observables are then given by projection operations P(B) with eigenstates $|B\rangle$, $|\neg B\rangle$ and P(A) with eigenstates $|A\rangle$, $|\neg A\rangle$, respectively. The projection operators P,P',... of a two-dimensional Hilbert space H_2 can be represented by the points on the surface of a three-dimensional unit sphere, the Poincaré (or Stokes) sphere $\mathcal{P}$[5]. Projection operators $P(X) = |X\rangle\langle X|$ and $1-P(X) = P(\neg X) = |\neg X\rangle\langle\neg X|$ which project onto orthogonal states $|X\rangle$ and $|\neg X\rangle$ correspond to diametrical points. The operators P(B) and P(A) of the path and of the interference pattern are then given by orthogonal unit vectors of $\mathcal{P}$. (Fig. 2). The complementarity between path and interference pattern means in this description that for all states, i.e. for all points of the Poincaré sphere the operators P(A) and P(B) do not commute, i.e. $[P(B),P(A)]_- \neq 0$.

In order to measure unsharply the wave- and particle nature of a photon we will consider the following experiment: An incoming photon in the state $|\varphi(\delta)\rangle = 1/\sqrt{2}\,(|B\rangle + e^{i}\,|\neg B\rangle)$ is split into two different components $|B\rangle$ and $|\neg B\rangle$ with probabilities 1/2 at the partly reflecting mirror (beam splitter) (BS_1) and then brought together with a phase shift at the partly reflecting mirror (beam recombiner) $(BS)_2$ to be detected in one of the detectors (Fig. 1). Theoretically we need only one of the detectors because the detection of a photon in one of them excludes it in the other. In the following we consider therefore D_1, when we deal with expectation values. (However, for the practical realization of this experiment two detectors D_1 and D_2 are used for reasons which will be explained below.)

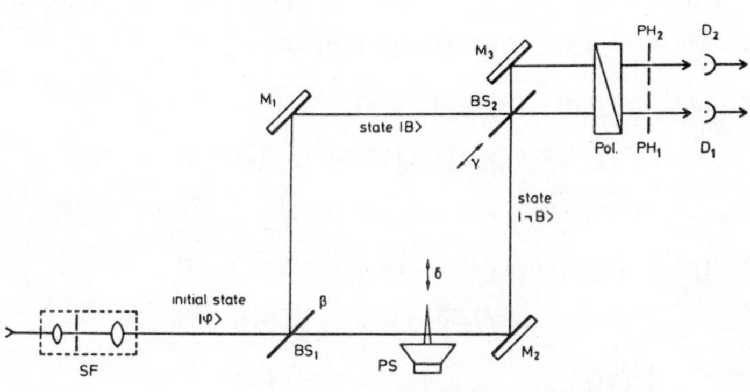

Fig. 1: Schematic diagram of a photon split-beam experiment with beam splitters $(BS)_1$ and $(SBS)_2$ (both partly reflecting mirrors), two fully reflecting mirrors M_1 and M_2, a phase shifter PS and two photon detectors D_1 and D_2.

The transparency γ of the beam recombiner $(BS)_2$ is variable, i.e. $0 \leq \gamma \leq 1$. Each value of γ determines an observable $P(C(\gamma))$ which is sharply measured with the apparatus. In particular, if $\gamma = 0$ one measures the observable

$$P(C(0)) = P(B) = |B\rangle\langle B|$$

of the <u>path</u> with eigenstates $|B\rangle, |\neg B\rangle$ and eigenvalues 1,0 respectively. The expectation value of P(B) is given by

$$\langle\varphi|P(B)|\varphi\rangle = |\langle\varphi|B\rangle|^2 = \frac{1}{2}.$$

If $\gamma = 1/2$ one measures the <u>interference</u> <u>observable</u>

$$P(C(\tfrac{1}{2})) = P(A) = \frac{1}{2}(1 + |B\rangle\langle\neg B| + |\neg B\rangle\langle B|)$$

with eigenstates

$$|A\rangle = \frac{1}{2}(|B\rangle + |\neg B\rangle)$$
$$|\neg A\rangle = \frac{1}{2}(|B\rangle - |\neg B\rangle)$$

which correspond to eigenvalues 1 and 0, respectively. For the expectation value of P(A) one obtains[6)]

$$\langle\varphi|P(A)|\varphi\rangle = |\langle\varphi|A\rangle|^2 = \frac{1}{2}(1 + \cos\delta)$$

In the general case $(0 \leq \gamma \leq 1)$ one measures the observable $P(C(\gamma))$ which can be expressed by P(A) and P(B) as:

$$P(C(\gamma)) = (\gamma - \sqrt{(1-\gamma)\gamma})\underline{1} + 2\sqrt{(1-\gamma)\gamma}\, P(A) + (1-2\gamma)P(B)$$
$$= (\gamma - 1/2\ S(\gamma))\underline{1} + S(\gamma)P(A) + (1-2\gamma)P(B)$$

with $S(\gamma) = 2\sqrt{(1-\gamma)\gamma}$. The eigenstates of

$P(C(\gamma) = |C(\gamma)\rangle\langle C(\gamma)|$ are given by

$$|C(\gamma)\rangle = \sqrt{1-\gamma}\,|B\rangle + \sqrt{\gamma}\,|\neg B\rangle$$
$$|\neg C(\gamma)\rangle = \sqrt{\gamma}\,|B\rangle - \sqrt{1-\gamma}\,|\neg B\rangle$$

and the expectation value of $P(C(\gamma)$ reads

$$\langle\varphi|P(C(\gamma)|\varphi\rangle = \frac{1}{2}(1+2\sqrt{\gamma(1-\gamma})\cos\delta)$$

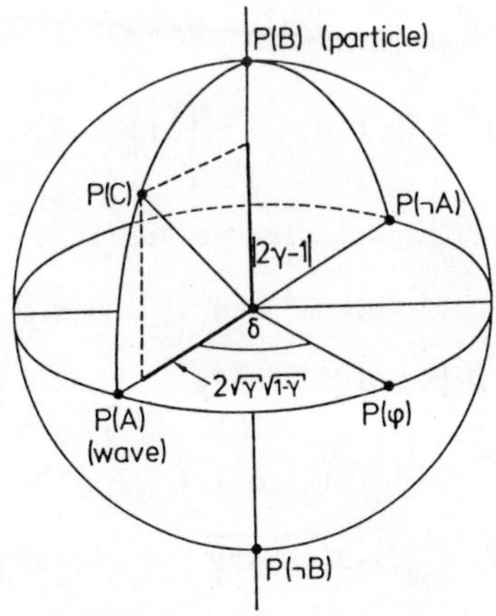

Fig. 2: Projection operators of a two-dimensional Hilbert space of the photon represented by the Poincaré-sphere. Initial preparation $P(\varphi(\delta))$, phase shift δ , path observable P(B) and interference-observable P(A). The intermediate observable P(C) depends on γ. Arbitrary projection operators depend on γ and δ .

The observables P(B), P(A) and $P(C(\gamma))$ are projection operators of H_2 and can thus be represented by points on the surface of the Poincaré sphere (Fig. 2). $P(C(\gamma))$ lies between P(A) and P(B) and is moved along the meridian circle through P(A) by varying γ . The angle α beween P(C) and P(B) is given by $\alpha = 2 \text{ arc sin}\sqrt{\gamma}$ and the components of P(C) in the direction of P(A) and P(B) are $2\sqrt{(1-\gamma)\gamma}$ and $|1-2\gamma|$ respectively. The coefficient $S(\gamma) = 2\sqrt{(1-\gamma)\gamma}$ of P(A) is the visibility of the interference pattern, which reaches its maximum for $\gamma = 1/2$ (when we do not know anything about the photon's path) and vanishes for $\gamma = 0$ (when we know exactly which way the photon came). $P(\varphi(\delta))$ projects onto the initial state $|\varphi(\delta)\rangle$ and is orthogonal to P(B), whereas δ is the angle between $P(\varphi(\delta))$ and P(A).

Obviously the transparency γ of the mirror and with it the visibility $S(\gamma)$ of the interference pattern give some kind of information about the photon's particle- and wave nature. A satisfactory formula for the deficiencies of information $H(\gamma,A)$ and $H(\gamma,B)$ about further measurements of P(A) and P(B) in an eigenstate $|C(\gamma)\rangle$ of $P(C(\gamma))$ can be expressed by means of information theory (Cf. [3]).

$$H(\gamma,A) = -|\langle A|C\rangle|^2 \ln(|\langle A|C\rangle|^2) - |\langle \neg A|C\rangle|^2 \ln(|\langle \neg A|C\rangle|^2)$$

$$H(\gamma,B) = -|\langle B|C\rangle|^2 \ln(|\langle B|C\rangle|^2) - |\langle \neg B|C\rangle|^2 \ln(|\langle \neg B|C\rangle|^2)$$

With the corresponding expectation values:

$$|\langle B|C\rangle|^2 = 1-\gamma, \quad |\langle \neg B|C\rangle|^2 = \gamma, \qquad 0 \le \gamma \le 1/2$$

$$|\langle B|C\rangle|^2 = \gamma, \quad |\langle \neg B|C\rangle|^2 = 1-\gamma, \qquad 1/2 \le \gamma \le 1$$

$$|\langle A|C\rangle|^2 = \frac{1}{2}(1+2\sqrt{(1-\gamma)\gamma}\,) = \frac{1}{2}(1+S(\gamma))$$

$$|\langle \neg A|C\rangle|^2 = \frac{1}{2}(1-2\sqrt{(1-\gamma)\gamma}\,) = \frac{1}{2}(1-S(\gamma))$$

one obtains

$$H(\gamma,B) = -\gamma\ln\gamma-(1-\gamma)\ln(1-\gamma)$$

$$H(\gamma,A) = -\frac{1}{2}(1+S(\gamma))\ln(\frac{1}{2}+S(\gamma)))-\frac{1}{2}(1-S(\gamma))\ln(\frac{1}{2}(1-S(\gamma)))$$

These functions are shown in the following figure.

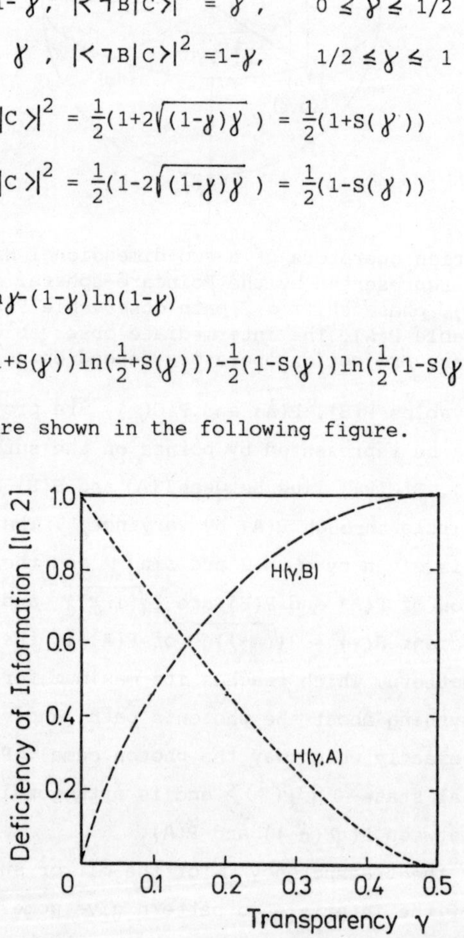

Fig. 3: The deficiencies of information $H(\gamma,A)$ and $H(\gamma,B)$ about P(A) and P(B) in the state $|C(\gamma)\rangle$ as functions of the transparency γ.

3. THE PHOTON SPLIT BEAM EXPERIMENT

The practical realization of the experiment differs slightly from the described arrangement. Since the transparencies β and γ of the partly reflecting mirrors $(BS)_1$ and $(BS)_2$ respectively are not exactly known, these transparencies must also be determined experimentally. If one considers the more general situation with arbitrary values β and γ for the transparencies of $(BS)_1$ and $(BS)_2$, the decomposition of the initial state reads

$$|\varphi(\beta,\delta)\rangle = \sqrt{1-\beta}\,|B\rangle + e^{i\delta}\sqrt{\beta}\,|\neg B\rangle$$

for the special value $\beta = 1/2$ this state agrees with the state mentioned above. In general the angle χ between $P(\varphi(\beta,\delta))$ and $P(B)$ is given by $\chi = 2 \arcsin\sqrt{\beta}$. The intensities

$$I_1(\beta,\gamma) = |\langle C\,|\,\varphi\rangle|^2 \quad \text{and} \quad I_2(\beta,\gamma) = |\langle\neg C\,|\,\varphi\rangle|^2$$

in the detectors D_1 and D_2 respectively are then given by

$$I_1 = (1-\gamma)(1-\beta) + \gamma\beta + 2\sqrt{(1-\gamma)\gamma}\sqrt{(1-\beta)\beta}\cos\delta$$

$$I_2 = (1-\gamma)\beta + \gamma(1-\beta) - 2\sqrt{(1-\gamma)\gamma}\sqrt{(1-\beta)\beta}\cos\delta$$

with $I_1 + I_2 = 1$. These intensitities $I_1(\delta)$ and $I_2(\delta)$ can also be written in the form

$$I_1(\delta) = I_1(\pi/2)(1+S_1\cos\delta),\; I_2(\delta) = I_2(\pi/2)(1-S_2\cos\delta)$$

where the (positive) coefficients S_1 and S_2 are the visibilities

$$S_1 = \frac{2\sqrt{\beta(1-\beta)}\sqrt{\gamma(1-\gamma)}}{1-(\gamma+\beta-2\gamma\beta)}, \qquad S_2 = \frac{2\sqrt{\beta(1-\beta)}\sqrt{\gamma(1-\gamma)}}{\gamma+\beta-2\gamma\beta}$$

of the interference pattern on the detectors D_1 and D_2 respectively.

In order to make use of the results of both detectors D_1 and D_2 in the following we will consider the mean "effective" intensity $I(\delta) := \frac{1}{2}(I_1(\delta)) + I_2(\delta+\pi)$ of the detectors and the corresponding mean "effective" visibility S, which is given by

$$S = 2\left(\frac{1}{S_1} + \frac{1}{S_2}\right)^{-1} = 4\sqrt{\beta(1-\beta)}\sqrt{\gamma(1-\gamma)}$$

Furthermore we define an "effective" transparency γ' by

$$S = 4\sqrt{\beta(1-\beta)}\sqrt{\gamma(1-\gamma)} =: 2\sqrt{\gamma'(1-\gamma')}$$

The effective transparency corresponds to an experimental set-up with transparencies 1/2 and γ' for the beam-splitters $(BS)_1$ and $(BS)_2$, respectively, which leads to the same mean visibility S as the original arrangement.

Since the values of the transparencies β and γ are not exactly known, these values must be measured independently of the measurement of the interference pattern. This can be done by changing the experimental set-up such that one of the ways (B or $\neg B$) is closed. If, e.g., the B-path is closed the intensities in the detectors D_1 and D_2 read $I_1(\neg B) = \gamma\beta$, $I_2(\neg B) = \beta - \gamma\beta$ and the transparencies β and γ can easily be obtained from the measured intensities $I_1(\neg B)$ and $I_2(\neg B)$. If once the values of β and γ are determined in this way, by the formulas mentioned above one can calculate the values of the visibilities $S_1(\beta,\gamma)$ and $S_2(\beta,\gamma)$ which will be considered here as "theoretical" visibilities.

4. EXPERIMENTAL REALISATION

A scheme of the experimental set-up is shown in Fig. 1. A 1mW He-Ne-laser is used as a light source. The incoming laser beam is prepared by a spatial filter telescope SF in order to reduce unwanted diffraction effects. The Mach-Zehnder interferometer consisting of the beam splitters BS_1 (transmission β) and BS_2 (transmission γ), the mirrors M_1 and M_2 and the phase-shifter PS is mounted on a rigid Invar base so that thermal drifts of the optical pathlength become negligible. The influence of atmospheric pressure fluctuations is minimized by positioning the whole interferometer into a plastic box. BS_1 is a 50% metallic layer beamsplitter and BS_2 is a variable dielectric beamsplitter. The transmission can be varied by a horizontal displacement as indicated by an arrow in Fig. 1. In order to avoid unwanted variation of the transmission across the laser beam profile the spatial filter telescope is adjusted to focus precisely on the beamsplitter BS_2.

The phaseshifter PS consists of a wedged glassplate (wedge angle 3°) mounted on the membrane of an ordinary loudspeaker. Supplying a current to the loudspeaker moves the wedge in or out of the beam $\neg B$. This method has the advantage that the parallel displacement of the laser beam is practically zero while tuning. In addition the very small optical pathlength variations are transformed to macroscopic movements according to the small wedge angle. Thus the tuning mechanism is very insensitive to mechanical perturbations and the requirements for the control electronics are very low. A second wedge is placed into the beam B to compensate for a difference in the optical pathlength.

The two output beams are fed through one common polarizer. This guarantees that depolarization of the beams does not influence the interference differently. Two pinholes PH_1 and PH_2 are placed in front of the photomultipliers D_1 and D_2. They select only a small fraction of the beamprofiles so that the interference fringes observed are not deterioriated by averaging over a nonuniform interference pattern across the laser beam profile.

The output signals of the detectors are recorded on a two-channel chart recorder (instead of being counted). This does not influence the interpretation of the experiment according to the following arguments: The maximum current measured at the anode of the photomultipliers corresponded to about 1.5×10^8 photoelectrons/sec. With a quantum efficiency of 8% we had about 2×10^9 photons/sec. on the photocathode. Thus the photon density of the beams reaching the detectors was about 0.07 photons/cm. Since the beam path difference was made very small ($\leq$ 1mm) we can conclude that the probability to detect interference caused by more than one photon simultaneously is negligible.

For adjustment the interference of both output beams is observed. First the interferometer is aligned to show as little spatial variation of the interference fringes as possible. After that the signals of both detectors are monitored on an oscilloscope while supplying the loudspeaker with a ramp. Since the two signals should be sinewaves precisely 180° out of phase, the relative position of the two pinholes

PH_1 and PH_2 can be accurately adjusted to transmit the same portion of the laser beam.

The next step is to measure the currents of both detectors while blocking one of the arms in the interferometer and then the other. One gets four signals which are

	Detector 1	Detector 2
$\neg B$ closed; B open:	$d_1(1-\beta)(1-\gamma)$	$d_2(1-\beta)\gamma$
$\neg B$ open; B closed:	$d_1\beta\gamma$	$d_2\beta(1-\gamma)$

The factors d_1 and d_2 take into account the different sensitivities of both detectors. From these measurements one can calculate the transmissions β and γ. β should have a value of 1/2 but it varies around this value with ± 0.03. This represents the fact that the beams B and $\neg B$ did not overlap identically for the different measurements because the interferometer had to be realigned for every new value of γ.

After this determination of the transmission coefficents β and γ the interference pattern of both detectors is recorded simultaneously while the phase shifter is tuned supplying the loudspeaker with a ramp of about 5 sec. duration. Two examples are shown in Figure 4. The upper part shows the interference with almost equal intensity of the interfering beams, i.e. with an effecitve transparency $\gamma' \approx 0.5$ and an almost vanishing particle property. The lower part of Figure 4 demonstrates a surprisingly strong interference signal although one of the beams contributes only about 6% of the intensity on the detectors. The value of the effective transparency is given here by $\gamma' = 0.006$ which means that there is only less than 2% wave property left. This case demonstrates very clearly how the wave nature is still obvious while the path through the interferometer is already well determined.

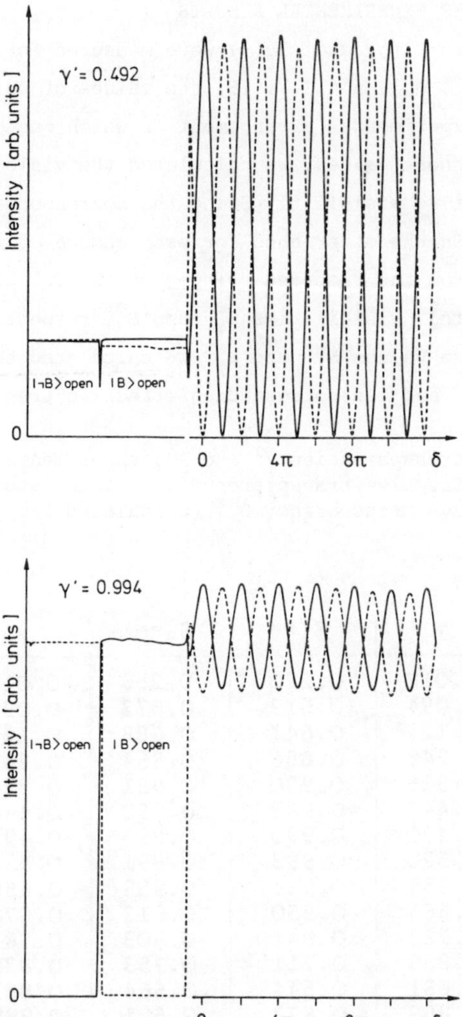

Fig. 4: Interference pattern for two effective transparencies γ'
Upper part: γ' = 0.492, i.e. 0.02% particle property, 99.98% wave property and an effective visibility S^{ex} = 0.992.
Lower part: γ' = 0.006, i.e. 98.2% particle property, 1.8% wave property and an effecitve visibility S^{ex} = 0.145.

5. DISCUSSION OF THE EXPERIMENTAL RESULTS

The intensities $I_1(\delta)$ and $I_2(\delta)$ were measured for 18 different transparencies (β_n, γ_n), n = 1,...18, the values of which were first determined by an experimental arrangement in which one of the two ways was closed. Using these values, we calculated the visibilities S_1 and S_2, the mean effective visibilitiy S and the corresponding effective transparency γ'. On the other hand for each choice of the transparencies (β_n, γ_n) from the recorded fringes at the detectors D_1 and D_2 the two "experimental" visibilities S_1^{ex} and S_2^{ex} respectively could be obtained. From these values S_1^{ex} and S_2^{ex} we calculated the mean effective visibility S^{ex} and the corresonding effective transparency γ'_{ex}.

Table: Calculated transparencies β and γ; experimental visibilities S_1^{ex}, S_2^{ex}; effective transprarency γ' (calculated from β and γ); mean effective transparency S^{ex} (calculated from S_1^{ex} and S_2^{ex}).

β	γ	S_1^{ex}	S_2^{ex}	γ'	S^{ex}
0.512	0.018	0.270	0.258	0.018	0.270
0.521	0.094	0.619	0.573	0.094	0.595
0.477	0.127	0.647	0.688	0.126	0.667
0.487	0.248	0.856	0.864	0.248	0.860
0.469	0.325	0.920	0.951	0.323	0.935
0.480	0.448	0.982	0.990	0.444	0.986
0.505	0.490	0.990	0.993	0.492	0.992
0.477	0.520	0.993	0.991	0.531	0.992
0.524	0.656	0.934	0.958	0.658	0.946
0.460	0.666	0.950	0.913	0.670	0.931
0.484	0.783	0.841	0.803	0.783	0.822
0.528	0.838	0.711	0.753	0.839	0.732
0.511	0.881	0.644	0.664	0.881	0.654
0.475	0.889	0.673	0.621	0.889	0.646
0.472	0.940	0.515	0.466	0.940	0.489
0.485	0.953	0.459	0.423	0.953	0.440
0.479	0.990	0.224	0.181	0.990	0.200
0.490	0.994	0.147	0.144	0.994	0.145

The experimental visibility S_1^{ex} and S_2^{ex} are to be compared with the visibilities S_1 and S_2 calculated from β and γ. They agree very well within an RMS experimental error of 0.008. The error is due to the reading from the recorded fringes, instabilities of the laser power and - this seems to be the most important contribution - due to additional small fringes resulting from reflection on the various surfaces the beams have to cross in the interferometer. Part of these fringes are removed with additional apertures in and outside the interferometer since most of the reflections are displaced with respect to the main beam. With a more sophisticated set-up it is very likely that an accuracy for β and γ about two orders of magnitude better than derived here can be obtained. Therefore this technique could be a very useful method for measuring reflectivities of mirrors, index of refraction of unknown materials etc.

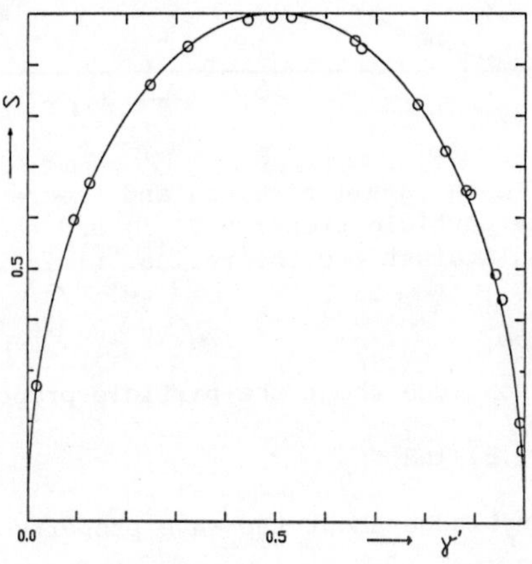

Fig. 5: The mean effective visibilities S^{ex} (small crosses) and S (solid line) plotted against the theoretical effective transparency γ'. The RMS experimental error is 0.007.

In Fig. 5 the theoretical and the experimental mean effecive visibilities S and S^{ex} are plotted against the theoretical effective transparency γ'. The experimental values of S^{ex} are drawn by small circles, the theoretical function $S = S(\gamma')$ by a solid line. The values of S and S^{ex} agree very well within an RMS experimental error of 0.007.

The most important problem which is treated in this paper is the question whether the particle and the wave property of a photon can be measured simultaneously.

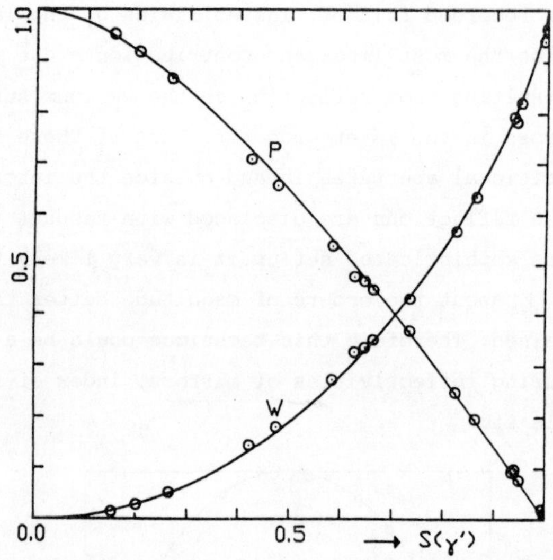

Fig. 6: Experimental data (small circles) and theoretical values (solid lines) of the particle property $P(\gamma')$ and the wave property $W(\gamma')$ plotted against the theoretical visibility $S(\gamma')$.

In Fig. 6 we show

(1) the normalized knowledge about the particle property of the photon

$$P(\gamma') = 1 - H(\gamma',B)/\ln 2$$

(2) the normalized knowledge about the wave property of the photon

$$W(\gamma') = 1 - H(\gamma',A)/\ln 2$$

plotted against the mean effective visibility $S(\gamma')$. The experimental values P^{ex} and W^{ex} (small circles) are calculated from the corresponding values of γ'_{ex}, the theoretical values $P(\gamma')$ and $W(\gamma')$ (solid lines) are obtained from the values of γ'. - Theoretical and experimental results are again in very good agreement and show the expected behaviour: A photon possesses simultaneously particle properties and wave properties, the degrees of which are given by the values of $P(\gamma')$ and $W(\gamma')$, respectively. Even for a very high value of the particle property $P(\gamma')$ there is still a nonvanishing amount of the wave property $W(\gamma')$ which leads to an observable and significant interference pattern.

This statement is illustrated very well by the two examples for the interference patterns which are shown in Fig. 4 :

1) For $\gamma' = 0.492$ (above) we have $P(\gamma') = 1.85\cdot 10^{-4}$ and $W(\gamma') = 0.999$, i.e. a photon with 0.02% particle property and 99.98% wave property. The mean visibility $S^{ex} = 0.992 \simeq S$ corresponds to a very significant interference pattern.
2) For $\gamma' = 0.994$ (below) we have $P(\gamma') = 0.947$ and $W(\gamma') = 0.017$, i.e. a photon with 98.2% particle property and only 1.8% wave property. Even in this extreme situation the visibility $S^{ex} = 0.145 \simeq S$ is still sufficiently high for a significant interference pattern.

These results demonstrate in a very convincing way that the measurement of the observable $C(\gamma)$ provides an unsharp simultaneous measurement of the photon's particle property as well as of its wave property. In accordance with previous theoretical expectations[3] it follows that even if the path of the photon is known to a degree of 98.2%, its wave character still manifests itself in a surprisingly significant interference pattern. Hence we find that there is in general only an approximate duality between the particle property and the wave property of the photon, except when one of these properties has been measured exactly.

REFERENCES

1) Bohr, N., "Discussion with Einstein on epistemological problems in atomic physics", in: P.A. Schilpp (ed.), Albert Einstein: Philosopher-Scientist, p. 199, Library of Living Philosophers, Evanston, Ill., 1949.
2) Bohr, N., "The Quantum postulate and the recent development of atomic theory", Nature 121, 580 (1928).
3) Wootters, W.K. and W.H. Zurek, "Complementarity in the double slit experiment: Quantum nonseparability and a quantitative statement of Bohr's principle", Phys. Rev. D 19, 473 (1979).

4) Busch, P., "Some Realizable Joint Measurements of Complementary Observables", Foundations of Physics 1987 (in print).

5) Beltrametti, E. and G. Casinelli, The Logic of Quantum Mechanics, p. 31, Addison-Wesley Publishing Co., Reading, Mass. (1981).

6) Mittelstaedt, P., "Quantum Logical Analysis of Delayed-Choice Experiments" in: Namiki et al. (eds.), The Proceedings of the 2nd Symposium on Foundations of Quantum Mechanics - in the Light of New Technology, Tokyo 1986, Physical Society of Japan (1987).

IS BOHM'S THEORY A NONLOCAL HIDDEN VARIABLES THEORY?

Willem M. de Muynck
Department of Theoretical Physics,
Eindhoven University of Technology
Eindhoven, The Netherlands

ABSTRACT

Three different kinds of hidden variables theories are discussed and compared with Bohm's theory. It is demonstrated that Bohm's theory does not satisfy the requirement that a reasonable account of the measurement process is given. The relation between the nonlocality issue of the Bell inequalities and Bohm's nonlocal quantum potential is discussed.

1. INTRODUCTION

One reason for Bell to doubt the validity of von Neumann's impossibility proof of hidden variables may have been the idea that Bohm's theory[1)] is yielding a counterexample to this impossibility thesis. As is well known Bohm's theory is completely equivalent with quantum mechanics, at least as far as experiments are concerned that fall within the domain of quantum mechanics. Hence, if Bohm's theory is a hidden variables theory von Neumann's proof cannot be correct. Once thus far, it was not difficult for Bell[2)] to prove the failure of von Neumann's proof.

The nonlocality of Bohm's quantum potential subsequently suggested a direction for Bell to look into in order to obtain his own impossibility proof. If Bohm's theory is a *nonlocal* hidden variables theory, such theories seem capable of being equivalent with quantum mechanics. Hence, at most *local* hidden variables theories seem to be excluded. This idea must have led Bell[3)] to the construction of the inequalities, named after him, which, in consonance with Bohm's ideas, are interpreted by the vast majority of physicists as a proof of the nonlocality of the quantum world.

The ideas of Bohm and Bell supplement each other in a very suggestive way. Consequently, when we resist the interpretation of the Bell inequalities as being due to a locality assumption[4], it seems necessary to contemplate also Bohm's theory from the same point of view.

2. BOHM'S THEORY

Point of departure of Bohm's theory is the possibility to write the Schrödinger equation

$$i\hbar \frac{\partial \psi}{\partial t} = -\frac{\hbar^2}{2m} \Delta\psi + V\psi \tag{1}$$

in a way that suggests a classical interpretation. If we put

$$\psi = R\, e^{iS/\hbar}, \ R \text{ and } S \text{ real}, \tag{2}$$

then the following equations for R and S can be derived:

$$\frac{\partial P}{\partial t} + \operatorname{div}\left(P \frac{\vec{\nabla} S}{m}\right) = 0, \qquad P = R^2 \tag{3}$$

$$\frac{\partial S}{\partial t} + \frac{(\vec{\nabla} S)^2}{2m} + V + Q = 0, \qquad Q = -\frac{\hbar^2}{2m}\frac{\Delta R}{R} \tag{4}$$

Equation (3) is a continuity equation for the transport of the quantity $P = |\psi|^2$ if $\vec{\nabla}S/m$ can be interpreted as the local transport velocity. In the usual quantum mechanical interpretation this relation is interpreted as a law of probability conservation and expressed in terms of the probability flux

$$\vec{J} = \frac{i\hbar}{2m}(\psi^* \overleftrightarrow{\nabla} \psi - \psi \vec{\nabla} \psi^*) = P\frac{\vec{\nabla} S}{m}, \tag{5}$$

P being interpreted as a probability density.

It was the peculiar form of equation (4) that induced Bohm to propose a different interpretation, in which

$$\vec{p} = \vec{\nabla} S(\vec{x}) \tag{6}$$

is the "classical" momentum of the particle if the latter is situated at position $\vec{x}$. Indeed, equation (4) is the "classical" Hamilton-Jacobi equation of a particle acted upon by a potential V+Q. The extra quantum potential Q represents the deviation from purely classical behavior. It, hence, should describe the quantum mechanical fluctuations as embodied by the wave function $\psi(\vec{x})$. In Bohm's view the wave function is to be interpreted as a description of a real field acting on the particle. In later work[5)] Q is indicated as a quantum *information* potential in order to point out that the quantum potential does not represent a mechanical force but should rather be considered as something like an "information content", to be compared with "radar waves that guide a ship".

The quantum potential has a nonlocal character. This becomes especially evident if the procedure of equation (2) - (4) is applied to the Schrödinger equation of two particles. Unless the two-particle wave function is a product of single-particle wave functions (which obtains if the particles are completely uncorrelated), the two-particle quantum potential seems to describe an interaction that does *not* vanish if the interparticle distance becomes infinite. The particles seem to influence each other however big their distance. Since, as is well known, it is impossible to use this nonlocality for superluminal signalling, as a first point of criticism it should be remarked that it is not explained by Bohm how it is possible that such nonlocal interactions do not lead to a violation of relativistic causality.

3. SOME OBJECTIONS AGAINST BOHM'S THEORY

An objection against Bohm's theory, already raised by Pauli[6)] against a theory by de Broglie[7)], is formulated by Einstein[8)] as follows:

Consider a one-dimensional particle in an eigenstate

$$\psi(x) = A \cos kx \qquad (7)$$

of an infinite well potential. Then,

$$p = \frac{\partial S}{\partial x} = 0$$

contrary to the values $\pm \hbar k$ that follow from developing $\psi(x)$ into plane waves. More generally, it is clear that $\vec{\nabla} S = 0$ for any wave function $\psi(\vec{x})$ that is real-valued. An analogous objection was also raised by Rosen[9], who, for this reason, deemed a similar theory "hard to accept".

As a reaction to Einstein's objection Bohm[10] refers to his theory of measurement which "takes into account the disturbance of the observed system by its interaction with the measuring apparatus". The momentum measurement, then, should transform the wave function $\psi(x)$ "into a wave packet, with momentum approximately $+\hbar k$ or $-\hbar k$, and with equal probability of either result". Hence, the momentum measurement should transform the pre-measurement value $\vec{p} = 0$ into one of the post-measurement values $\pm \hbar\vec{k}$. The pre-measurement value $\vec{p} = 0$ cannot be measured within the quantum mechanical domain. All quantum mechanical measurements have this disturbing character, except, perhaps, position measurement, which seems to be interpretable as measuring the particle position without disturbing it.

The distinction between pre- and post-measurement values has been a severe reason for criticizing Bohm's theory because of the alleged metaphysical character of pre-measurement quantities[11]. I do not think that such criticisms are decisive, since, as also done in a reply[12] by Bohm, rightly can be referred to the possibility of future measurements falling outside the domain of quantum mechanics and yielding values that are closer to the pre-measurement ones than the quantum mechanical measurement results. For me, a more important point of concern is the fact that, in order to justify his theory of measurement, Bohm relies on a *quantum mechanical* theory of measurements of the first kind[1],[13], in which the initial state Ψ_i of the combined system of object and measuring instrument

$$\Psi_i = \sum_n c_n \psi_n(\vec{x}) f_o(\vec{y}) \tag{8}$$

is transformed by the measuring process into the final state

$$\Psi_f = \sum_n c_n \psi_n(\vec{x}) f_n(\vec{y}), \tag{9}$$

$f_o(\vec{y})$ and $f_n(\vec{y})$ being initial and final "pointer" states of the measuring instrument, each final state corresponding with an eigenstate $\psi_n(\vec{x})$ of the measured observable. In the conventional interpretation this quantum mechanical theory of measurement is viewed upon as a reflection of Heisenberg's ideas[14)] with regard to quantum mechanical measurement, to the effect that the measurement process disturbs quantities that are incompatible with the observable that is actually measured, but the measured observable itself is not disturbed. This picture of what is going on in a measurement evidently is very different, and even incompatible, with Bohm's views. This difference is explicitly acknowledged by Bohm (ref. 1, p. 183). It is remarkable, then, and also a bit disturbing, that the relation between the pre-measurement value $\vec{p} = \vec{\nabla}S$ of momentum and the momentum value that is actually obtained in a momentum measurement, should be explained by means of a measurement theory that is developed according to a completely different, and even incompatible, picture of what is happening in a measurement process. Also, the pre-measurement value $\vec{p} = \vec{\nabla}S$ does not have a counterpart in the quantum mechanical theory of measurement processes of the first kind. What is clearly missing from Bohm's theory is a detailed account of the way the pre-measurement value of $\vec{p}$ is transformed into the quantum mechanical measurement result.

4. DIFFERENT INTERPRETATIONS OF QUANTUM MECHANICAL PROBABILITIES

In interpreting quantum mechanics we try to answer the question of the physical significance of the mathematical quantities obtaining in the theory. In this section I want to discuss possible meanings of the arguments of quantum mechanical probability distributions $W(a_i)$, in which the mathematical meaning of a_i is that of an eigenvalue of a selfadjoint operator representing a quantum mechanical observable. The question is: "What do we mean by saying that $W(a_i)$ is the relative frequency that observable A has the value a_i?", or analogous questions for the joint probabilities of several quantum mechanical observables.

In discussing *interpretations* of quantum mechanics we need not go beyond quantum mechanics. Any quantity to be encountered will, for this reason, be considered to be *a quantum mechanical quantity*, that is, some expectation value of a selfadjoint operator A:

$$\langle A\rangle = \mathrm{Tr}\, \rho\, A. \tag{10}$$

Here ρ is the density operator, representing a preparation procedure which, by definition, is in the domain of quantum mechanics. The expectation value is a linear functional of ρ. By the same token A represents a quantum mechanical measurement procedure. If

$$A = \sum_i a_i P_i,\ P_i P_j = P_i \delta_{ij}, \tag{11}$$

then the probability distribution $W(a_i)$ is given by a linear functional of the density operator,

$$W(a_i) = \mathrm{Tr}\, \rho\, P_i. \tag{12}$$

If $\{R_i\}$ is a set of positive operators satisfying

$$0 \leq R_i \leq I, \tag{13}$$

$$\sum_i R_i = I,$$

then the linear quantities

$$q_i = \mathrm{Tr}\, \rho\, R_i \tag{14}$$

are quantum mechanical expectations obeying

$$0 \leq q_i \leq 1,$$

$$\sum_i q_i = 1, \tag{15}$$

and , hence, are interpretable as probabilities. The set $\{R_i\}$ is called a positive operator valued (POV) measure, a projection valued measure $\{P_i\}$ being a special case. By means of POV-measures it is possible to define, *within the domain of quantum mechanics*, more general joint probability distributions

$$q_{ij} = \mathrm{Tr}\, \rho\, R_{ij}, \tag{16}$$

$$0 \leq R_{ij} \leq I,$$

$$\sum_{ij} R_{ij} = I \tag{17}$$

than the ones traditionally encountered in quantum mechanics, the latter generally being represented as

$$W_{ij} = \text{Tr}\, \rho\, P_i Q_j, \tag{18}$$

in which $\{P_i\}$ and $\{Q_j\}$ are projection valued measures satisfying $[P_i, Q_j] = 0$.

I shall now give a short discussion of several possibilities of interpreting the arguments of these probability distributions. In the following I shall distinguish two kinds of *realistic* interpretations and a *phenomanalist/instrumentalist* one.

The language of realistic interpretations is, like the language of classical mechanics, a language of properties. Properties refer to quantities that may be intrinsic (i.e. independent of the state of the object, like e.g. the rest mass of a particle), or extrinsic (i.e. dependent on the state of the object, like e.g. position and momentum of the particle, both liable to change with time). In classical mechanics a property generally is an *objective* property, that is, a property that can be attributed to the object independently of its surroundings. This, essentially, implies that the object is considered to be isolated from the rest of the world (apart perhaps from certain fields that are taken into account as parameters in the Hamiltonian).

Sometimes properties seem to be attributable to the object only if this object is situated in certain environments or contexts. For this reason it seems necessary to take into account also the possibility of *contextual* properties the value, or even the existence, of which depends on the environment. In quantum mechanics this environment is determined by the preparation and measurement set up. A contextual quantity does not have an unambiguous meaning, but depends on the measurement procedure. Not only its measured value is dependent on this procedure, but even the *definition* of the measured quantity may depend on it. Thus, the measured value may be associated with either the *final* value of the contextual quantity (i.e. *after* completion of the measurement) or with some average value obtained by averaging over the measuring process. What is essential, here, is that the value a_i

of the observable A can be attributed to the object only as far as the object is influenced by the measurement arrangement.

The language of instrumentalist/phenomenalist interpretations is a language of phenomena. Phenomena are often introduced[15], in the context of empiricism, as "observable phenomena" or "appearances" in order to represent the *empirical* substructures of a theory. For the present purpose it suffices to characterize a phenomenon as an *observable* part of reality. Within the domain of classical mechanics, any property can be considered to be observable. Hence, within the domain of classical mechanics every phenomenon can be taken to correspond to a property of the object itself: the classical world is observed "as it is". The event of a macroscopic particle having velocity $\vec{v}(t)$ at time t is both a property of the particle, *and* a phenomenon in case this property is actually observed.

The situation is different in microphysics because here a macroscopic measuring instrument acts as an intermediary between object and observer. The observer has only access in a direct way to the measuring instrument, not to the object system. Therefore the phenomenon should be attributed as a property to the measuring instrument rather than to the object. Thus, a track in a Wilson or bubble chamber is a phenomenon playing a role in the detection of an elementary particle. The track consists of a great number of small droplets or bubbles, the development of which is caused by a particle crossing the chamber. For this reason the track provides us with information about the particle. However, it is not a property of the particle itself, but should be considered, - in the sense of classical mechanics - , as a property of the chamber. Although the existence of a particle trajectory is made highly plausible by the existence of the track, the track may not be identified with the trajectory. Actually, the same track could be caused by many different trajectories!

A pure phenomenalist/instrumentalist interpretation considers quantum mechanics as a theory about measurement results of quantum mechanical measurements, without any necessity to attribute the measurement result to the object itself. The theory just yields predictions about the reaction of certain macroscopic measuring instruments if certain macroscopic preparation procedures are performed. Then, the quantum mechanical probabilities are just probabilities of measurement results, i.e. of properties of the measuring instruments.

In retrospect, we can distinguish three possible interpretations of the argument a_i in a quantum mechanical probability distribution $W(a_i)$:

1) a_i is an objective property of the object,

2) a_i is a contextual property of the object,

3) a_i is a phenomenon, i.e. a property of the measuring instrument.

5. PROBABILITY DISTRIBUTIONS IN HIDDEN VARIABLES THEORIES

If it is assumed that quantum mechanics can be reproduced by means of a realistic (hidden variables) theory, this classification yields different representations of quantum mechanical probabilities. Let Λ denote the configuration space of hidden variable λ, and $\rho(\lambda)$ its probability distribution. Then, in the objectivistic case 1 we can write

$$W(a_i) = \int_{\Lambda_i} d\lambda\, \rho(\lambda) = \int_{\Lambda} d\lambda\, \chi_{\Lambda_i}(\lambda)\, \rho(\lambda), \tag{19}$$

in which Λ_i is the subset of Λ for which $A(\lambda) = a_i$ and $\chi_{\Lambda_i}(\lambda)$ is the indicator function of Λ_i. It should be noticed that, as far as the hidden variables theory is intended to reproduce quantum mechanics and $\rho(\lambda)$ is thought to represent a quantum mechanical state, the probability $W(a_i)$ should also be expressible as a linear functional (12) or (14) of a density operator. A general hidden variables theory, however, might have a wider scope, and contain expressions of the form

(19) that are not expressible as linear functionals of a density operator. Moreover, (19) might refer to properties that do not correspond with quantum mechanical observables.

In the contextualistic case (case 2) we take into account the possible influence of the detection process on the resulting probability. This can be done by introducing a stochastic operator[16] $\mathcal{T}_\lambda$ which transforms $\rho(\lambda)$ into a new probability distribution, thus yielding

$$W(a_i) = \int_{\Lambda_i} d\lambda \, \mathcal{T}_\lambda \rho(\lambda) = \int_{\Lambda} d\lambda \, \rho(\lambda) \, \mathcal{T}^*_\lambda \, \chi_{\Lambda_i}(\lambda), \tag{20}$$

$\mathcal{T}^*_\lambda$ being the adjoint of the stochastic operator $\mathcal{T}_\lambda$. If we write

$$p(a_i|\lambda) = \mathcal{T}^*_\lambda \chi_{\Lambda_i}(\lambda), \tag{21}$$

we can interpret $p(a_i|\lambda)$ as the conditional probability of a_i for given λ. This quantity, evidently, depends on the measuring process. Sometimes it is assumed that $W(a_i)$, nevertheless, can be attributed to the object as an objective property. This would be reasonable if there would exist only one single procedure to measure observable A. We might call this the *quasi-objectivistic* case. Even then, however, it would not be allowed to attribute $W(a_i)$ to the object in a context that is different from this single measurement context.

If we want to take into account explicitly the interaction between object and measuring instrument we should extend the description to include the space M_A of the hidden variables μ of the measuring apparatus for observable A. Introducing the stochastic operator $\mathcal{T}^A_{\lambda\mu}$ describing this interaction, we obtain for the contextualistic case

$$W(a_i) = \int_{\Lambda_i} d\lambda \int_{M_A} d\mu \, \mathcal{T}^A_{\lambda\mu} \, \rho(\lambda) \, \rho_A(\mu), \tag{22}$$

$\rho_A(\mu)$ being the initial probability distribution of the hidden variable of the measuring instrument. This expression can be reduced to

$$W(a_i) = \int_\Lambda d\lambda \;\; p(a_i|\lambda)\, \rho(\lambda), \tag{23}$$

with

$$p(a_i|\lambda) = \int_{M_A} d\mu \;\rho_A(\mu)\, \mathcal{T}^{A*}_{\lambda\mu}\, \chi_{A_i}(\lambda). \tag{24}$$

So, introducing the dynamics of the measurement interaction does not yield much news over the case of equation (20) in which the dynamics of the measuring apparatus is not taken into account explicitly.

The instrumentalist/phenomenalist case, finally, is represented by the expression

$$W(a_i) = \int_\Lambda d\lambda \int_{M_{A_i}} d\mu \;\mathcal{T}^{A}_{\lambda\mu}\, \rho(\lambda)\, \rho_A(\mu) \tag{25}$$

which differs from (22) in the boundary of the domain of integration, M_{A_i} being the subset of M_A the apparatus finally is in if the final state is to be interpreted as the measurement result a_i. Although, also here, it is possible to write (25) in the form (23), with

$$p(a_i|\lambda) = \int_{M_A} d\mu \;\rho_A(\mu)\, \mathcal{T}^{A*}_{\lambda\mu}\, \chi_{A_i}(\mu) \tag{26}$$

($\chi_{A_i}(\mu)$ being the indicator function of M_{A_i}), there is a fundamental difference with the other cases: it is now impossible to attribute the probability $W(a_i)$ to the object system as an objective property, since the probability evidently only exists if the measuring instrument is present.

6. BOHM'S INTERPRETATION

I shall now try to answer the question which of the three interpretations is applicable to Bohm's theory. The problem here is that not all assertions in the articles by Bohm and cooperators are equally clear. Some even appear to be inconsistent. Thus, in a recent review article[5b)] we read the following: "We feel that these examples are sufficiently broad in their applications to indicate how the ontology works out more generally, and to enable us to understand individual quantum processes *without having to bring in measurement*" (p.324, my italics). This suggests that the quantities should be considered as objective properties of the particle (case 1). This is also apparent from p. 325 where the probability density of position is interpreted as "the probability density for the particle to *be* at a certain position" (their emphasis), and where is stated explicitly that "here we differ from Born who supposed that it was the probability of *finding* the particle there in a suitable measurement" (their emphasis). This, in any case, seems to exclude the third, instrumentalist/phenomenalist interpretation.

Bohm is well aware of the important role the measuring apparatus plays in the measurement process. As he puts it in the second part of ref. 1: " Observables are not properties belonging to the observed system alone, but instead potentialities whose precise development depends just as much on the observing apparatus as on the observed system". This points into the direction of the second, contextualistic interpretation. Indeed, Bohm distinguishes the pre-measurement value $\vec{p} = \vec{\nabla}S$ of momentum (he calls this the "actual" value) from the value obtained in a quantum mechanical measurement. As already discussed before these are different in general. Whereas the probability distribution of the measured values is given by the usual quantum mechanical expression

$$W(\vec{p}) = |a(\vec{k})|^2 = \mathrm{Tr}\ \rho\ P_{\vec{p}}\ , \quad P_{\vec{p}} = |\vec{k}\rangle\langle\vec{k}|,\ \rho = |\psi\rangle\langle\psi|,\ \vec{p} = \hbar\vec{k} \tag{27}$$

$a(\vec{k})$ being the Fourier transform of $\psi(\vec{x})$, the probability of the actual value of momentum seems to be derivable from $\psi(\vec{x})$, using (6), as

$$W_{act}(\vec{p}) = \int_{\vec{\nabla}S(\vec{x})=\vec{p}} d\vec{x}\,|\psi(\vec{x})|^2 \tag{28}$$

the integration being restricted to that part of configuration space for which $\vec{\nabla}S(\vec{x}) = \vec{p}$. Indeed, for real-valued $\psi(\vec{x})$ this part amounts to the whole space for $\vec{p} = 0$, and vanishes for $p \neq 0$, in accordance with previous results. On the other hand, for real $\psi(\vec{x})$ the probability distribution (27) is only restricted by the relation

$$a(\vec{k}) = a^{*}(-\vec{k}). \tag{29}$$

This, evidently, does not imply $W(\vec{p})$ to vanish for $\vec{p} = \hbar\vec{k} \neq 0$.

It, incidentally, should be remarked that the probability distribution $W_{act}(\vec{p})$ does not have a quantum mechanical representation of the form (12). Formally, (28) can be written as an expectation value of a projection operator, viz. the one projecting a wave function $\varphi(\vec{x})$ according to

$$P'_{\vec{p}}\ \varphi(\vec{x}) = \chi_{\vec{p}}(\vec{x})\ \varphi(\vec{x}), \tag{30}$$

$\chi_{\vec{p}}(\vec{x})$ being the indicator function of the region of configuration space for which $\vec{\nabla}S(\vec{x}) = \vec{p}$. However, now $P'_{\vec{p}}$ depends on the wave function $\psi(\vec{x})$, thus causing (28) to deviate from linearity. Evidently, if the "actual" value $\vec{p} = \vec{\nabla}S(\vec{x})$ is interpreted in the way Bohm proposes, then it can not be viewed as a quantum mechanical quantity. This need not bother us too much, because it is possible to deny the status of a quantum mechanical quantity to quantities having values that can be interpreted as *pre*-measurement values. This would be tantamount to saying that not the objectivistic probabilities (19) but the contextualistic ones (23) should reproduce the quantum mechanical probabilities in case the preparation and measurement procedures are within the domain of quantum mechanics. This is a perfectly reasonable position. On the other hand, as stated before, without a measurement theory relating "actual" and measured values of a quantity, it is rather arbitrary to declare the result of any algorithm, however suggestive this algorithm may be, as the "actual" value of a quantity.

One of Bohr's leading principles in guessing the generalisation of classical mechanics into the domain of quantum mechanics was the correspondence principle. The validity of this principle induces certain similarities between the different theories. Bohm's representation of the Schrödinger equation is a clear manifestation of this similarity. Another manifestation is Ehrenfest's theorem yielding still another way to represent quantum mechanics in a manner that is very suggestive of classical mechanics. Both representations are mathematically equivalent with the Schrödinger equation. Since nobody seems to deny that the latter equation describes *only* statistical properties of the quantum world, we should have solid arguments if we want to interpret essentially the same mathematics in a different way. In fact, we have not. On the contrary, as indicated in equation (5), the quantity $\vec{\nabla}S$ has, in the "usual" interpretation, a *statistical* meaning which, in principle, lends itself to experimental verification. The equality $\vec{p} = 0$, obtaining for real-valued wave functions, can be interpreted as the, in principle observable, *statistical* result that the probability flux $\vec{J}$ vanishes, consistent with the vanishing of the quantum mechanical expectation value $\langle\vec{p}\rangle$ resulting from (29). It does not seem to be very reasonable to suppose that the same mathematical quantity has still another, completely different meaning, viz. that of a pre-measurement value. This is not to deny the possibility that a particle may have an "actual" momentum value. It, however, appears to be very improbable that this value can be derived from an approvedly statistical theory like quantum mechanics. For this reason an interpretation of Bohm's theory as yielding "actual" pre-measurement values is hardly plausible.

Analogous remarks can be made with respect to the nonlocality of the quantum potential. If this is interpreted as a nonlocal interaction, we are confronted with the problem of explaining the principle of local commutativity in quantum mechanics which is an expression of the mutual nondisturbance of quantum mechanical measurements performed at a space-like distance from each other. It remains completely unclear why such a nonlocal interaction would be contrived in such a way that in EPR-like experiments the probability distribution of one observable is not disturbed by the other measurement. Why does this nonlocal interaction hide itself from detection?

Such problems disappear if it is realized that the quantum potential formalism is equivalent with the wave function formalism and, hence, should encompass the same experimental information, i.e. probabilities like the ones given in equations (12)-(18) and the expectation values calculated from them. More specifically, the joint probability distributions of EPR-like measurements are described by (18) if $\{P_i\}$ and $\{Q_j\}$ are two spectral representations of observables of each of the particles separately. Unless the density operator ρ can be written as a product $\rho_1\rho_2$ of two single-particle density operators the particles are said to be statistically correlated. It is interesting to observe that precisely under this circumstance of statistical correlation the nonlocal character of the quantum potential becomes manifest. For this reason it is rather natural to view the nonlocality of the quantum potential as an expression of the statistical correlation between distant particles. Since such correlations can always be understood from a common preparation process there is no nonlocal interaction involved. It is even misleading to speak about *nonlocal* correlations because the correlation can readily be explained in terms of local interactions being effective when the particles were near each other.

Since the Bell inequalities are often invoked to disprove the possibility of the picture presented here, this issue is discussed in the next section.

7. RELATION WITH THE BELL INEQUALITIES

After the derivation of the Bell inequalities EPR-like experiments generally have been analyzed in terms of these inequalities. Although it is doubtfull whether this is the most appropriate way to do such an analysis[17),18)], in the present context I will follow the same route. It is a widespread belief that the Bell inequalities are due to a locality assumption.

Consider EPR-like measurements performed on a system of two particles, A_1 and B_1 being incompatible observables of particle 1 and analogously A_2 and B_2 for particle 2. By Rastall[19)] it was demonstrated that, under the assumption of macrolocality[17], for the validity of the Bell inequalities it is necessary and sufficient that a joint probability distribution (jpd) exists of the four observables that are involved. If such a jpd exists it is impossible that the measurement results of all EPR-like experiments can be derived from it by taking marginals. Evidently, such a jpd does not describe the quantum world.

By Dieks[20] the nonexistence of such a jpd is interpreted, once again, as a consequence of nonlocality: since the value of observable A_2 may, because of nonlocality, depend on the circumstance whether we measure A_1 or B_1 on the other particle, it is impossible to attribute simultaneously a well-defined value to A_1,A_2 and B_1.

Although this reasoning, taken by itself, is not incorrect, it is not very relevant because also in case of locality such a jpd cannot exist. This is so, because the existence of the jpd of the four observables does not only imply the existence of jpd's of pairs of *compatible* observables (which can be compared with the outcomes of EPR-like experiments), but also of pairs of *incompatible* observables, like, for instance, $p(A_1 = a_i^1, B_1 = b_j^1)$. According to Heisenberg's disturbance theory of measurement this is impossible if both marginals of this expression have to yield undisturbed probabilities. For this reason, it is possible to attribute the nonexistence of Rastall's jpd, to the Heisenberg measurement disturbance of the incompatible observable B_1 if A_1 is measured (and vice versa). Since A_1 and B_1 are observables of the same particle, no nonlocality is involved.

In comparing the implications of the three possible interpretations considered in section 5 with respect to the existence of a jpd, it is clear that for EPR-like measurements the instrumentalist/phenomenalist interpretation does not lead to such a jpd because of the necessity, in this interpretation, to measure the four observables jointly. Also, since in the objectivistic realistic case the quantities are attributed to the object independently of the measurement procedure, this interpretation cannot lead to the quantum mechanical results. For this reason the contextualistic realistic case remains as the only realistic possibility. Bohm and Hiley[21)] arrive at a comparable conclusion in their attempt "to explain quantum mechanical nonlocality along Einstein's lines, i.e. in terms of a local field". They come to the conclusion that Bell's inequalities cannot be derived if the hidden variables probability distribution $\rho(\lambda)$ is dependent on the measurement setup.

From the foregoing it is, indeed, clear that contextualism is necessary for a realistic understanding of quantum mechanics. However, an unqualified contextualism may not be sufficient. As a matter of fact, also in the contextualistic case, in which $p(a_i|\lambda)$ depends on the measurement procedure, it is possible to write down a jpd according to

$$p(a_i^1, a_j^2, b_k^1, b_l^2) = \int d\lambda\ \rho(\lambda)\ p(a_i^1|\lambda)\ p(a_j^2|\lambda)\ p(b_k^1|\lambda)\ p(b_l^2|\lambda) \tag{31}$$

from which the single-observable probabilities (23) can be obtained by marginalization, and which satisfies macrolocality. Hence, the Bell inequalities are satisfied even in this contextualistic case.

However, the jpd (31) does not seem to represent the physical situation of EPR-like experiments because it does not allow for the possibility that e.g. the probability distribution of A_1 might differ in the context of a B_1-measurement from the distribution of A_1 in the context of an A_1-measurement. The construction of the jpd (31) is, evidently, based on an unwarranted quasi-objectivism (cf. sect 5). The lack of a realistic theory of measurement disturbance makes it impossible to estimate whether Bohm's theory excludes such quasi-objectivistic jpd's. In any case, they are excluded by the traditional Heisenberg theory of measurement, which, therefore, seems to me to have a certain advantage over Bohm's theory, even if we try to come to a realistic understanding of quantum mechanics.

8. CONCLUSION

My conclusion is that Bohm's theory, as it stands, should not be considered as a hidden variables theory, but as a suggestive way to write down the Schrödinger equation. For this reason, Bohm's theory should not be interpreted as yielding information transcending the quantum mechanical one. Far from representing a nonlocal interaction the nonlocality of the quantum potential can be interpreted as descriptive of a correlation between the particles, which may be caused by the preparation process and which need not vanish when the particles separate.

The basic idea of Bohm's theory is, to consider quantum mechanical states as a kind of equilibrium states of some subquantum dynamics. This idea is very attractive and is worth to be developed further, both as an explanation of the quantum riddle and as a source of inspiration in finding ways to transcend the experimental domain of quantum mechanics. It is questionable, however, whether this will be possible by sticking to a disguised form of the Schrödinger equation and a rather arbitrary interpretation of the mathematical formalism. Admittedly, it is impossible to present a falsification of the interpretation, mainly because it does not offer a sufficiently developed theory of measurement to be able to draw even the most general conclusions like the ones necessary to investigate whether the Bell inequalities are satisfied or not. As Heisenberg's theory of measurement is developed not much further, it remains possible to prefer Bohm's measurement scheme over Heisenberg's one. For this reason it is, at this moment, not strictly impossible to view Bohm's theory as a hidden variables theory.

On the other hand, several arguments were advanced in the foregoing which make such a choice unattractive. Among these arguments the nonlocality issue is not the least important one. Indeed, we should have strong indications of the correctness of the theory, if we are to accept the consequence of a theoretical nonlocality that has no counterpart in experimental physics. Such indications we do not have, at least not if we resist the temptation to justify the nonlocality of Bohm's theory by pointing to the Bell inequalities, while referring to Bohm's theory in justifying the role of the locality assumption in the derivation of the Bell inequalities. Hence, unless we accept, without any necessity, Bohm's theory as a hidden variables theory, we are not constrained to believe that we live in a nonlocal world.

By renouncing an interpretation of Bohm's theory as a hidden variables theory I do not intend to deny the possibility of a realistic interpretation of quantum mechanics. On the contrary, if we want to transcend the experimental domain of application of quantum mechanics, such a realistic interpretation may be very helpful in finding new theories and new phenomena. It seems to me that a nonlocal theory like Bohm's one could only contribute to this goal if it would predict experimental evidence of this nonlocality. As long as the theory is completely equivalent with quantum mechanics this, however, will fail because of the quantum mechanical principle of local commutativity. For this latter reason Bohm's theory has to provide for an explanation why the nonlocality on the subquantum level does not manifest itself on the level of quantum phenomena. Far from providing an explanation of macrolocality it introduces a nonlocality that is of a completely metaphysical nature. For this reason it seems to be more of a hindrance to the attempt of finding a realistic interpretation of quantum mechanics, than a step forward into this direction, even though it contains quite a lot of seminal ideas. As demonstrated above, the locality issue is not the relevant issue in assessing the importance of the Bell inequalities. If the disturbance of the object system by the measurement interaction is duly taken into account it does not seem impossible to give an interpretation of quantum mechanics in realistic terms that is basically local. As far as Bohm's theory hampers attempts to develop such a theory its influence seems to be that of a degenerative rather than a progressive research program. This contribution has the intention to help to try to improve on this state of affairs.

REFERENCES

1. Bohm D., Phys. Rev. 85, 166,180 (1952).
2. Bell J.S., Rev. Mod. Phys. 38, 417 (1966).
3. Bell J.S., Physics 1, 195 (1964).

4. de Muynck W.M. and Abu-Zeid O., Phys. Lett. 100A, 485 (1984); de Muynck W.M., Phys. Lett. 114A, 65 (1986).

5.a)Bohm D, and Hiley B.J., Found. of Phys. 14, 255 (1984),

b)Bohm D, Hiley B.J and Kaloyerou P.N., Physics Reports 144, 321 (1987).

6. Pauli W., in Reports on the Solvay Congress (Gauthiers-Villars et Cie., Paris, 1928), p. 280.

7. de Broglie L., Comptes Rendues 183, 447 (1926); 184, 273 (1927); 185, 380 (1927).

8. Einstein A.,"Elementare Ueberlegungen zur Interpretation der Grundlagen der Quantenmechanik", in Scientific papers presented to Max Born, Edinburgh, Oliver and Boyd, 1953), p. 33.

9. Rosen N.; Journ. of the Elisha Mitchell Scientific Society 61,67 (1945); See also M. Jammer, The philosophy of quantum mechanics (John Wiley & Sons, 1974), p. 285.

10. Bohm D., p. 13 of Scientific papers (l.c. ref. 8).

11. Takabayasi T., Progr. Theor. Phys. 8, 143 (1952).

12. Bohm D., Progr. Theor. Phys. 9, 273 (1953).

13. Bohm D.and Hiley B.J., Found. of Phys. 14, 255 (1984).

14. Heisenberg W., Physikalische Prinzipien der Quantentheorie (Bibliografisches Institut Mannheim/Wien/Zürich, 1958), sect. II.2.

15 van Fraassen B.C., The scientific image, Clarendon Press, 1980, p. 64.

16. Primas H., in Quantum dynamics of molecules, ed. Woolley R.G., Plenum Press, 1980, p. 39.

17. de Muynck W.M., Found. of Phys. 14, 111,119 (1984).

18. de Muynck W.M., Found. of Phys. 16, 973 (1986); de Muynck W.M. and van Stekelenborg J.T.P.M., Phys. Lett. 116A, 420 (1986).

19. Rastall, P., Found. of Phys. 13, 555 (1983).

20 Dieks D., Phys. Lett. 117A, 433 (1986).

21. Bohm D. and Hiley B.J., Found. of Phys. 11, 529 (1981).

MANY-HILBERT-SPACES THEORY OF QUANTUM MEASUREMENTS

Mikio NAMIKI and Ichiro OHBA
Department of Physics, Waseda University
Tokyo 160, Japan

ABSTRACT

Based on the macroscopic nature of local systems of measuring apparatus to be described by a continuous direct sum of many Hilbert spaces, we explicitly show that quantum mechanics can derive the wave packet reduction by measurement within its theoretical framework. In this report, we study implications of the criterion to examine any detector as to whether it can work well or not in deriving the wave packet reduction. Besides general aspects of the theory, we present numerical simulations of a few quantum measurement processes having a simple model detector, which will be demonstrated with a movie film produced by Y. Murayama, H. Nakazato and ourselves.

1. INTRODUCTION

The quantum measurement problem is undoubtedly one of the most controversial topics in modern physics. There we must deal with a wide class of basic problems, including not only theoretical analyses of measurement processes but also serious questions against the Copenhagen Interpretation itself. In this report, however, we restrict ourselves to a rather narrow question of whether the <u>wave packet reduction</u> in the measurement of an observable can be described by quantum mechanics itself.

Several years ago, S. Machida and one of the present authors (M.N.)[1] gave an affirmative answer to the question, taking into account the macroscopic nature of measuring apparatus by means of a continuous direct sum of many Hilbert spaces. The direct sum is usually called the continuous-super-selection-rule space, of which mathematical basis was already discussed by H. Araki.[2] We call the theory simply the many-Hilbert-spaces theory. This can give a definite criterion to examine any instrument as to whether it works well or not toward the wave packet reduction as a detector. In this report, we review and develop the theory mainly in numerical simulations of measurement processes based on a simple model.

In section 2, we explain physical implications of two successive steps of a typical quantum measurement consisting of spectral decomposition and detection. In section 3, we discuss physics of the detection process leading us the wave packet reduction in some detail, using numerical simulations based on a simple model. In section 4, we review a general theory of quantum measurements by making explicit use of a continuous direct sum of many Hilbert spaces. Section 4 is devoted to concluding remarks.

2. MEASURING APPARATUS AND WAVE PACKET REDUCTION

Speaking in general, we perform every quantum measurement by making use of a measuring apparatus composed of two successive steps, the first being responsible for spectral decomposition and the second for detection. In typical momentum measurement, each object particle Q is deflected, as $p_i \rightarrow q_i$, by a magnetic field, correspondingly to its initial momentum p_i, and then is guided to detectors, $D_1, D_2, \ldots, D_i \ldots$, located far from the field. This is the spectral decomposition step. See. Fig. 1. If one detector, say D_i, generates a signal of particle arrival, then we know that the particle has momentum q_i (and had p_i before entering into the field). This is the detection step, by which the measurement process is completed. Note that the wave packet reduction takes place at the detection step but not at the spectral decomposition step.

In this case, each detector has only to possess the ability of reporting the particle arrival. Its location itself informs us about the particle momentum through the above arrangement of instruments. One

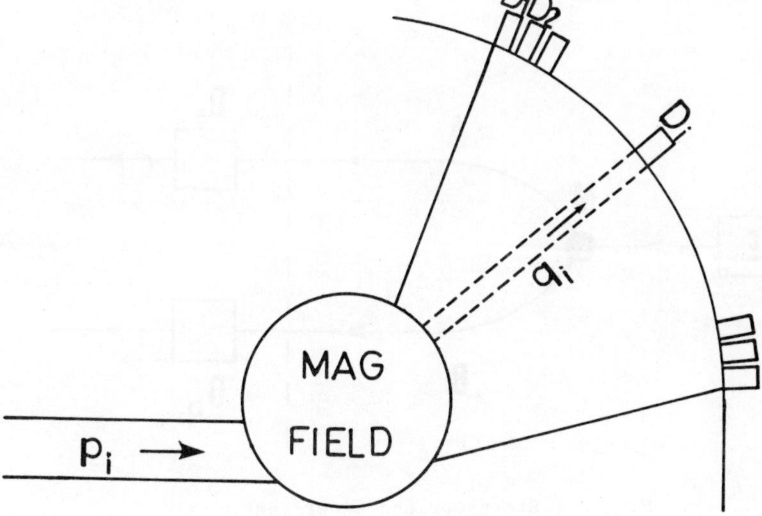

Fig. 1: Momentum measurement.

detector, say D_i, is responsible only for a yes-no experiment in which its triggering, called "yes", means p_i (or q_i) for particle momentum, while no triggering, called "no", is corresponding to other values. The whole experiment can be decomposed into a set of such yes-no experiments, as in most of practical experiments.

The spin measurement by Stern-Gerlach experiment is a typical yes-no experiment. See Fig. 2. In the spectral decomposition step, a stationary beam of object particles with spin one-half, coming from the emitter E, is spatially divided by a magnetic field in upward and downward directions in such a way that the initial wave function changes as

$$\psi_0^Q = (c_a u_a + c_b u_b)\,\phi \rightarrow \psi^Q = c_a u_a \phi_a + c_b u_b \phi_b, \tag{2.1}$$

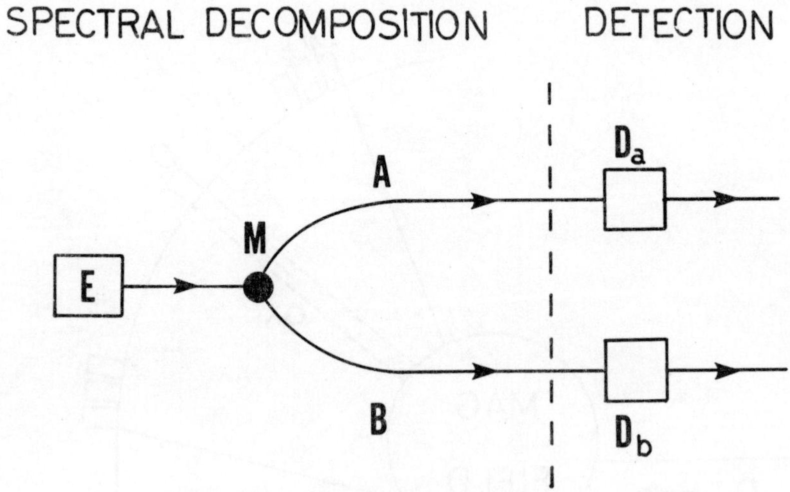

Fig. 2.: Stern-Gerlach experiment.

where c_a and c_b are constants, u_a and u_b represent up-spin and down-spin eigenfunctions, respectively, and ϕ , ϕ_a, and ϕ_b stand for position wave functions. We denote the object system by Q. In the standard experiment, ϕ_a and ϕ_b are, respectively, wave packets going to different detectors, say counters D_a and D_b (see Fig. 2). Each detection by D_a or D_b means that the particle has up-spin or down-spin, respectively, and that the measurement has been completed.

However, we need not use two detectors for the spin measurement. Remove D_b retaining D_a. Triggering of D_a (case "yes") means again that the particle has up-spin, as in the above case. On the other hand, an anti-coincidence experiment between E with signal and D_a with no signal (case "no") tells us that the particle went downward and hence had down-spin. This is called the <u>negative-result measurement</u>, in which the down-spin measurement is completed, and consequently, the wave packet reduction took place without resort to the "real firing" of D_a. For this reason, this is sometimes called the negative-result-measurement <u>paradox</u> against the ergodic amplification theory in which the wave packet reduction is supposed to originate in thermally irreversible processes, such as the real firing of counters.

In order to examine physical effects of the change in the particle wave function, $\psi_a \rightarrow \psi'_a$, given by D_a, we introduce the next-step experiment of the Young-interferometry type (super-script Q being suppressed here). See the right-hand side of Fig. 3. Both the wave,

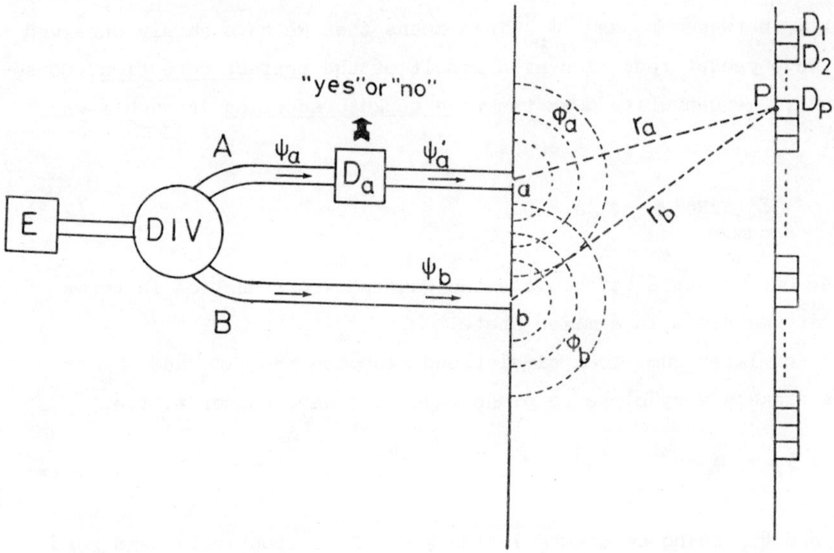

Fig. 3: Yes-no experiment and examination of its effects.

ψ'_a and ψ_b, are guided to small slits a and b, respectively, to make spherical waves, ϕ'_a and ϕ_b, diverging toward the screen. As is well known, one particle makes only one spot on the screen, but we can obtain the distribution proportional to

$$|\phi'_a + \phi_b|^2 = |\phi'_a|^2 + |\phi_b|^2 + 2\mathrm{Re}(\phi_b^* \phi'_a) \tag{2.2}$$

on the screen, if we accumulate many spots brought by a stationary beam. Suppose that the detector D_a is <u>perfect</u>, then we have only ϕ'_a or only ϕ_b corresponding to its signal "yes" or "no", respectively, because of the wave packet reduction. Gathering particle spots in the case of only "yes" or "no", we must obtain the distribution proprotional to $|\phi'_a|^2$ or $|\phi_b|^2$, respectively. In the case of "yes" and "no", we should have the distribution proportional to $|\phi'_a|^2 + |\phi_b|^2$. Here it is noted that the interference term disappears, in other words, that there is no coherence between ϕ'_a and ϕ_b, or equivalently, between ψ'_a and ψ_b. This means that we have surely observed the wave packet reduction as a result of the <u>perfect detection</u>. Consequently, we can write down the <u>wave packet reduction</u> in such a way that

$$\sum_{\text{accum.}} 2\mathrm{Re}(\phi_b^* \phi'_a) = 0. \tag{2.3}$$

Later on, we shall reformulate the wave packet reduction in terms of density matrices in a mixed state.

For later numerical simulations, suppose that ψ_a and ψ_b are wave packets very close to plane wave with wave number k, i.e.,

$$\psi_a = \theta_a e^{ikx} \quad \text{and} \quad \psi_b = \theta_b e^{ikx}, \tag{2.4}$$

θ_a and θ_b being constants inside A and B, respectively, and zero otherwise, and x standing for the position variable along the channel axes. Then we have diverging waves given by

$$\phi'_a = CSr_a^{-1}\exp[ikr_a] \quad \text{and} \quad \phi_b = Cr_b^{-1}\exp[ikr_b], \tag{2.5}$$

where r_a and r_b are distances from the slits, a and b, to an observation point, P, on the screen, respectively, and C is a constant. S stands for the S-matrix element defined by

$$\psi'_a = \theta_a S e^{ikx}. \tag{2.6}$$

3. PHYSICS OF THE DETECTION PROCESS

Here we discuss physics of the detection process to give the wave packet reduction by measurement, that is Eq. (2.3) in a yes-no experiment, as shown in Fig. 3. Our problem is to find what kind of physical processes in the detector can erase the coherence or correlations between ψ_a and ψ_b. Needless to say, the detector is a macroscopic system having a huge number of degrees of freedom and a finite size on the macroscopic scale. Remember that a macroscopic system has usually a large number of local systems, one of which is still a microscopically very large system with the macroscopic nature though it is macroscopically very small.

In most cases, it is unnatural to consider that an object particle interacts with the whole detector system, because its wave packet spread is macroscopically very small. We mean that the object particle only interacts with one local system of the detector and then produces the wave packet reduction. In some cases, the wave packet reduction triggers secondary processes due to interactions only among constituents of the detector, such as counter discharge phenomena, which are brought in for the purpose of amplification or energy supply to make displays easy. We have strictly distinguished the wave packet reduction from the secondary thermally irreversible-processes for displays. The distinction is the key to solve the negative-result-measurement paradox mentioned in section 2. Really, the amplification process is not necessary to yield the very wave packet reduction.

In this report, we exclusively discuss the case of detectors of the non-destructive type to keep the same object particle in the detection process, but we can also formulate our theory even in the case of the destructive type, for example, of photon and neutron counters to absorb object particles.[3)]

First of all, we have to solve the collision problem of an object particle with a local system of the detector. Following the theory of nuclear reactions,[4)] the S-matrix for collisions of a particle with a finite-size system can be written as

$$S = e^{i\Delta}\frac{1 + iK}{1 - 1K}e^{i\Delta} \qquad (3.1)$$

where Δ is diagonal and K off-diagonal in the channel representation. The finite-size effect is reflected in the diagonal element of Δ (in the relevant channel, for example, the elastic one) proportional to a size parameter of the target, say ℓ, apart from others slowly depending on it. Here ℓ is supposed to be a linear size, but we can use other size parameters such as volume, electric current, and magnetic flux, etc.

One of the simplest models for local systems of a detector may be a one-dimensional lattice with n+1 scatterers represented by fixed potentials, $V_1, V_2, \ldots, V_{N+1}$, and with spacings, $a_1, a_2, ---, a_N$. See Fig. 4. The S-matrix element (see Eqs. (2.6) and (3.1)) for the elastic

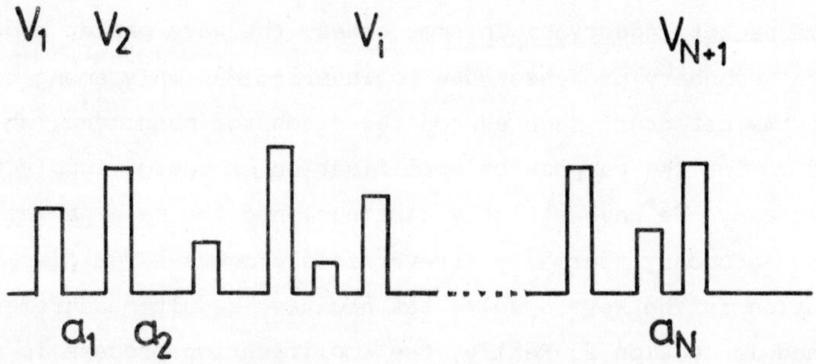

Fig. 4: One-dimensional detector model.

collision of an object particle (with wave number k) with the lattice is generally written as[5)]

$$S(k) = \varepsilon(k)e^{2i\Delta(k)} \qquad (3.2)$$

where $\varepsilon(k)$ is a complex function of k and

$$2\Delta = -\bar{k}\ell \quad ; \quad \ell = \sum_{i=1}^{N} a_i \quad , \tag{3.3}$$

$\bar{k}$ being the effective wave number. ε and $\bar{k}$ are supposed to be slowly varying functions of ℓ in the detection process. Corresponding to Eq. (3.2), we have

$$I(\ell) = 2\,\frac{|\varepsilon C|^2}{r_a r_b}\cos\,[k(r_a - r_b) + 2\Delta + \delta] \tag{3.4}$$

for the interference term of Eq. (2.2), where $\delta = \arg\ \varepsilon$.

In the present interference experiment, we separately send particles, one by one, into the apparatus from the emitter, and then accumulate independent particle spots on the screen. If we observed an interference pattern on the accumulated distribution even though it were a little distorted, the coherence between two waves, ϕ_a' and ϕ_b, (or equivalently, between ψ_a' and ψ_b) still remains, thereby telling us no wave packet reduction. If the pattern disappears, then we can conclude that the wave packet reduction took place. The wave packet reduction occurs or the coherence remains, all depending on whether the interference term vanishes or not, as anticipated in section 2.

Suppose that we had a frozen lattice with fixed N and a_i's, then we should only observe the distribution proportional to the formula Eq. (2.2) with Eq. (3.4). This means that we have no wave packet reduction. However, such a situation is usually not realistic. After one object particle passed through, constituents of the detector continue to change their states. Therefore, each object particle meets, particle by particle, a different local system with different N, a_i's, and V_i's. Such a difference is reflected in the S-matrix element mainly through its phase shift. Strictly speaking, the dependence of the phase shift on N, a_i's, and V_i's is to be obtained by solving a dynamical collision problem, based on detailed structures and motions of the detector constituents. Nevertheless, if we take the huge number of N into account, we can surely replace details of such a dynamics with a probabilistic law, for example, a Gaussian distribution given by

$$W(\Delta) = \sqrt{[2\pi(\delta\Delta)^2]^{-1}}\exp[-\frac{(\Delta - \langle\Delta\rangle)^2}{2(\delta\Delta)^2}]. \tag{3.5}$$

In case we can discard the dependence of $\bar{k}$ on random variations of V_i's, we have only to use

$$W(\ell) = \sqrt{[2\pi(\delta\ell)^2]^{-1}}\exp[-\frac{(\ell - \langle\ell\rangle)^2}{2(\delta\ell)^2}] \tag{3.6}$$

instead of Eq. (3.6). In Eqs. (3.5) and (3.6) we have followed the conventional notations for averages and variances of Δ and ℓ. Here we have introduced a limiting procedure to give continuous variables Δ and ℓ from the original discrete ones. For simplicity, in this report, we proceed to study of the measurement problem with Eq. (3.6), ignoring the dependence of $\bar{k}$ on random variations of V_i's, even though it cannot always be neglected.[6)]

For $\langle\ell\rangle$ and $\delta\ell$, it is easy to derive the useful formulas

$$\langle\ell\rangle = \langle N\rangle\langle a\rangle, \quad (\delta\ell)^2 = \langle N\rangle(\delta a)^2 + (\delta N)^2\langle a\rangle^2, \tag{3.7}$$

where we have assumed that every spacing has the same average and variance. In the case of fixed N, Eq. (3.7) becomes

$$\langle\ell\rangle = N\langle a\rangle, \quad \delta\ell = \sqrt{N}(\delta a). \tag{3.8}$$

It is natural to consider $\langle\ell\rangle$ to have a value of the order of the wave packet length of the object particle. On the other hand, estimation of $\delta\ell$ requires us to dynamically analyze the inner structure and motion of the detector. It may be necessary to give 100 to $\langle N\rangle$, at least, for numerical simulations. Anyway, an assembly of the local systems participates in interactions with the incident beam of object particles, as a whole. Such an assembly is mathematically represented by a direct sum of many Hilbert spaces but not by a single Hilbert space. Assuming that $\bar{k} = k$ and $\varepsilon = 1$ for simplicity, we perform numerical simulations of the measurement process in the case of fixed N (equal to 1, 25, 50, 75, and 100) and of $\lambda (= 2\pi k^{-1}) = 2.8\times10^{-8}$cm, $\langle a\rangle = 10^{-8}$cm, and $\delta a = \langle a\rangle\times10^{-1}$. Choosing one value of ℓ at random, according to

Eq. (3.6), we determine one particle spot on the screen, according to the distribution formula Eq. (2.2) with Eq. (3.4) having its value of ℓ. Figure 5 shows the accumulated spot distributions on the screen,

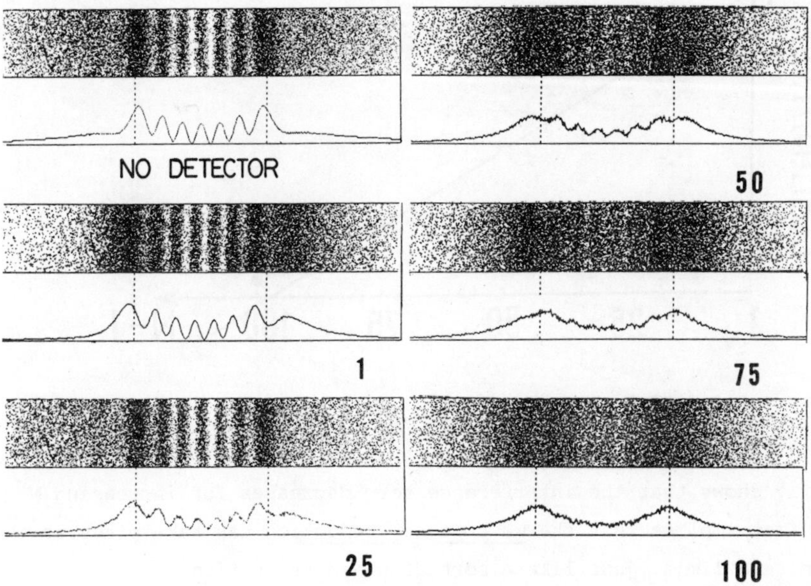

Fig. 5: Results of numerical simulations. The integers show the number of spacings. We can clearly observe that the interference pattern is going to disappear for increasing N.

in which the integers describe the number of spacings to N. We can clearly observe that the spot distribution gradually changes from the interference-type to the no coherence-type for increasing N. The spot distribution with N = 100 is very close to $|\phi_a'|^2 + |\phi_b|^2$, as was expected in section 2. In order to explicitly show the vanishing of the interference term, we identify the area of the center peak, $\bar{I}$, with the interference term and then plot it versus N in Fig. 6. The plot also

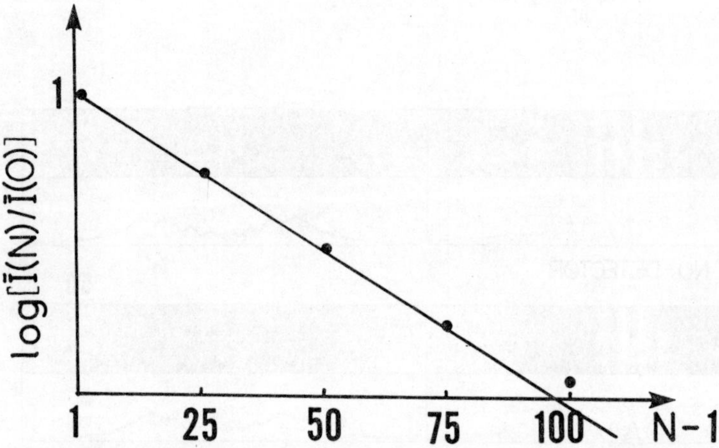

Fig. 6

clearly shows that the interference term decreases for increasing N. This fact implies that the wave packet reduction can take place at the infinite N limit, just like a sort of phase transition.

The result of the numerical simulations can be theoretically reproduced in the following way. Averaging the interference term, Eq. (3.4), over ℓ with the weight function given by Eq. (3.6), we obtain

$$\langle I \rangle \equiv \int I(\ell) W(\ell) d\ell$$

$$= 2 \frac{|\varepsilon C|^2}{r_a r_b} \cos[k(r_a - r_b) - \bar{k}\langle \ell \rangle + \delta] \exp[-\frac{1}{2}(\bar{k}\delta\ell)^2], \qquad (3.9)$$

where we neglected the dependences of ε and $\bar{k}$ on ℓ. Hence, we know that the interference term vansihes if

$$(\bar{k}\delta\ell)^2 \gg 1, \text{ or } N(\bar{k}\delta a)^2 \gg 1, \qquad (3.10)$$

where Eq. (3.8) has been used in the second one. This is the criterion to examine the detector as to whether it can work well or not toward the wave packet reduction. Equation (3.9) gives the same plot versus N as that in Fig. 6 due to the last damping factor tending to zero as N goes to infinity. In the above simulations with N = 100, we have $\frac{1}{2}N(\bar{k}\delta a)^2 \sim 2.5$, which is nearly enough to diminish the last damping factor.

If random variations of the potential strength are taken into account, the criterion must be replaced with $\max[(\bar{k}\delta\ell)^2, (\delta\bar{k}\langle\ell\rangle)^2] \gg 1$, V standing for the strength parameter and $\delta\bar{k}$ for the variance of $\bar{k}$ coming from V, respectively. Detailed discussions[6)] were given and will be reported on other occasions.

Now, we remark that the lattice with N = 1 in the above simulations is to be compared with a one-atom detector. In this case we cannot have the wave packet reduction in agreement with the arguments of Wootters and Zurek.[7)] However, this does not mean that any detector can never work, but simply tells us that one-atom detector cannot do. As for the cases with medium N's, we can say that we have performed a sort of <u>imperfect</u> measurement to give a half-coherence and half-mixed state. Recent development of the microfabrication technique enables us to produce devices, by which we can observe <u>mesoscopic</u> behaviors.[8)] Such devices will show us imperfect measurements, when used as detectors.

4. GENERAL THEORY OF QUANTUM MEASUREMENT PROCESSES

Consider a measurement of an observable, F, of an object system, Q, in a superposed state $\psi^Q = \Sigma_i c_i u_i$, u_i being an eigenstate of F belonging to an eigenvalue λ_i and $c_i = (u_i, \psi^Q)$. When we found one eigenvalue, say λ_k, in one measurement (of the first kind), we should have the change, $\psi^Q \to u_k$, called the <u>wave packet reduction</u>. However, this expresion does not faithfully correspond to the wave packet reduction being a probabilistic and acausal event as a whole. Following J. von Neumann, many authors write down the measurement process as

$$\rho^Q = |\psi^Q\rangle\langle\psi^Q| \to \bar{\rho}^Q = \Sigma_i |c_i|^2 \xi^Q(u_i) \tag{4.1}$$

in terms of density matrices only of system Q, in which $\xi^{Q}(u_i) = |u_i\rangle\langle u_i|$. However, this is also not complete. Here it is inevitable to bring states of the apparatus system, say A, into the mathematical expression of the wave packet reduction by measurement (of the first kind) in the following way:

$$\Xi^{tot} = \rho^{Q} \times \sigma^{A} \rightarrow \bar{\Xi}^{tot} = \Sigma_i |c_i|^2 \xi^{Q}(u_i) \times \sigma^{A}_{F(i)} \tag{4.2}$$

where Ξ^{tot} and σ^{A} stand for the initial-state density matrices of the total system and of system A, respectively, $\bar{\Xi}^{tot}$ for the final-state density matrix of the total system, and $\sigma^{A}_{F(i)}$ for the final-state density matrix of system A showing that F had the i-th eigenvalue. This is the precise expression of the <u>wave</u> <u>packet</u> <u>reduction</u>, which was provisionally described by Eq. (2.3) and shown by numerical simulation in section 3.

We are now going to derive Eq. (4.2) for a quantum measurement process (of an observable F) consisting of the two successive steps as shown in Figs. 1 or 2. The first step prepares the following spectral decomposition:

$$\psi^{Q} = \{\Sigma_i c_i u_i\}\phi \rightarrow \psi^{Q} = \Sigma_i c_i u_i \phi_i , \tag{4.3}$$

where ϕ and ϕ_i are the wave packet functions before and after the step, respectively. The latter goes to the i-th detector.

As was mentioned in section 3, the beam of object particle will interact, as a whole, with an assembly of local systems of a detector (say D), subject to a statistical law. Such an assembly of the local systems can be described in terms of the density matrix given by

$$\rho^{D} = \Sigma_N W_N \rho^{D}_N ; \tag{4.4a}$$

$$\rho^{D}_N = \Sigma_N |\phi^{N}_n\rangle w_n \langle\phi^{N}_n| , \tag{4.4b}$$

where ρ^{D}_N stands for the denstiy matrix of a local system with N constituents and W_N for the weight factor distributing around $\langle N \rangle$ with width δN. ϕ^{N}_n and w_n are, respectively, the n-th eigen-state of

Hamiltonian H_N of the local system and Boltzmann factor. Since ρ_N^D is an operator on a single Hilbert space $\mathcal{H}_N$, ρ^D is belonging to a direct sum of the Hilbert spaces, $\Sigma_N \mathcal{H}_N$.

Needless to say, the macroscopic nature requires $\langle N \rangle$ and δN to be very large but $\delta N/\langle N \rangle$ very small. In the infinite $\langle N \rangle$ limit, as performed in section 3, we can replace a discrete parameter $\ell_N = \Sigma_i a_i$ (a_i being a length of the order of atomic size) with a continuous size parameter ℓ, through the limiting procedure $\ell_n \to \ell$ as $N \to \infty$ and $a_i \to 0$. Correspondingly, the discrete sum of Eq. (4.4) is replaced by the following integral:

$$\rho^D = \int d\ell\, W(\ell)\rho^D(\ell) \equiv \mathcal{W} \cdot \rho^D(\ell) \,, \tag{4.5}$$

where $W(\ell)$ is a weight function distributing around $\langle \ell \rangle$ with width $\delta\ell$ subject to the normalization condition $\int W(\ell)d\ell = 1$. We have introduced the symbol $\mathcal{W}$ for the averaging procedure over ℓ with the weight function $W(\ell)$. $W(\ell)$, or especially $\delta\ell$ is a representative of the sturcture of the local system of D. Following discussions in section 3, we can use Eq. (3.6) for $W(\ell)$ in which $\langle \ell \rangle$ and $\delta\ell$ are estimated by Eq. (3.7) or Eq. (3.8).

Now we can describe the whole detection process by the time-dependent density matrix of the total system as follows:

$$\Xi_t^{tot} = \mathcal{W} \cdot e^{-iHt/\hbar} \rho^Q \times \rho^D(\ell)\, e^{iHt/\hbar}$$

$$\xrightarrow[t\to\infty]{} \mathcal{W} \cdot e^{-iH_0 t/\hbar} S\rho^Q \times \rho^D(\ell)\, S^\dagger e^{iH_0 t/\hbar} \,, \tag{4.7}$$

where we have introduced the S-matrix defined by $\exp[-iHt/\hbar] \xrightarrow[t\to\infty]{} \exp[-iH_0 t/\hbar]S$, H and H_0 being the total and free Hamiltonians of the total system Q+D, respectively. ρ^Q and $\rho^D(\ell)$ represent, respectively, systems Q and one of local systems of D immediately before the detection process, that is,

$$\rho^Q = \sum_{i,j} c_i c_j^* |i\rangle\langle j| \quad ; \quad |i\rangle = u_i \phi_i \,, \tag{4.8}$$

$$\rho^D(\ell) = \prod_\nu \rho_\nu^D(\ell_\nu) , \tag{4.9}$$

where $\rho_\nu^D(\ell_\nu)$ is the initial-state density matrix of the ν-th detector D_ν. Our task is to explicitly show that the phase correlations in ρ^Q are erased out in the asymptotic behavior of Ξ_t^{tot}.

Putting Eqs. (4.8) and Eq. (4.9) into Eq. (4.7), we obtain

$$\Xi_t^{tot} = \sum_{i,j} c_i c_j^* \Xi_t^{ij} ; \tag{4.10a}$$

$$\Xi_t^{ij} = \prod_\nu \mathcal{W}_\nu \cdot e^{-iHt/\hbar}[|i\rangle\langle j| \times \rho_\nu^D(\ell_\nu)]e^{iHt/\hbar}, \tag{4.10b}$$

and then

$$\Xi_t^{ij} \xrightarrow[t\to\infty]{} \prod_\nu \mathcal{W}_\nu \cdot e^{-iH_0^\nu t/\hbar} S_\nu[|i\rangle\langle j| \times \rho_\nu^D(\ell_\nu)]S_\nu^\dagger e^{iH_0^\nu t/\hbar}, \tag{4.11}$$

where we have put $H_0 = \Sigma_\nu H_0^\nu$ and $S = \Pi_\nu S_\nu$. Hamiltonian H_0^ν and S-matrix S_ν describe free motions of D_ν and its interactions with the ν-th partial wave ($|\nu\rangle = u_\nu \phi_\nu$) of Q, respectively. S_ν brings in interactions between $\rho_\nu^D(\ell_\nu)$ and $|i\rangle$ or $\langle j|$ only when $\nu = i,j$, and then yields the phase shift factor $\exp[-i\bar{k}_i \ell_i]$ or $\exp[i\bar{k}_j l_j]$, respectively, as was seen in section 3. Hence the averaging procedures, $\Pi_\nu W_\nu$, in Eq. (4.11) give the ℓ-integrals

$$[\int d\ell_i W(\ell_i)e^{-i\bar{k}_i \ell_i} e^{i\bar{k}_i \ell_i} \ldots] \prod_{\nu\neq i} [\int d\ell_\nu W(\ell_\nu) \ldots] \tag{4.12a}$$

in Ξ_t^{ii}, where the phase shift factors cancel each other out in the ℓ_i-integral, and also

$$[\int d\ell_i W(\ell_i)e^{-i\bar{k}_i \ell_i} \ldots][\int d\ell_j W(\ell_j)e^{i\bar{k}_j \ell_j} \ldots] \times \prod_{\nu\neq i,j} [\int d\ell_\nu W(\ell_\nu) \ldots] \tag{4.12b}$$

in Ξ_t^{ij} ($i\neq j$). The remaining integrands other than W and the phase shift factors are smooth functions of ℓ_ν, so that Eq. (4.12b) vanishes for the condition given by

$$\bar{k}_i \delta\ell_i \gg 1, \text{ and/or } \bar{k}_j \delta\ell_j \gg 1, \tag{4.13}$$

because of the Riemann-Lebesgue lemma, while Eq. (4.12a) never vanishes due to the lack of the phase shift factors. Consequently, we are led to the following asymptotic behavior:

$$\Xi_t^{tot} \xrightarrow[t\to\infty]{} \Sigma_i \; |c_i|^2 \; \xi_F^Q(i) \times \sigma_{i,F}^D \times \Pi_{\nu\neq i} \sigma_{\nu,I}^D , \tag{4.14}$$

where $\xi_F^Q(i)$ and $\sigma_{i,F}^D$ are, respectively, the final density matrices of systems Q and D_i, which are brought by

$$\xi_F^Q(i) \times \sigma_{i,F}^D$$
$$= \mathcal{W}_i \, e^{-iH_0 t/\hbar} \, S[\,|i\rangle\langle j| \times \rho_i^D(\ell_i)]\, S^\dagger \, e^{iH_0 t/\hbar} \tag{4.15}$$

Here we have suppressed their t-dependences. $\sigma_{\nu,I}^D$ denotes the initial density matrix of D_ν. Equation (4.14) has no phase correlations among $|i\rangle$'s. Thus we achieved the derivation of the wave packet reduction. Compare $\xi_F^Q(i)$ and $\sigma_{i,F}^D \times \Pi_{\nu\neq i} \sigma_{\nu,I}^D$ in Eq. (4.14), respectively.

In the case of the standard Stern-Gerlach experiment, Eq. (4.14) becomes

$$\Xi_t^{tot} \xrightarrow[t\to\infty]{} |c_a|^2 \xi_F^Q(a)\times\sigma_{a,F}^D\times\sigma_{b,I}^D + |c_b|^2 \xi_F^Q(b)\times\sigma_{a,I}^D\times\sigma_{b,F}^D. \tag{4.16}$$

For the negative-result measurement, Eq. (4.12b) does not have the ℓ_b-integral but still retains the ℓ_a-integral which vanishes, if $\bar{k}_a \delta\ell_a \gg 1$. Thus we obtain

$$\Xi_t^{tot} \xrightarrow[t\to\infty]{} |c_a|^2 \xi_F^Q(a)\times\sigma_{a,F}^D + |c_b|^2 \xi_I^Q(b)\times\sigma_{a,I}^D , \tag{4.17}$$

which is the wave packet reduction in this case.

Here we have derived the wave packet reduction, under the condition given by Eq. (4.13). However, we still keep the phase correlations and hence have no wave packet reduction, if the condition is not satisfied. In fact, we have observed interesting examples of such cases in recent experiments on neutron interference.[9] The condition is surely a definite criterion for the wave packet reduction.

5. CONCLUDING REMARKS

In this report we have formulated the many-Hilbert-spaces theory of quantum measurements which enables us to explicitly derive the wave packet reduction by measurement within the framework of quantum mechanics, based on the following two properties: The first is the macroscopic nature of a local system of detectors which is to be represented in a continuous direct sum of large number of Hilbert spaces. The second is the phase shift of S-matrix for collisions of an object particle with a local system which is proportional to the finite size of the local system. We should notice that the wave packet reduction is really shown even in the case of the negative-result measurement and that our theory is free from the Wigner theorem.

We stress that our theory offers a definite criterion to examine detectors through the condition, Eqs. (3.10) or (4.13). For detectors characterized by Eq. (3.8), we learned that the criterion becomes $N(\bar{k}\delta a)^2 \gg 1$ and then the wave packet reduction exactly holds at the infinite N limit. However, when we have $(\bar{k}\delta\ell)^2 \ll 1$, no wave packet reduction occurs. For medium values of $(\bar{k}\delta\ell)^2$, we shall perhaps have an imperfect measurement or may meet with a mesoscopic behavior. Even the same detector can or cannot make the wave packet reduction, depending on whether the condition is satisfied or not. Furthermore, note that N in our theory is proportional to the time in Araki's limiting procedure,[2] because the latter means the passage time in a detector.

The case with $(\bar{k}\delta\ell)^2 \lesssim 1$ is also interesting to us, because every measurement theory must be examined as to whether it really gives no wave packet reduction in this case. One of the nice examples has been presented by recent experiments on neutron interference,[9] from which we learned that a macroscopic system does not always work well as a measuring apparatus. They have observed beautiful coherence between two channels A and B of an apparatus just like that shown in Fig. 3, one of which has an Al-phase shifter or an oscillating magnetic field in itself. The interference experiment with the oscillating field is of special interest from the measurement-theoretical point of view, because it gives a question why such a dynamical change as spin-flipping in one channel causes no wave packet reduction to determine the particle path. In this case,[9] both the Al-phase shifter and the oscillating field give $D_\lambda = 2\pi/\bar{k}$ larger than $\delta\ell$. Hence, we can understand the reason why the interference phenomena are observed despite of spin-flipping.

ACKNOWLEDGMENT

One of the authors (M.N.) is grateful to S. Machida for stimulating discussions and for many collaborations. The authors are also indebted to Y. Murayama, H. Nakazato, T. Ohya, and other members of our institute for discussions and for their help in numerical simulations.

REFERENCES

1) S. Machida and M. Namiki: Prog. Theor. Phys. 63 (1980) 1457-1473, 1833-1847; Proc. of the International Symposium on Foundation of Quantum Mechanics - In the Light of New Technology (ISQM-Tokyo '83), S. Kamefuch et al., Eds., 127-135, 136-139, (Phys. Soc. Japan, Tokyo, 1984); Fundamental Questions in Quantum Mechanics, L.M. Roth and A. Inomata, Eds., 77-92, (Gordon and Breach, New York, 1986); M. Namiki: Ann. New York Acad. 480 (1986) 78-88; Proc. of the International Conference on Microphysical Reality and Quantum Formalism, van der Merwe and Tarozzi, Eds., to be published, (Reidel, Dordrecht); Quantum Mechanics of Macroscopic Systems and Measurement Problem (in Japanese), Y. Otsuki, Ed., 139-219, (Kyoritsu, Tokyo, 1985).
2) H. Araki: Prog. Theor. Phys. 64 (1980) 719-730; Fundamental Aspects

of Quantum Theory, V. Gorini and A. Frigerio, Eds., 23-33, (Plenum, New York, 1986).

3) See earlier papers of Machida and Namiki.
4) See, for example, K.M. McVoy: Ann. of Phys. 54 (1969) 552-565.
5) See ref. 1).
6) Y. Murayama, H. Nakazato, M. Namiki, and I. Ohba: Movie on "Numerical Simulations of Reduction of Wave Packet in Quantum Measurement Processes", (Cinesell, Tokyo, 1986); Y. Murayama and M. Namiki: In preparation.
7) W.K. Wootters and W.H. Zurek: Phys. Rev. D19 (1979) 443-456.
8) See, for example, many papers in Proc. of ISQM-Tokyo '86, M. Namiki et al., Eds., (Phys. Soc. Japan, Tokyo, 1987).
9) See, for example, H. Rauch: Proc. of ISQM-Tokyo '83, 277-288; Proc. of ISQM-Tokyo '86, 3-17; M. Namiki and H. Soshi: Prog. Theor. Phys. 77 (1987) 508-513.

VARIETIES OF REALISM

ILKKA NIINILUOTO
Department of Philosophy
University of Helsinki
00170 Helsinki, Finland

ABSTRACT

Twentieth-century epistemology and philosophy of science have been battle-fields between phenomenalist, pragmatist, and realist approaches. The demise of logical positivism was followed by the rise of scientific realism, but a new pragmatist turn in the 1970s has made anti-realist attitudes fashionable again. Yet, Arthur Fine's recent declaration that "realism is dead" seems overhasty. In this paper, I argue that realism is well and alive: it exists today in many different variants. After defending critical scientific realism against its most important current alternatives (descriptivism, instrumentalism, constructive empiricism, internal realism), I proceed to evaluate some of the most interesting recent commentaries on the challenge Quantum Theory gives to realism: Fine's natural ontological attitude, Folse's account of Bohr's philosophy of complementarity, d'Espagnat's conception of veiled reality, and the idea of interactive realism.

1. REALISM IN ONTOLOGY

As an ontological thesis, realism is the doctrine that there exists a mind-independent reality.[1] This simple formulation is liable to several kinds of misunderstanding, since ontological realism in fact turns out

---

[1] In a more specific (medieval) sense, realism asserts the mind-independent existence of universals or properties (like redness or roundness), denying thus the nominalist thesis that the world consists of individuals only. See D. Armstrong, Universals and Scientific Realism, vols. 1 - 2, Cambridge University Press, Cambridge, 1978.

to be compatible with a surprising variety of philosophical positions.

Materialism, in its different variants, claims that the whole reality is identical with, or reducible to, or emergent from, the realm of physical things and processes ("nature" or Popper's World 1). While this view entails realism, the converse does not hold. Realism need not claim that minds (Popper's World 2) are unreal, or that only mind-independent things are real. Emergent materialism may accept the existence of genetically mind-dependent, man-made concrete and abstract cultural objects (Popper's World 3). Also dualism, which accepts the independent existence of both matter and mind, asserts the existence of "mind-independent reality". Furthermore, ontological realism may also be regarded as compatible with objective idealism which claims that matter and "subjective" human minds both are somehow generated from the ultimate spiritual reality (God, Plato's ideas, Plotinos' One, Spinoza's substance, Hegel's absolute spirit).

Ontological realism should, therefore, be understood as the thesis that nature is independent of human minds. It is thus incompatible with subjective idealism - such as solipsism ("the world equals the contents of my thought"), Bishop Berkeley's doctrine esse est percipi ("to exist is to be perceived"), and Ernst Mach's phenomenalism ("the world consists of sensations").[2]

The concept of mind-independence in our definition of realism should be understood in the sense of ontological presupposition: A is dependent on B if A could not exist without B existing. This concept should be clearly distinguished from causal independence: even though nature is ontologically mind-independent, minds and physical objects are constantly

---

[2] A phenomenalist is an anti-realist in the sense that he denies reality behind appearances: for Mach, physical objects are not external causes of sensations, but rather complexes of sensations. Like Rudolf Carnap in Aufbau (1928), Mach denies that his position is idealistic: sensations are not mind-dependent, since mind (ego) itself is a bundle of sensations. But what on earth are sensations without a perceiving subject? See I. Niiniluoto, "Ernst Mach as a Philosopher of Science", in Essays in Honour of Kalervo Vihtori Laurikainen, Annales Academiae Scientiarum Fennicae, Series A.VI. Physica 431, Helsinki, 1986, pp. 190 - 203.

in causal interaction. For example, a tree as an object of my visual perception brings about a mental event, a sensation, in my mind; I can plan in my mind and execute with my body an action of moving and shaping a stone. The existence of such interactions is not an argument against realism - not even against materialism (as, e.g., Karl Marx clearly understood).[3]

The thesis of ontological realism alone does not say anything about the nature of mind-independent reality (except its existence). It is thus compatible with various theories which identify the elements of reality as things[4], events, processes, fields, or systems[5]. In particular, this means that realism does not entail the world picture (classical realism) usually associated with classical mechanics - the world consisting of mass points, and their systems, which have a sharply defined location, mass, and momentum, and obey deterministic laws of motion.

Ontological realism need not deny that physical objects may have non-physical, mind-involving properties, which ontologically presuppose relations to consciousness or consciously created cultural and social institutions (e.g., beauty, popularity, economic value). On the other hand, the extreme claim that all properties are mind-involving has been defended by the assumption that reality remains a kind of "noumenal jam" without any structure before man introduces concepts for describing

---

[3] The situation is not essentially different, if realism is defined partly in epistemic terms - e.g., requiring "reality that exists independently of our knowledge of it". See M. Dummett, "Realism", Synthese 52, 55 - 112 (1982). The fact that empirical knowledge presupposes causal interaction with nature does not imply that nature ontologically depends on our knowledge of it.

[4] Things may be understood as (i) "bare" or propertyless particulars with some predicates applying to them (nominalism), (ii) pieces of material substance with attributes (substance theory), or (iii) sets of properties (bundle theory). See M. Loux, Substance and Attribute, D. Reidel, Dordrecht, 1978.

[5] For a system ontology, see M. Bunge, Treatise on Basic Philosophy, vols. 3 - 4, D. Reidel, Dordrecht, 1977 - 78.

it. This doctrine - called internal realism by Putnam[6], transcendental nominalism by Hacking[7], and conceptual idealism by Rescher[8] - is untenable as an ontological thesis, since it fails to give a reasonable answer to the classical query: what did exist in the world before the evolution of man and the first conceptual systems?

This criticism of internal realism does not imply the position known as essentialism or metaphysical realism (Putnam), i.e., the claim that the world has a unique "ready-made" structure, and there is a privileged conceptual framework (the nature's own language as it were) for describing this structure. Against essentialism, ontological realism may admit that the world can be categorized or "carved into pieces" (things, properties, systems, etc.) in several alternative ways - but it should insist that the world (rather than we) decides the results of such categorizations. In this sense, an ontological realist is willing to refer to the mind-independent world as "objective reality".

Finally, an ontological realist need not be an actualist who identifies the real with the actual. For example, fire has the capability of burning, even if it does not actually burn anyone anywhere now. The atoms of transuranic substances belong to reality, even if they may fail to be actualized at a given moment of time. Following Aristotle, the domain of reality can thus be taken to consist of the actual and the potential things and properties: dispositional properties, causal powers, and objective physical probabilities ("propensities") could be included among real potencies.

---

[6] See H. Putnam, Reason, Truth, and History, Cambridge University Press, Cambridge, 1981.

[7] See I. Hacking, Representing and Intervening, Cambridge University Press, Cambridge, 1983. Note that Hacking's aim is to criticize internal realism.

[8] See N. Rescher, Conceptual Idealism, Blackwell, Oxford, 1973. Rescher himself denies an ontological interpretation of his conceptual idealism: what is mind-dependent is "not reality itself (whatever that might be), but reality-as-we-picture-it".

2. REALISM IN EPISTEMOLOGY

As an epistemological thesis, realism claims that it is possible to obtain knowledge about the mind-independent reality.

Modern philosophy has been dominated by Descartes' formulation of the problem of knowledge. Classical empirism and rationalism had assumed that certain knowledge about reality is obtainable by sense experience or by reason. Against this _naive realism_, Descartes urged that the thinking subject (or "I") can have "clear and distinct" knowledge only of its own states, so that knowledge about the "external world" (matter and other minds) becomes problematic.

Kant's "Copernican revolution" solves this problem by making a distinction between the "noumenal" realm of things-in-themselves (_Dinge an sich_) and the "phenomenal" realm of things-for-us. Of the things-in-themselves we know only that they exist and cause our sensations.[9] The proper object of our knowledge is the world of phenomena - this world is cognizable for us, since it is partly mind-dependent, constituted by the specifically human mental faculty (space and time as forms of our sensation, causality as one of the categories of our understanding). The epistemological subject is for Kant a _Ding an sich_, and it has to be distinguished from the "empirical self" in the world of phenomena.

Kant's _critical idealism_ is still realistic in the ontological sense, since it accepts things-in-themselves. German _idealism_, in its subjective and objective forms, expelled _Dinge an sich_ from the Kantian system. _Phenomenalism_ takes the further step of eliminating the epistemological subject, the transcendental ego, leaving only the phenomenal world, including a phenomenal ego as a bundle of sensations. Husserl's _phenomenology_ puts the objective world into "brackets", i.e., suspends judgment about its existence, and concentrates on the analysis of the ways in which out theoretical and practical interests constitute our "life-world". Similarly, _pragmatism_ excludes all talk about the world-in-

---

[9] This is a contradiction in Kant's system: things-in-themselves cause phenomena, but still the category of causality is applicable only to the world of phenomena.

itself as meaningless: it is senseless to speak of reality as divorced of the conceptual and cognitive practices of the (temporally unlimited) scientific community. Hence, also the objective concept of truth (correspondence with the world-in-itself) has to be replaced with truth-for-us (coherence, verifiability, or assertability relative to the world-for-us).

The logical positivism of the Vienna Circle in the late 1920s was a sophisticated form of anti-metaphysical pragmatism. Both Moritz Schlick and Rudolf Carnap argued that ontological realism and its negation (i.e., the "old positivism" of Mach) are meaningless: "The denial of the existence of a transcendental external world would be just as much a metaphysical statement as its affirmation."[10] New influential versions of this doctrine have emerged in the 1970s in Michael Dummett's anti-realism, Nicholas Rescher's conceptual idealism, Hilary Putnam's internal realism, and Nelson Goodman's theory of worldmaking[11].

Another line of epistemological thought, critical realism, can also be motivated in Kantian terms: Kant's phenomena could be interpreted as expressions of our partial knowledge of the things as they are "in themselves" in the mind-independent reality. This idea accepts "direct" realism and the correspondence theory of truth, but it is not "naive", since our knowledge of reality is always laden with some conceptual framework and is never assumed to be completely and exactly true. Still, this view rejects relativism and scepticism. If we recognize that Kant's categories are not fixed, but rather historically changing (Hegel) and bound to man-made languages (Peirce), then each conceptual framework provides us with a perspective on objective reality. By employing and enriching such frameworks, and by empirically testing scientific theories formulated in these frameworks, we have a rational method of approaching to the truth, i.e., finding deeper and deeper partial descriptions of the world.

---

[10] See M. Schlick, "Positivism and Realism", in A. Ayer (ed.), Logical Positivism, The Free Press, New York, pp. 32 - 33.

[11] See N. Goodman, Ways of Worldmaking, Harvester Press, Hassocks, 1978.

On the basis of this idea, it is natural to regard independence of perspectives as a good indicator of reality. Charles Peirce distinguished "real things" from fictional entities by saying that their "characters are independent of what anybody may think them to be".[12] Elaborating a similar idea from Leibniz, the Finnish philosopher Eino Kaila defined in the 1930s reality in terms of _invariance_. He distinguished from each other the perceptual or phenomenal _p-objects_, the macroscopic physical _f-objects_ of our everyday experience, and the scientific _s-objects_ postulated by theories of physics. For Kaila, the sequence from p-objects (e.g., a sense-impression of red) to f-objects (e.g., a stone) and to s-objects (e.g., elementary particles or fields) exhibits increase of invariance and, hence, increase in their character of reality.[13]

In this respect, Kaila's view is similar to the later formulation of scientific realism by Wilfrid Sellars: the objects of the "scientific image" (s-objects) are ontologically but not epistemically prior to the objects of the "manifest image" (f-objects).[14] However, Kaila does not accept the implication of the Sellarsian realism that p-objects and f-objects are unreal and only the ultimate s-objects are real.

---

[12] See C. Peirce, _Collected Papers_, Harvard University Press, Cambridge, Mass., 1931 - 35, vol. 5, para. 405. The famous Einstein-Podolsky-Rosen criterion of physical reality may be regarded as a special case of this idea: a physical quantity is real if its value can be predicted with certainty.

[13] See E. Kaila, _Experience and Reality_, D. Reidel, Dordrecht, 1979.

[14] Sellars keeps clearly separate our current s-objects and the ultimate s-objects obtainable in the ideal limit of the opinion of the scientific community. See J. Pitt, _Pictures, Images, and Conceptual Change: An Analysis of Wilfrid Sellars' Philosophy of Science_, D. Reidel, Dordrecht, 1981.

3. CRITICAL SCIENTIFIC REALISM

Critical scientific realism[15] can be distinguished from its alternatives by the following theses (cf. Fig. 1):

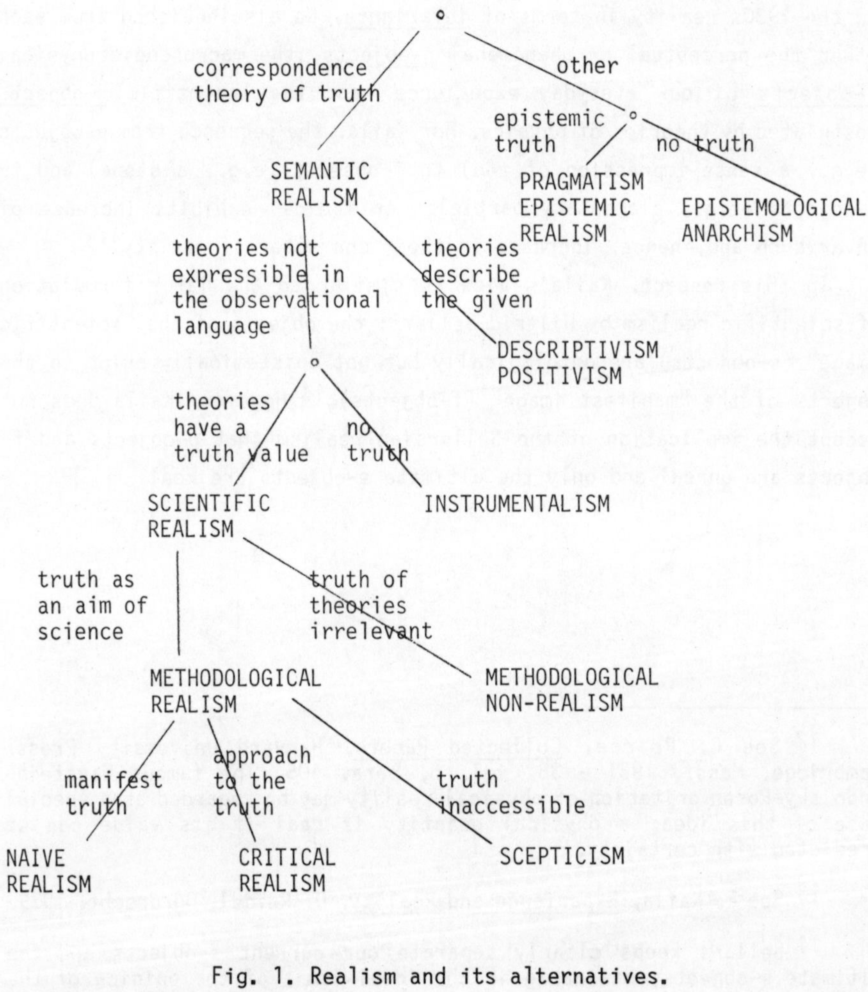

Fig. 1. Realism and its alternatives.

---

[15] For discussion on scientific realism, see J. Leplin (ed.), Scientific Realism, University of California Press, Berkeley, 1984; I. Niiniluoto, Is Science Progressive?, D. Reidel, Dordrecht, 1984.

(R0) At least part of reality is ontologically independent of human minds.

(R1) Truth is a semantical relation between language and reality. Its meaning is given by a modern (Tarskian) version of the correspondence theory, and its best indicator is given by systematic inquiry using the methods of science.

(R2) The concepts of truth and falsity are in principle applicable to all linguistic products of scientific inquiry, including observation reports, laws, and theories. In particular, claims about the existence of theoretical entities have a truth value.

(R3) Truth (together with some other epistemic utilities) is an essential aim of science.

(R4) Truth is not easily accessible or recognizable, and even our best theories can fail to be true. Nevertheless, it is possible to approach to the truth, and to make rational assessments of such cognitive progress.

(R5) The best explanation for the practical success of science is the assumption that scientific theories in fact are approximately true or sufficiently close to the truth in the relevant respects. Hence, it is rational to believe that the use of the methods of science in the long run has been, and will be, progressive in the cognitive sense.

Thesis (R0) is the minimum assumption which distinguishes ontological realists from subjective idealists, phenomenalists, pragmatists, and logical positivists (cf. Section 1).

Thesis (R1) separates semantic realists, who define truth as correspondence between language and reality, from the pragmatists, who replace this realist concept of truth with some epistemic surrogate (such as verified, proved, coherent with our knowledge, warrantedly assertable, limit of inquiry, consensus of the scientific community)[16], and from the epistemological anarchists, who deny the existence of any reasonable

---

[16] This view is represented, among others, by James, Dewey, Rescher, Dummett, Putnam, Rorty, and Habermas. Tuomela's "causal internal realism" also employs an epistemic notion of truth, but he reintroduces a language-world connection through a causal account of knowledge formation. See R. Tuomela, Science, Action, and Reality, D. Reidel, Dordrecht, 1985. For a good criticism of internal realism as a form of anti-realism, see J.J.C. Smart, "Realism vs. Idealism", Philosophy 61, 295 – 312 (1986).

concept of truth.

Thesis (R2) distinguishes scientific realists from the descriptivists, who regard theories as economical descriptions of the observationally given empirical world, and from the instrumentalists, who deny that theoretical statements have a truth value. Descriptivism and instrumentalism are typical philosophies of science for radical empiricists who identify reality with the observable world - either with the p-objects (phenomenalism) or the f-objects (physicalism). Descriptivists assume that all terms in the language of science are explicitly definable in the observational language. Instrumentalists, who usually give special emphasis to the idealizational character of scientific theories (Mach, Duhem, Poincaré), admit "auxiliary" theoretical concepts for the purpose of prediction, but treat them as symbolic tools without interpretation. Theoretical entities are thus at best useful fictions. Scientific realists instead take theories seriously as attempted descriptions of reality: a theory may go beyond the edge of direct observability by postulating theoretical entities, if it yields predictions testable by public observation. (See Fig. 2.)

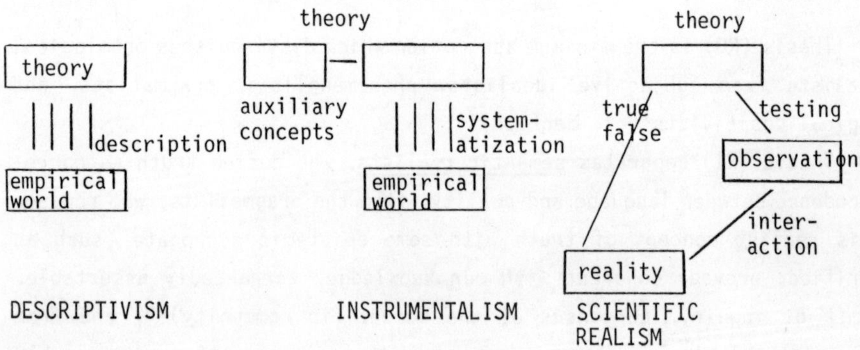

Fig. 2.

Thesis (R3) separates methodological realists from the methodological non-realists, who admit that theories have a truth value, but regard it as irrelevant to the aims of science. This kind of non-realism replaces truth as an aim of science with some methodological surrogate - such as successful prediction, simplicity, or problem-solving ability[17]. Bas van Fraassen's constructive empiricism requires that a theory should be "empirically adequate": what the theory says about the observable should be true (see Fig. 3).[18]

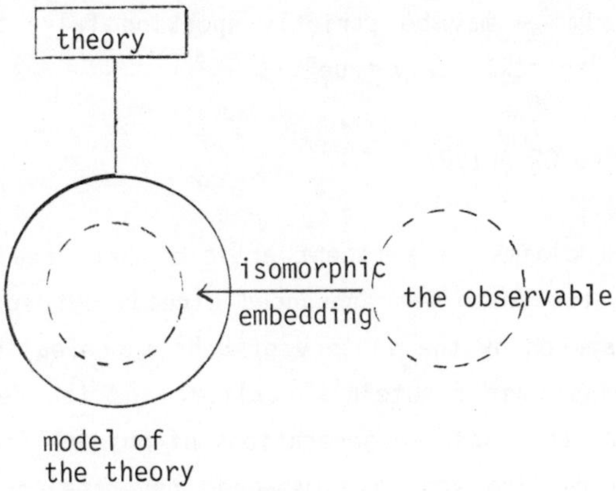

Fig. 3.

Finally, theses (R4) and (R5) distinguish critical realists from the naive realists, who believe that certified truth is easily accessible, and from the sceptics, who deny the possibility of true knowledge or progress towards it. Critical realism has been represented by the "fallibilist" tradition in epistemology - with such diverse thinkers as Cusanus, Engels, Peirce, Popper, Bunge, and Sellars.

Philosophers have recently developed positions that might be characterized as critical "half-realism". McMullin defends scientific realism, but suggests that theoretical laws are metaphorical statements which

---

17 See T. Kuhn, The Structure of Scientific Revolutions, 2nd ed., The University of Chicago Press, Chicago, 1970; L. Laudan, Progress and Its Problems, Routledge and Kegan Paul, London, 1977.

18 See B. van Fraassen, The Scientific Image, Oxford University Press, Oxford, 1980.

are not "literally true or false".[19] Cartwright defends an entity realism, accepting theoretical entities which play a role in causal explanations, but denies realism about theoretical laws.[20] My own favourite form of realism does not need this distinction between entity and theory realism: it appeals to the idea that theoretical statements in science - both universal ones expressing laws and existential ones expressing ontological claims - may be strictly speaking false but nevertheless "truthlike" or "approximately true".[21]

4. REALISM: DEAD OR ALIVE?

Arthur Fine claims in a recent article that "realism is well and truly dead".[22] Its death was announced already by neopositivists, the process was hastened by the victory of Bohr's nonrealist philosophy of quantum mechanics over Einstein's realism, and the death was finally certified, when "the last two generations of physical scientists turned their backs on realism and have managed, nevertheless, to do science successfully without it".

Fine - well-known for his earlier sustained attempts to defend a realist interpretation of quantum theory - is not advocating anti-realism, either: anti-realism is not "the winner in the philosophical debate that realism has lost".[23] What he suggests instead as "a third way" is the natural ontological attitude (NOA): accept the results of science

---

19 See E. McMullin, "A Case for Scientific Realism", in Leplin (ref. 15), 8 - 40.

20 See N. Cartwright, How the Laws of Physics Lie, Oxford University Press, Oxford, 1983. Cf. also Hacking, ref. 7, and R. Harré, Varieties of Realism, Blackwell, Oxford, 1986.

21 These concepts are explicated in precise terms in I. Niiniluoto, Truthlikeness, D. Reidel, Dordrecht, 1987.

22 See A. Fine, "The Natural Ontological Attitude", in Leplin (ref. 15), 83 - 107.

23 See A. Fine, "Unnatural Attitudes: Realist and Instrumentalist Attachments to Science", Mind 95, 149 - 179 (1986)

in the same way as the evidence of our senses, but resist the impulse to ask any questions or to propose any additions that go beyond the history and practice of science itself. Thus, NOA claims to be neither realist nor anti-realist, since it refuses to talk of "the external world" or to propose any particular analysis of the concept of truth.

Fine's NOA can be regarded as an expression of philosophical despair. He in effect suggests that we suspend judgment about the positions in Fig. 1 - not because they are meaningless (as the Vienna Circle urged), not because they are irrelevant for the actual practice of science (Fine in fact argues that a non-realist attitude has been important for the success of the quantum theory[24]), but rather because there are no resources for settling such philosophical disputes. Perhaps this is a "natural attitude" towards philosophy - but not one that we philosophers are willing to take, in spite of the fact that battles between such wholesale philosophical orientations as realism and anti-realism will never be finally settled.

What is more, it seems to me that NOA after all is a variant of realism, even if it wishes to avoid the customary realist jargon and its refinements (such as the concept of approximate truth). According to Fine, NOA treats truth in "the usual referential way" and so "commits us, via truth, to the existence of the individuals, properties, relations, processes, and so forth referred to by the scientific statements that we accept as true". Unless 'existence' has a very unnatural meaning here, this is a realist position - for example, a statement about the existence of electrons could not be scientifically acceptable, if electrons existed only in A. Fine's mind, or if for each electron e there

24 Only some years earlier, Fine argued for the opposite claim that "contemporary science as a whole is struggling to free itself" from the "regressive, if still somewhat fashionable, scientific ideology", viz. "the deeply positivist legacy of Bohr and Heisenberg". See A. Fine, "How to Count Frequencies: A Primer for Quantum Realists", Synthese 42, 145 - 154 (1979).

existed one person p such that e exists in p's mind.[25]

In my own evaluation, realism is well alive. It is only recently that realists have developed adequate logical tools for defending theses (R4) and (R5), i.e., for showing in what sense even idealizational theories may be truthlike, how the truthlikeness of scientific statements can be estimated on evidence, and how the approximate truth of a theory explains its empirical success.[26]

Besides this positive development of his own philosophical programme, the realist can also point out serious problems in the rival views.

Descriptivism (and the related doctrine of operationalism) was convincingly refuted already in 1936 by Carnap, who showed that dispositional concepts cannot be explicitly defined in observational terms.[27]

Against instrumentalism, it can be shown that theoretical terms are logically indispensable for many important methodological functions of scientific theories,[28] and their introduction has been a fallible but effective strategy in promoting theoretical progress in science. The instrumentalist device of cutting down the cognitive content of theories loses its main motivation, if the realist can account (with his concept of truthlikeness) for the fact that theoretical laws are not strictly speaking true. Further, the whole doctrine runs into trouble with the problem that it does not seem possible to find any semantic way of

---

[25] This view is sometimes called "microphenomenalism": electrons exist if and only if they are perceived. If NOA accepts the scientific statement that tables are constituted by microparticles, then it should exclude microphenomenalism: the table in front of me exists, but no one perceives any electrons in it.

[26] These challenges to realism are forcefully presented by L. Laudan, "A Confutation of Convergent Realism", Philosophy of Science 48, 19 – 49 (1981). (Also in Leplin (ref. 15).) For replies, see Niiniluoto (ref. 15, 21), and I. Niiniluoto, "Theories, Approximations, and Idealizations", in Logic, Methodology, and Philosophy of Science VII, North-Holland, Amsterdam, 1986.

[27] See R. Carnap, "Testability and Meaning", Philosophy of Science 3, 419 – 471 (1936); 4, 1 – 40 (1937).

[28] See R. Tuomela, Theoretical Concepts, Springer-Verlag, Berlin, 1973.

dividing scientific statements into observational (with a truth value) and theoretical (without a truth value).

Against methodological non-realism, we may also ask why theoretical truth should be irrelevant: if theories have a truth value, why is the truth "about the observable" the only relevant feature of acceptable theories (as van Fraassen's constructive empiricism requires)?[29] Further, any textbook of physics reminds us that physical theories typically have intended applications that are idealized or counterfactual - even if such applications could not occur in the observable empirical reality, it is important that theories give a correct description of them.

The strategy of pragmatism or internal realism to define reality and truth in epistemic terms, as the ideal limit of scientific inquiry, faces also several problems. This kind <u>scientia mensura</u> doctrine puts epistemology before ontology: science ultimately decides what is true and what exists. This view thus guarantees only by a verbal definition that the whole "reality" is accessible to science. However, there is no general guarantee that such a limit of inquiry would always be well-defined and unique, or that the ultimate opinion of the scientific community would decide all meaningful questions about reality. Moreover, internal realism fails to give any explanation for the cognitive success of science, since it transforms the thesis that science approaches to the truth into a mere tautology.[30]

5. REALISM AND QUANTUM THEORY

We have seen that there are general philosophical reasons for regarding realism as an attractive position in ontology and epistemology. But is it also able to survive the challenge of the quantum theory? I can approach this broad and deep question here only briefly, indicating why

---

[29] For discussions of van Fraassen's constructive empiricism, see P. Churchland and C.A. Hooker (eds.), <u>Images of Science</u>, The University of Chicago Press, Chicago, 1985.

[30] Cf. Niiniluoto, ref. 15. Fine, in ref. 23, does not notice that this difficulty disproves his "Metatheorem 1".

I fail to be impressed by the frequently voiced claim that the "paradoxes" in the interpretation of quantum mechanics inevitably lead to anti-realist conclusions.[31]

The following discussion aims to illustrate how one can locate some well-known and important positions about the nature of the quantum theory in the conceptual net developed in Sections 1 – 4. This method has its obvious limitations: the philosophical views of real-life physicists are usually "mixed" cases of different influences and doctrines, and they cannot be reduced to the "pure" states defined by the abstract distinctions of the philosophers. But this fact also makes the continuing re-evaluation of the foundational discussions of physics philosophically rewarding. I believe that, in particular, the elaboration of various kinds of realist and non-realist approaches in the recent philosophy of science gives us tools for understanding better the positions taken by the generation of such great physicists as Einstein, Bohr, Heisenberg, Schrödinger, and Pauli.[32]

5.1. Folse's Bohr

Niels Bohr's philosophy of complementarity, now celebrating its sixtieth anniversary, continues to be an important object of various interpretations. As a realist, I believe there was a real Bohr who can be distinguished from different conceptions of Bohr (including Bohr's own conception of himself). Here I comment only on the very interesting

---

[31] For an excellect survey, see M. Jammer, The Philosophy of Quantum Mechanics, John Wiley and Sons, New York, 1974.

[32] Fine has argued that, apart from a weak "motivational" commitment to realism, Einstein's philosophical position could be understood as representing constructive empiricism. See A. Fine, "Einstein's Realism", in J.T. Cushing, C.F. Delaney, and G.M. Gutting (eds.), Science and Reality, University of Notre Dame Press, Notre Dame, 1984. I find this implausible: Einstein gives a strong "Platonist" emphasis on man's ability to grasp the mathematical structure of the universe by inventing physical theories, which distinguishes him from van Fraassen's demand that theories should only "save the appearances".

recent interpretation presented by Henry J. Folse.[33]

Folse's Bohr is a physicist who wishes to avoid ontological discussion - and thus favours a position similar to Fine's NOA. However, Folse argues, Bohr is committed to a form of realism, i.e., to "upholding the view that it is the task of science to understand the nature of physical reality as it exists independently of the phenomena through which we experience it". More precisely, Bohr must accept that the same "atomic object" causes, via physical interactions, different complementary "phenomenal objects". The classical description - with the wave and particle pictures - refers to properties belonging only to a phenomenal object. Indeed, no pictorial description of the atomic object is available. While the complementary phenomena are not representations of the physical object that produces them, still the combination of these complementary descriptions exhausts all that can be known about this independent reality.

As Folse's Bohr accepts the existence of quantum-mechanical objects behind the phenomena, he is an entity realist. Hence, he cannot be a phenomenalist. While there is a strong Kantian flavour in the distinction between atomic and phenomenal objects, where the former are causes of the latter,[34] Bohr is not an epistemological anti-realist: quantum theory in the framework of complementarity gives us knowledge about the atomic objects in terms of their potential for causal interaction.

How could Bohr defend the idea that his atomic objects are not merely unknowable Kantian *Dinge an sich*, that it is possible to have

---

[33] See H. Folse, *The Philosophy of Niels Bohr: The Framework of Complementarity*, North-Holland, Amsterdam, 1985. Other philosophical interpretations of Bohr are given by C.A. Hooker, "The Nature of Quantum Mechanical Reality: Einstein Versus Bohr", in R.G. Colodny (ed.), *Paradigms and Paradoxes: The Philosophical Challenge of the Quantum Domain*, University of Pittsburgh Press, Pittsburgh, 1972, pp. 67 - 302; P. Teller, "The Projection Postulate and Bohr's Interpretation of Quantum Mechanics", in *PSA 1980*, vol. 2, Philosophy of Science Association, East Lansing, 1981, pp. 201 - 223.

[34] Folse points out also important differences between Kant and Bohr. In particular, Bohr does not accept that we are *a priori* committed to using classical concepts for the description of phenomena.

some knowledge of them, in spite of the fact that he did not interpret the ψ-function as describing the state of quantum-mechanical objects?[35] In his paper for this symposium, Folse suggests that Bohr not only rejected classical realism (or the substance/property ontology) but the spectator account of knowledge and the correspondence theory of truth as well.[36] Thus, Bohr turns out to be a sort of pragmatist or epistemic realist.

While Bohr rejected classical realism, he also insisted against Einstein that Quantum Mechanics provides a "complete" framework for describing nature. This means that quantum theory does not assign any non-relational _intrinsic properties_ to atomic objects, but only what might be called relational or _interactive properties_, i.e., properties which the object possesses or fails to possess only consequent upon the occurrence of a suitable interaction.[37] Moreover, for Bohr these interactions are _observational_ interactions, i.e., they involve "preparation" or "measurement" by macroscopic devices.

If the claim of completeness were an epistemic thesis, then atomic objects might have intrinsic properties, but they would remain unknown or it would be meaningless to speak about them. This kind of position might be attractive to some logical empiricists with a physicalist outlook or to pragmatists with a verificationist theory of meaning. However, it would be compatible with local hidden variable theories - and thus vulnerable to the objections from Bell's Theorem. I don't believe that this is the view of Folse's Bohr. Rather, his claim of

---

[35] In this sense, Bohr treats the ψ-function in an instrumentalist way. Teller (ref. 33) expresses Bohr's view as follows: "The state function must be taken as a purely symbolic device for calculating the statistics of classically or commonly described experimental outcomes in collections of phenomena grouped by shared specifications of experimental conditions". This interpretation also excludes the possibility that Bohr would be a constructive empiricist.

[36] See H. Folse, "Niels Bohr's Concept of Reality", this volume.

[37] This term is used in R. Healey, "Comments on Kochen's Specification of Measurement Interactions", in _PSA 1978_, vol. 2, Philosophy of Science Association, East Lansing, 1981, pp. 277 - 294.

completeness is an ontological thesis: there are no "pictures" of atomic objects as they exist in isolation from any observation, since such objects don't have any properties at all. It is only after observational interaction that we can attribute to an atomic object x the relational property 'x caused this particle-type phenomenon' or 'x caused this wave-type phenomenon'.

This conclusion suggests that Folse's Bohr is virtually identical with Feyerabend's Bohr, who maintains that "all state descriptions of quantum-mechanical systems are relations between the systems and measuring devices in action".[38] On this view, a micro system isolated from measuring devices is a "naked individual", a bare particular without properties.[39]

If this interpretation is correct, Bohr is a radical nominalist with respect to the micro world. He is also a representative of a special kind of internal realism: the independently existing naked individuals will be dressed with relational properties only through observational interactions.[40]

5.2. Interactive Realism

Bohr's complementarity, as interpreted by Folse, is a weak form of realism: only a naked atomic object exists independently of observations, and our knowledge of such objects concerns their phenomenal manifestations.

This view rejects the claims that atomic objects are fictions (in-

---

38 See P. Feyerabend, "Problems of Microphysics", in R.G. Colodny (ed.), Frontiers of Science and Philosophy, University of Pittsburgh Press, Pittsburgh, 1962.

39 Hooker (ref. 33) argues that the idea of such "propertyless ghosts" is incoherent.

40 This version of internal realism is different from Tuomela's (ref. 16), since the latter gives strong emphasis to man's conceptualizing activity and takes the best-explaining theories to be epistemically true in a sense which involves Sellarsian "picturing".

strumentalism), mere potentialities (Heisenberg[41]), exist only during observational interactions (microphenomenalism), or are created in observational interactions.[42] It also gives up attempts to restore classical realism within deterministic (Bohm) or indeterministic (Popper[43]) frameworks.

However, there are objections to Folse's Bohr which suggest that it is reasonable to search for a version of quantum realism which is stronger than Bohr's but weaker than classical realism.

First, it is not plausible that an atomic object does not have any permanent intrinsic properties. It is true that the state of a quantum-mechanical system need not be independent of our measurements: such a system does not have a definite or sharply defined position and momentum, but our measurement may have the effect that the system is "localized". But while position and momentum are "non-classical" interactive properties[44], the same does not hold of mass and electric charge. Without

---

41 "In the experiments about atomic events we have to do with things and facts, with phenomena that are just as real as any phenomena in daily life. But the atoms or the elementary particles themselves are not as real; they form a world of potentialities or possibilities rather than one of things or facts." W. Heisenberg, Physics and Philosophy, Harper Torchbooks, New York, 1962, p. 186. Heisenberg adds that the transition from the "possible" to the "actual" takes place during the act of observation (p. 54). Even if Heisenberg says that atoms as potentialities are not "real", he also speaks of "objective" tendencies and possibilities (p. 53, 180). Thus, it is not clear whether his position about atoms is anti-realist or not.

42 Of course some micro particles are produced through man-made experiments within high-energy accelerators.

43 For Popper's most recent formulation of his position, see his "Realism in Quantum Mechanics and a New Version of the EPR Experiment", in G. Tarozzi and A. van der Merwe (eds.), Open Questions in Quantum Physics, Reidel, Dordrecht, 1985, pp. 3 - 25.

44 Constantin Piron calls such properties non-classical. In classical physics all properties are classical (i.e., either P or non-P is actual), but all properties in quantum physics need not be non-classical. See C. Piron, "Quantum Mechanics: Fifty Years Later", in P. Lahti and P. Mittelstaedt (eds.), Symposium on the Foundations of Modern Physics, World Scientific Publishers, Singapore, 1985, pp. 207 - 211. Piron says that a property is actual, "if it is certain that the positive result

such properties, we could not identify atomic objects - it hardly would make sense to say with Bohr that the same object produced two different phenomenal objects.

Secondly, if it is true to say that an atomic object x caused a particular phenomenon y, then it is likewise true that x has a causal power or potency to produce phenomena of type y. The object x can then be said to possess a dispositional property - and such a disposition is real even when x is isolated from measuring devices or from observations.[45] The idea that the isolated x is propertyless, and cannot be described by true statements, seems to be incoherent.[46]

Hence, it is possible to give quantum mechanics such a realist interpretation that atomic objects and some of their properties exist independently of observations.[47] If this is correct, we may also have descriptions of atomic objects - about their properties, relations, and structure - which are true in the correspondence-theoretical sense. We can thus resist the pragmatist move suggested by Folse in this volume. With non-classical properties we need a non-standard logic, and perhaps an idea of unsharp reality as well[48], but not a non-realist theory of

---

would be obtained if the experimentalist would decide to perform the corresponding experiment". This is not adequate, since no probabilistic disposition or propensity could be actual in this sense.

45 A glass window is fragile even at those moments when we don't test whether it breaks or not. For articles on the nature of dispositions, see R. Tuomela (ed.), Dispositions, D. Reidel, Dordrecht, 1978.

46 This remark is of course related to the idea that the square of the ψ-function can be interpreted to express an objective probability or propensity for a quantum-mechanical system. Cf. K. Popper, The Open Universe: An Argument for Indeterminism, Rowman and Littlefield, Totowa, N.J., 1982.

47 Cf. P. Lahti, "Kvanttiteoria ja realismi", Ajatus 42, 69 - 105 (1985).

48 This conception refers to the idea that reality itself may be unsharp or indefinite relative to some quantities. As shown in Niiniluoto (ref. 21), the realist concept of truthlikeness can be applied in such cases, too. Schrödinger understood clearly the distinction between blurred reality and blurred knowledge. See F. Rorlich, "Schrödinger's

truth.

Thirdly, we may accept the existence of interactive properties without restricting our attention on observational interactions, as Bohr without warrant does. For example, an electron may fail to have a sharply defined location in a box; it may become localized by our manipulating activity with measuring devices, but equally well by interactions with other macroscopic or microscopic entities. As Simon Kochen remarks, there is no reason to give a "privileged role to the measurement process among all interactions": quantum theory can be interpreted as "decribing an objective world of individual interacting systems".[49]

Thus, for an <u>interactive realist</u>, the main content of quantum mechanics is expressible by <u>interaction conditionals</u> of the form

(1) If a system x is in a particular state, and if a certain interaction occurs for x, then x will subsequently possess a specified property with a certain probability.[50]

The conditionals need not have anything to do with observations or measurements - but they are true in the quantum world, i.e., in the independent reality which quantum theory is primarily about. (For example, they hold of all unobserved micro processes inside the table in front of me at this moment.)

On the other hand, quantum mechanics also entails manipulative measurement conditionals of the form

(2) If a system x is prepared in a particular state, and if a certain measurement occurs, then the measurement gives a specified value with a certain probability.[51]

It is typical to descriptivism and instrumentalism to think that such

---

Criticism of Quantum Mechanics - Fifty Years Later", in Lahti and Mittelstaedt (ref. 44), pp. 555 - 572. Unsharp observables are discussed in the same book by P. Mittelstaedt, P. Busch, and E. Prugovečki.

49 S. Kochen, "A New Interpretation of Quantum Mechanics", in Lahti and Mittelstaedt (ref. 44), pp. 151 - 169.

50 See Healey (ref. 37), p. 281.

51 Cf., e.g., Heisenberg (ref. 44), p. 46.

statements exhaust the content of a scientific theory; for a realist, they are test statements which can be used in the empirical appraisal of the theory (see Fig. 2).[52] In brief, (1) explains (2), and (2) gives observational evidence for (1).

In this view, Bohr's philosophy of complementarity is a valuable and interesting contribution to discussions about the peculiar subject-object relations within experimental testing of quantum mechanics. But this philosophy does not imply that we (or our minds) are somehow involved in the unobserved object-object relations described by interactive conditionals of type (1).

Bohr made the well-known remark, in a Deweyan pragmatist spirit, that we are not only spectators but also actors in the drama of life. It is natural that quantum-theorists are impressed by our ability to manipulate the micro world - by influencing, transforming, and even "creating" atomic objects and some of their properties. But then they start to speak as if our use of this ability, which could be understood as a proof of realism,[53] would somehow be responsible for all the novel exciting features of the world that quantum theory has revealed - such as wave-particle dualism, transition from potentiality to actuality, and indeterminism. For a realist, this confuses the directions of explanation and evidence: quantum measurements and observations, as described by (2), are indeterministic because the ontologically mind-independent quantum world, as described by (1), is indeterministic.[54]

---

[52] Cf. M. Bunge, Philosophy of Physics, D. Reidel, Dordrecht, 1973.

[53] Cf. Hacking (ref. 7).

[54] Pauli considered "the impredictable change of the state by a single observation" to be "an abandonment of the idea of the isolation (detachment) of the observer from the course of physical events outside himself". He compares the role of the observer in the quantum theory with that of "a person, who by its freely chosen experimental arrangements and recordings brings forth a considerable 'trouble' in nature, without being able to influence its unpredictable outcome and results which afterwards can be objectively checked by everyone" (my italics). See K.V. Laurikainen, "Wolfgang Pauli and the Copenhagen Philosophy", in Lahti and Mittelstaedt (ref. 44), p. 282. The italicized part of the quotation shows, however, that the observer is "detached" even from

5.3. Veiled and Unveiled Reality

I conclude with brief remarks on Bernard d'Espagnat's concept of "veiled reality".[55]

d'Espagnat starts from a critical discussion of the problems that "forced" quantum mechanics to give up the idea that atoms and their constituents "possess a reality that is completely independent of our means of observing it". Less critically he then asserts that "consciousness exists" and "cannot be reduced to the notions of physics". In the next step, he postulates the existence of "independent" or "nonphysical reality" which is "beyond the frames of space and time and cannot be described by our current concepts". This independent reality is "veiled", since it is "intrinsically impossible" to describe it "as it really is". Still, "empirical reality" (the set of all phenomena extended in space and time) and "consciousness" are two complementary "reflections" of this independent reality.

There is again a strong Kantian flavour in d'Espagnat's distinction between an unknowable independent reality and its reflection as a mind-dependent empirical reality.[56] There is also a clear influence from objective idealism, since the "nonphysical" reality is referred to by such concepts as Spinoza's substance and "God".

---

those changes which he himself has initiated (in addition to being detached from those changes that no one observes): the single case satisfies an objective probabilistic law. Note also that, if we reject the idealizing assumption that the quantities in classical mechanics have sharp values, then a person who tosses a coin is in the same situation as the observer in quantum theory according to Pauli. See J. Ford, "How Random is a Coin Toss?", Physics Today, 40 - 47 (1983).

[55] See B. d'Espagnat, In Search of Reality, Springer-Verlag, New York, 1983.

[56] K.V. Laurikainen (this volume) also accepts the idea of "veiled reality". He suggests, as an interesting interpretation of Pauli's view, that the veil between empirical and "irrational" reality arises from the unconscious functioning of our psyche. However, if unconsciousness is somehow connected with irrationality, then should not the mind-dependent empirical reality (rather than the world behind the veil) be the "irrational" one?

Apart from the other traditional difficulties in Spinozistic and Kantian ontologies, d'Espagnat's theory faces the problem that he seems to multiply domains of reality beyond necessity. His need for saving the possibility of objective knowledge by postulating an unknowable nonphysical reality arises from the fact that he accepts a too mind-dependent account of empirical reality.[57] The critical realist will, therefore, reply to d'Espagnat that the concept of empirical reality is unnecessary in the ontological sense (see Section 2): it refers only to our partial knowledge about objective reality.[58]

With this conclusion, the idea of veil in front of intrinsically unknowable reality becomes unnecessary. Science is a search of reality - according to realism, the aim of science is informative truth about the world. The Greek word for truth, aletheia, means literally "to uncover" or "to unveil". There is only a veil between the known and so far unknown parts of reality. The task of science is to increase the former part and reduce the latter by moving this veil. Our aim as scientists and philosophers is not to veil reality but to unveil it.

---

[57] For example, to say that "our eyes contribute to the creation of atoms", by allowing the particles to emerge from the potential into the actual (ref. 55, p. 97), has a misplaced emphasis on our visual cognitive activity.

[58] Laurikainen (this volume) claims that "scientism", which is "very generally accepted" in Finland "as a kind of modern religion in the academic world", "does not acknowledge any conception of reality independent of empirical reality". I am glad to see that I am not a supporter of "scientism" in this sense: for a critical realist, nothing "depends" on empirical reality, since that reality does not even exist in an ontological sense.

Quantum Chaos and Irreversibility*

Asher Peres

Department of Physics

Technion—Israel Institute of Technology

32 000 Haifa, Israel

ABSTRACT

In classical physics, irreversibility is explained by invoking the notions of mixing and coarse graining. The quantitative criterion for chaotic evolution is the following: A system having *deterministic* dynamics is called chaotic if, in order to predict its motion over long periods of time, the demand for precision of the initial data increases *exponentially* with the time separation. In classical mechanics this indeed happens because neighboring trajectories diverge exponentially when we numerically integrate the equations of motion. There is no corresponding property in quantum theory and therefore there can be no chaos in a closed, finite quantum system. However, when a system is large enough, it cannot be perfectly isolated from its "environment," namely from external (or even internal) degrees of freedom which are not fully accounted for in the Hamiltonian of that system. As a consequence, the long range evolution of such a quantum system is essentially unpredictable—it is effectively chaotic.

*Talk presented at the Symposium on the Foundations of Modern Physics: The Copenhagen Interpretation 60 Years After the Como Lecture (Joensuu, Finland, 6-8 August 1987).

1. IRREVERSIBILITY IN CLASSICAL PHYSICS

The intuitive meaning of "irreversibility" is that there are some dynamical evolutions which can easily be prepared, but it is extremely difficult (we say "impossible") to prepare the time-reversed evolutions. For example, it is easy to get a cup of lukewarm water by placing, one hour earlier, a cube of ice in a cup of boiling water. It is impossible to prepare a cup of lukewarm water in such a way that, one hour later, it will turn into an ice cube floating in boiling water.[1]

The familiar *classical* explanation of irreversibility is based on the notions of *mixing* and *coarse graining*. Mixing is a property of chaotic systems which can be proved rigorously, under suitable assumptions, in the limit $t \to \infty$.[2] It essentially means the following: Consider two finite (and *fixed*) subsets of phase space, say V_1 and V_2, whose measures are fractions μ_1 and μ_2 of the total phase space. Suppose that the distribution $f_1(\mathbf{p}, \mathbf{q})$ is uniform in V_1 at time t_1, with $\int f_1 dV_1 = 1$. The dynamical evolution causes this distribution to become $f_2(\mathbf{p}, \mathbf{q})$ at time t_2. Then, for any t_2 *sufficiently remote* from t_1 (in the future *or* the past) and for sufficiently large μ_1 and μ_2, we have $|\int f_2 dV_2 - \mu_2| < \delta$, with arbitrarily small δ, irrespective of where V_1 is. The system thus appears to have no memory of its origin. This is the property called *mixing*. Notice that it is *time symmetric*. By itself it cannot explain irreversibility. Notice also that the smaller δ, or μ_1, or μ_2, the larger the time $|t_1 - t_2|$ needed for mixing. (In this paper I consider only *finite* times, not the unattainable mathematical limit of infinite time.)

In the ice-cube paradigm we have μ_1(ice cube plus boiling water)$\ll$ μ_2(lukewarm water) and μ_2 is of order 1. Therefore, with a suitable value of the total energy, *almost* every evolution will lead to a cup of lukewarm water, with only extremely small inhomogeneities. Indeed, the probabilities of finding the final state in V_1 and in V_2 are, respectively, $\mu_1/(\mu_1 + \mu_2) \ll 1$ and $\mu_2/(\mu_1 + \mu_2) \simeq 1$. Nevertheless, we *can*, conceptually, prepare the lukewarm water at time time t_2 so that, at a *later* time t_1 it will separate into an ice cube and boiling water. This "only" requires a *very* special preparation (not just any cup of lukewarm water, but one with delicate correlations between all the molecules) and this preparation has an *extremely* small μ_2, so small indeed that the mixing property, as defined above, will not yet be valid after the given *finite* time $t_1 - t_2$.

Now comes *coarse graining.* The standard textbook claims that we are unable, because of the coarseness of our instruments, to achieve such a preparation. We cannot coerce the initial state to reside within such a tiny V_2 (i.e., one with μ_2 so small that mixing has not yet occurred after a finite $t_1 - t_2$). In summary, we are incapable of preparing the system at time t_2 so that, after a finite time $t_1 - t_2$, it will be located with certainty in the small desired region V_1 of phase space. There are evolutions (e.g., from lukewarm water to an ice cube in boiling water) which cannot be made to proceed.

This argument deserves close scrutiny.[3] The claim that "we are unable to achieve such a preparation" ought to be *proved.* What is the degree of difficulty which is called "impossible?" (How, indeed, do we *measure* difficulty?) One should never underestimate the skill of experimental physicists! The well know spin echo[4,5] phenomenon shows that an apparently irreversible behavior can, under appropriate circumstances, be reversed. In that experiment, a large number of spins, initially aligned, acquire random phases under the influence of random, unknown, but *fixed* internal fields, acting during a *known* time t. A suitable rf pulse can then invert all these phases and, after the *same* time t, all the spins are again aligned. This is possible because all the spin precessions are *decoupled* from each other. Their equations of motion are integrable (the system is random, but not chaotic). The phases *appear* random to the untrained eye, but actually they are *correlated.* They are similar to a text which has been encrypted by a deterministic key, looks incomprehensible, but *can* be deciphered by anyone knowing the key. Information is not destroyed by encryption, it is only made less accessible. Likewise, in the spin echo case, the "key" is the strength of the rf pulse which inverts the phases and realigns the spins.

We must therefore proceed with extreme caution before claiming that it is impossible to achieve this, that, or the other preparation. Consider again the problem of motion in phase space which was discussed at the beginning of this Section. The argument is graphically represented in Fig. 1: Our task is to prepare initial data for a dynamical evolution *ending* at time $t = 0$ in a *designated area* of phase space (drawn as a circle). The experimental error box, due to the coarseness of our instruments, is represented by the small black rectangle (note that $\Delta\mathbf{p}$ and $\Delta\mathbf{q}$ are *uncorrelated*). Now, if we compute where the boundaries of the target area were located at $t = -1$ (in some arbitrary time units) we see that the error box

Figure 1: Dynamical evolution of a phase space domain, compared to the size of an experimental error box. Here **p** and **q** denote collectively all the coordinates and momenta.

could enter within these boundaries. This means that the required preparation was *possible*, if we performed it one unit of time in advance. However, if we want to perform this preparation two or more units of time in advance, we see that the shape of the bounday is so distorted that it is impossible to fit the error box inside it. We may conclude that we need a higher precision (a smaller error box). For a chaotic system—that is, for nearly all nonlinear systems—the demand in precision increases exponentially with the time separation (the coefficient is called the Lyapunov exponent[6]). Obviously, we may not be able to satisfy these more stringent precision requirements: They may not only be too costly in labor and parts, but also finally run afoul of the uncertainty principle (by this, I mean that the classical equations of motion would cease to be valid).

There is nevertheless a way to circumvent this difficulty, for some time at least. While the shape of the target area is distorted by the time evolution, its volume in phase space remains constant—this is Liouville's theorem. We may therefore try to likewise *adapt* the shape of the error box, without changing its size, by *correlating* $\Delta \mathbf{p}$ and $\Delta \mathbf{q}$. This can be done by preparing a *squeezed state*[7] of the measuring apparatus, as shown in Fig. 1 for time $t = -2$. Going to still earlier preparation times, we find that no similar squeezed state could fit into the required domain at $t = -3$ but perhaps, if we are able to pay the price, some very skilled experimental physicist could prepare a "twisted state" still fitting in it. After all, these different shapes of the target area are nothing but canonical transforms of each other. What appears complicated in one canonical coordinate system may be simple in another; all these systems are *mathematically* equivalent.

They are not, however, *physically* equivalent. The relevant question is: In a given *fixed* system, for example the one in which the observer appears to have a relatively simple shape, is there an end to this process of higher and higher refinement? It seems that there must be one, even if we ignore the uncertainty principle: The higher precision involves handling a larger amount of *information*, at the very least supplying more and more digits in the initial data. Getting this information is possible only at the cost of a corresponding increase of entropy in the rest of the world.[8] Now, if we assume that the Universe is finite, there must be an upper limit to the amount of entropy which may be generated in it.[9] Our wisdom may be large, but it is not infinite.

It thus appears that some of the laws of physics are subjective and may depend on the intellectual and experimental skills of physicists.[10] The spin echo example, discussed above, shows that it is indeed so: Some people may see only disorder (high entropy) where others may recognize a perfectly ordered—although encrypted— state, with zero entropy. This subjectivity should not be a matter of concern, as long as we admit that there is *some* limit to our skills. The laws of physics are almost indifferent to where this limit is set,[10] because the Lyapunov exponent of a *macroscopic system* is so huge that it will overcome, within an exceedingly brief time, even the most extravagant claims of reversibility.

This situation is often referred to as *chaos*, because the long term evolution of a physical system is *unpredictable*, except statistically, even when the basic dynamical law is deterministic. The final outcome appears to be *random*. The encryption key which we would need to decipher it is too long to be handled.

2. WHAT IS QUANTUM CHAOS?

The above argument thus explains classical irreversibility. The next question naturally is: What would be the corresponding argument in quantum theory? This is not at all a straightforward question because *a quantum state is not the analogue of a point in the classical phase space.*[11] If we consider two classical trajectories starting at neighboring points, their "distance" will usually increase, in a roughly exponential manner. Here, I put the word "distance" between quotes, because phase space has no metric structure (it has a symplectic structure) and the metric distance between two points is not canonically invariant. However, in any arbitrary but *fixed* metric, such as $ds^2 = d\mathbf{p}^2 + d\mathbf{q}^2$ say, an infinitesimal "distance" will ultimately grow exponentially if the system is chaotic.[12]

On the other hand, if we consider two quantum states which are initially almost identical (their scalar product is very close to 1) they will remain almost identical forever, because the Hamiltonian evolution is a unitary mapping which preserves scalar products. In fact, the same property appears in classical mechanics too: The analogue of a quantum state is a ***distribution*** in phase space[11] and, if we consider two such distributions initially centered around neighboring points, with some overlap, we likewise find that ***their overlap remains constant*** in time. Each one of these distributions may get distorted and grow "whorls and tendrils"[13] until all of phase space appears thoroughly mixed, when seen on a coarse scale. Yet, the overlap of the two distributions must remain constant, by virtue of Koopman's theorem.[14]

We must therefore carefully screen the questions which can legitimately be asked. In view of Koopman's theorem,[14] no one should be interested in how some distribution, which was initially given, will later overlap with the tortuous domain covered by the time evolution of another, initially given distribution. The legitimate question (the only one which is amenable to an experimental test) is how each one of these time evolving domains overlaps with a *fixed* domain of phase space, such as the error box drawn in Fig. 1. The answer was given at the beginning of Sect. 1: On a *fixed* coarse scale, the final distribution is homogeneous and roughly independent of the initial conditions. This is the property called ***mixing.***[2]

Considering again quantum theory, a natural question is: Does Wigner's distribution[15,16]

$$W(\mathbf{p},\mathbf{q}) = (\pi\hbar)^{-n} \int d\mathbf{y}\, \psi^*(\mathbf{q}+\mathbf{y})\, \psi(\mathbf{q}-\mathbf{y})\, e^{2i\mathbf{p}\cdot\mathbf{y}/\hbar}, \tag{1}$$

which is the quantum analogue of the Liouville distribution, also lead to mixing? The answer appears to be negative: $W(\mathbf{p},\mathbf{q})$ has a much smoother evolution in time than the Liouville distribution.[17] In particular, it cannot develop "whorls and tendrils" on scales smaller than $\hbar$. Therefore, $W(\mathbf{p},\mathbf{q})$ has not the mixing property, as defined above. (This does not conflict with the fact that, as $\hbar \to 0$, $W(\mathbf{p},\mathbf{q})$ does not tend to a limit, but rather has infinitely rapid oscillations.[18–20])

Indeed, it has been known empirically for a long time that quantum mechanics tends to suppress the appearance of chaos.[21] Quantum wave packets may remain localized, even though classical orbits are strongly chaotic, because cantori, associated with the breakup of invariant KAM surfaces, may effectively act as barriers to quantum wave packet evolution while permitting extensive classical flow. A similar phenomenon appears in models where the Hamiltonian includes a time-dependent external force.[17,22–24] In these models, which usually have a *single* degree of freedom, the physical system is prepared in a state involving only one, or at most a few energy levels of the unperturbed Hamiltonian. One then finds that, in its time evolution, the quantum system visits only a few more neighboring energy levels, so that the energy remains "localized" in a narrow domain (while in the corresponding classical evolution the energy would increase without bounds, in a diffusive way). As a consequence, the quantum motion is *almost periodic* and the initial state recurs repeatedly.[25] The same if of course true for any time-independent Hamiltonian with a discrete spectrum.[26] The problem here is not the recurrence itself, which has the same character as a Poincaré cycle[27] and is completed after an inordinately long time,[28] but the fact that the quantum state after an arbitrarily long time t can be computed accurately with a finite amount of work: The solution of the equations of motion has *null complexity.*[29] Everything is predictable. Nothing is left to chance.

Quantum chaos, if it exists, must therefore have a completely different meaning. To understand it, let us again examine the *operational* meaning of classical chaos: If we prepare a classically chaotic system in a small region of phase space and then let it evolve for *a long time*, the uncertainty in its position increases, first exponentially and then at a slower rate, until it finally reaches saturation when the entire accessible phase space is filled. If the size of the experimental error box is kept constant, the value of *every* nontrivial dynamical variable becomes uncertain. Only statistical predictions are possible. Now, quantum theory by itself is essentially probabilistic and, even for a well defined quantum state, the results of at least *some* measurements must be uncertain. One may thus say that quantum theory is intrisically chaotic.[30] There is, however, a profound difference: If ψ is known, there must be *some* nontrivial dynamical variables for which a measurement yields a definite value (namely, any variable A satisfying $A\psi = a\psi$). Therefore, the corresponding problem for a quantum system would be[29] whether or not there is randomness in its behavior *over and above the intrinsic randomness contained in the wavefunction* ψ. For example, one could conjecture that *no* nontrivial dynamical variable exists which, if measured after a long time t, would yield a definite and predictable outcome. This could perhaps be due to some unpredictability in the quantum eigenvalues or eigenfunctions, causing an unpredictable time evolution of the quantum state. It will be shown in the final section of this paper that this conjecture is indeed very plausible for quantum systems which are so large that they cannot be perfectly isolated from their environment.

The other ingredient needed for classical chaos is *coarse graining*, as symbolized by the "experimental error box" of Fig. 1. Now, coarse graining seems to make no sense in the quantum language, nor indeed in any language where dynamical variables occur with *discrete* values.[1,3] It was even suggested by Lamb[31] that it is in principle possible to prepare arbitrary, pure quantum states. In other words, there is no limit to our resolving power. This is a very bold conjecture. However, the opposite conjecture[32] would be equally bold and should be accompanied by some (possibly vague) specification of which states can—or cannot—be prepared (even the apparently trivial case of the 3-dimensional Hilbert space of a spin-1 particle has been a subject of controversy[33,34]). In this paper, I shall take the (admitedly optimistic) attitude that *any state which can be specified by means of a finite and explicit algorithm, can also be prepared experimentally.* The remaining problem thus is: Are there states which are *algorithmically* inaccessible?

3. THE PHYSICAL LIMITS OF COMPUTATION

In the realm of pure mathematics, a "proof of existence" is considered satisfactory, even if it does not show how to construct explicitly the "existing" object. Physics is more pragmatic: We want *operational* definitions of the concepts which we use. The millionth digit of π may "exist" and even be computable today, but it will *never* be possible to write 10^{100} digits of π, simply because there is not enough matter in the Universe to do the writing. It is therefore time to turn our attention to the "computational errox box." What can, or cannot, be computed?

One would like to have some objective measure for the amount of computational work needed to accomplish a given task. The notion of "logical depth" was introduced by Bennett.[35] It is, roughly speaking, the number of elementary logical steps required to compute a message from its minimal algorithmic description. Unfortunately, neither algorithmic information nor depth are effectively computable properties.[35] This limitation follows from the most basic result of computability theory, the unsolvability of the halting problem.[36,37] Moreover, the length of a computation can decrease dramatically if the minimal program which generates it is replaced by a slightly longer, but more efficient program. Nevertheless, if we impose some arbitrary rate of exchange between run time and program size, the resulting logical depth can be shown to be reasonably machine-independent.[35]

As a concrete example, consider again the time evolution of a nonintegrable dynamical system. As already stated, a system having *deterministic* dynamics can be called chaotic if, in order to predict the outcome of its motion, the *demand in precision of the initial data increases exponentially with the time separation.* For example, looking again at Fig. 1, if we want to be *sure* that the state at time $t = 0$ will be located in the given $2n$-dimensional ball, we must specify the initial data with a *number of digits which increases linearly* with the time separation. This is because neighboring trajectories diverge exponentially[38] when we numerically integrate the Hamilton equations of motion

$$\frac{d\mathbf{q}}{dt} = \frac{\partial H}{\partial \mathbf{p}} \qquad \text{and} \qquad \frac{d\mathbf{p}}{dt} = -\frac{\partial H}{\partial \mathbf{q}}. \tag{2}$$

If the complexity of that computation is beyond our means, we cannot predict what the final state will be (except statistically). Chance is supreme. The final state is said "random" and the evolution is "chaotic." In other words, chaos is *ineffectiveness*—specifically ineffectiveness in the compression of information.[39]

Let us now examine the corresponding problem in *quantum theory* and integrate the Schrödinger equation, starting with a given initial (or final) wavefunction $\psi_0(\mathbf{q})$. This could be, for instance, a wave packet located in the same region of phase space as before, if that region is much larger than $\hbar$. The standard procedure is to compute the eigenvalues and eigenfunctions of the Hamiltonian by solving

$$H\, u_n(\mathbf{q}) = \hbar\omega_n\, u_n(\mathbf{q}), \tag{3}$$

where it was assumed that the Hamiltonian has a discrete spectrum (see however below). The $u_n(\mathbf{q})$ are then normalized, and we can expand the initial state as

$$\psi_0(\mathbf{q}) = \sum c_n\, u_n(\mathbf{q}), \tag{4}$$

with

$$c_n = \int u_n^*(\mathbf{q})\, \psi_0(\mathbf{q})\, d\mathbf{q}. \tag{5}$$

Finally we obtain

$$\psi(\mathbf{q};t) = \sum c_n\, u_n(\mathbf{q})\, \exp(-i\omega_n t). \tag{6}$$

This result is essentially different from the classical one, because the computational complexity does *not* increase exponentially with time. It must increase somewhat, because in order to have meaningful phases in Eq. (6), the errors $\Delta\omega_n$ must be $\ll |t|^{-1}$. The higher precision required in ω_n means more numerical work, but the latter increases only as a polynomial in t, rather than exponentially. Actually, there is a swindle here, which is not difficult to find: The sum in Eq. (6) is *infinite*, so that this equation, taken literally, represents an infinite amount of numerical work, even if $t = 0$. This is already true for Eq. (3), unless the latter can be solved analytically, which would happen only for an integrable system. What is done in practice is to *truncate* the sums (4) and (6). Since these sums must converge, they can be truncated with an arbitrarily small, albeit finite, error. We thus replace $\psi_0(\mathbf{q})$ by

$$\psi_0'(\mathbf{q}) = \sum{}' c_n\, u_n(\mathbf{q}), \tag{7}$$

where $\sum'$ denotes the summation of a finite number of terms. Likewise we define

$$\psi'(\mathbf{q};t) = \sum{}' c_n\, u_n(\mathbf{q})\, \exp(-i\omega_n t). \tag{8}$$

Note that if $\psi_0'(\mathbf{q})$ is very close to the true initial distribution $\psi_0(\mathbf{q})$, then $\psi'(\mathbf{q};t)$ is very close to the true final distribution $\psi(\mathbf{q};t)$, by unitarity.

We thus reach the following rather curious conclusion: If we ask what are the initial conditions for a classical orbit so that, after a long time t, it will land with 100% certainty in a small region of phase space, the computational complexity of the answer increases exponentially with t. On the other hand, if we want to specify a *quantum state* such that, a time t later, it will become a small wave packet localized in the *same* small region of phase space, then the computational complexity, although formidable, will increase only as a polynomial in t. (Note however that in the quantum problem, the initial state from which we wish to obtain a small wave packet is *not* itself a small wave packet, but is spread throughout all accessible phase space.[40])

It follows that if we attach to the word "chaos" the meaning given above (the computational complexity increases exponentially with time) then *there can be no quantum chaos*, unless the Hamiltonian has a continuous spectrum (this could possibly happen in scattering problems[41]).

We are therefore still faced with our original problem, namely to explain how irreversibility occurs in quantum systems (for example, during a measurement). The inescapable cause of irreversibility was pointed out long ago by Blatt.[3] Considering Hahn's spin-echo[4] experiment, Blatt first remarked that it is not necessary to *measure* a velocity in order to *reverse* it: The process of *reversing* all velocities does not imply a *measurement* of all velocities (which would indeed be impossible for a macroscopic observer). In other words, *macroscopic observers are not restricted to coarse-grained experiments.* Therefore, it is not convincing to base fundamental arguments in statistical mechanics on coarse-graining. These remarks are in essential agreement with the conjecture proposed at the end of Sect. 2: Any state which can be specified by means of a finite and explicit algorithm, can also be prepared experimentally. There is no limit to our resolving power.

The fundamental limitation to an observer's ability is of a different and inescapable nature. He must restrict his observations to a *finite* part of the Universe. The concept of measurement involves an intrinsic dichotomy between the observer and the system being observed.[3,42,43] Someone trying to observe the whole Universe simultaneously would be completely confounded by *logical* as well as practical difficulties (he would then have to observe, among other things, the workings of his own brain).[3,42,43] Conversely, an omniscient Laplace demon would describe the world in terms incompatible with statistical mechanics, as there would be no random element involved in that description.[10] Statistical mechanics is not the mechanics of large, complicated systems; rather, it is the mechanics of limited, not completely isolated systems.[3]

It is sometimes claimed that physical systems can be isolated as well as we wish: It would then only be a matter of technological skill to reduce the influence of the external world to an arbitrarily low level. To show the fallacy of this claim, let us for a moment imagine that this external, unknown, world were completely known to us, and let us examine how it would affect the system under consideration. In the equations of motion, it would appear as additional degrees of freedom, very weakly coupled to the degrees of freedom of the original system. The coupling being very weak, one could legitimately ignore the *global* energy transfer between the internal and external degrees of freedom (superconducting currents have been shown to run for an "infinite" time, on a human scale). However, when we consider the *microscopic* behavior of a *chaotic* system, then the loss of memory of the details of its initial state proceeds much more quickly that its global approach to equilibrium with the external world.[3]

In these considerations, it should be clearly understood that the "external world" of a macroscopic system includes not only the walls of its container and everything beyond these walls. It also includes all the *internal* degrees of freedom which are not explicitly described in its Hamiltonian. For example, if we specify the state of a top by the coordinates of its center of mass and three Euler angles for its orientation, then all the remaining, unknown, "irrelevant" degrees of freedom, such as the displacements of individual atoms from their equilibrium positions, should be considered as the "external world," or the so-called "environment." Unless the "relevant" degrees of freedom are *rigorously* decoupled from the "irrelevant" ones (for example, by virtue of exact conservation laws), they are not isolated, and long range predictions are impossible, except statistically. As foreseen by Blatt,[3] the number of "relevant" degrees of freedom need not be large. We know today that *two* are enough, provided that they are nonlinearly coupled and form a nonintegrable system.[38]

To see formally how Eqs. (3)-(6) are affected by the unknown environment, let us try to use perturbation theory and write $H = H_0 + V$. It has been known empirically for a long time[44] that if the classical H_0 is chaotic, then the eigenvalues of the quantized H are extremely sensitive to the perturbation V. It can in fact be shown[45,46] that quantum perturbation theory must fail in the semiclassical limit $\hbar \to 0$. In general, semiclassical approximations may be valid in the limit $\hbar \to 0$ with t fixed, but break down if we make $t \to \infty$ with $\hbar$ fixed.[47]

There is thus a very close resemblance between the situation in quantum theory and that known to us from classical physics. The meaning of classical chaos is that, once enough time has elapsed, it is impossible to predict the value of *any* nontrivial dynamical variable. Only statistical predictions are possible. Likewise, in quantum physics, no nontrivial operator can be constructed such that its measurement would yield a predictable outcome. Indeed, if we want a dynamical variable A to have a well defined value a at time t, it must satisfy $A\psi(t) = a\psi(t)$. If we know $\psi(t)$ explicitly, it is trivial to construct an operator A satisfying this equation. But, if the system is sufficiently complex, we cannot specify $\psi(t)$ explicitly, and therefore it is also impossible to define A. In summary, a quantum system behaves irreversibly if it has some degrees of freedom which are nonlinealy coupled and interact, however weakly, with some degrees of freedom which are not explicitly accounted for in the Hamiltonian.

ACKNOWLEDGMENTS

Part of this research was done during visits to the University of Texas at Austin (Center for Theoretical Physics), the International Center for Theoretical Physics in Trieste, and Massachusetts Institute of Technology (Laboratory for Computer Science). I am grateful to all these institutions for their hospitality and support.

Research done at Technion was supported by the Gerard Swope Fund, the New York Metropolitan Research Fund, and the Fund for Encouragement of Research at Technion.

REFERENCES

1. Peres, A., Stability of Quantum Motion in Chaotic and Regular Systems, *Phys. Rev. A* **30**, 1610 (1984).
2. Arnold, V. I., and Avez, A., "Ergodic Problems in Classical Mechanics," Benjamin, New York (1968).
3. Blatt, J. M., An Alternative Approach to the Ergodic Problem, *Prog. Theor. Phys.* **22**, 745 (1959).
4. Hahn, E. L., Spin Echoes, *Phys. Rev.* **80**, 580 (1950).
5. Rothstein, J., Nuclear Spin Echo Experiments and the Foundations of Statistical Mechanics, *Am. J. Phys.* **25**, 510 (1957).
6. Reichl, L. E., "A Modern Course in Statistical Physics," Univ. of Texas Press, Austin (1980).
7. Mandel, L., Squeezed States and Sub-Poissonian Photon Statistics, *Phys. Rev. Lett.* **49**, 136 (1982).

8. Szilard, L., Über die Entropieverminderung in Einen Thermodynamischen System bei Eingriffen Intelligenter Wesen, *Z. Phys.* **53**, 840 (1929).

9. Peres, A., Can we Undo Quantum Measurements? *Phys. Rev. D* **22**, 879 (1980).

10. Peres, A., The Physicist's Role in Physical Laws, *Found. Phys.* **10**, 631 (1980).

11. Senitzky, I. R., Classical Statistics Inherent in Pure Quantum States, *Phys. Rev. Lett.* **47**, 1503 (1981).

12. Feingold, S. J. and Peres, A., Linear Stability Test for Hamiltonian Systems, *Phys. Rev. A* **26**, 2368 (1982).

13. Berry, M. V., Balazs, N. L., Tabor, M. and Voros, A., Quantum Maps, *Ann. Phys. (N. Y.)* **122**, 26 (1979).

14. Koopman, B. O., Hamiltonian Systems and Transformations in Hilbert Spaces, *Proc. Nat. Acad. Sc.* **17**, 315 (1931).

15. Wigner, E., Quantum Correction for Thermodynamic Equilibrium, *Phys. Rev.* **40**, 749 (1932).

16. Hillery, M., O'Connell, R. F., Scully, M. O. and Wigner, E. P., Distribution Functions in Physics: Fundamentals, *Phys. Reports* **106**, 121 (1984).

17. Korsch, H. J. and Berry, M. V., Evolution of Wigner's Phase-Space Density under a Nonintegrable Quantum Map, *Physica D* **3**, 627 (1981).

18. Ozorio de Almeida, A. M. and Hannay, J. H., Geometry of Two Dimensional Tori in Phase Space: Projections, Sections and the Wigner Function, *Ann. Phys. (N. Y.)* **138**, 115 (1982).

19. Ozorio de Almeida, A. M., The Wigner Function for Two Dimensional Tori: Uniform Approximation and Projections, *Ann. Phys. (N. Y.)* **145**, 100 (1983).

20. Peres, A., Semiclassical Properties of Wigner Functions, *Physica Scripta* **34**, 736 (1986).

21. Brown, R. C., and Wyatt, R. E., Quantum Mechanical Manifestation of Cantori: Wave-Packet Localization in Stochastic Regions, *Phys. Rev. Lett.* **57**, 1 (1986).

22. Fishman, S., Grempel, D. R. and Prange, R. E., Chaos, Quantum Recurrences, and Anderson Localization, *Phys. Rev. Lett.* **49**, 509 (1982).

23. Hogg T. and Huberman, B. A., Quantum Dynamics and Nonintegrability, *Phys. Rev. A* **28**, 22 (1983).

24. Casati, G., Chirikov, B. V., Guarneri, I. and Shepelyansky, D. L., Dynamical Stability of Quantum "Chaotic" Motion in a Hydrogen Atom, *Phys. Rev. Lett.* **56**, 2437 (1986).

25. Hogg T., and Huberman, B. A., Recurrence Phenomena in Quantum Dynamics, *Phys. Rev. Lett.* **48**, 711 (1982).

26. Percival, I. C., Almost Periodicity and the Quantal H Theorem, *J. Math. Phys.* **2**, 235 (1961).

27. Schulman, L. S. , Note on the Quantum Recurrence Theorem, *Phys. Rev. A* **18**, 2379 (1978).

28. Peres, A., Recurrence Phenomena in Quantum Dynamics, *Phys. Rev. Lett.* **49**, 1118 (1982).

29. Ford, J., Chaos: Solving the Unsolvable, Predicting the Unpredictable! *in*: "Chaotic Dynamics and Fractals," M. F. Barnsley and S. G. Demko, ed., Academic Press, New York (1986) pp. 1-52.

30. Lamb, W., Quantum Chaos and the Theory of Measurement, *in*: "Chaotic Behavior in Quantum Systems," G. Casati, ed., Plenum, New York (1985) pp. 353-361.

31. Lamb, W., An Operational Interpretation of Nonrelativistic Quantum Mechanics, *Phys. Today* **22**, 23 (April 1969).

32. Wigner, E. P., The Problem of Measurement, *Am. J. Phys.* **31**, 6 (1963).

33. Hultgren, B. O., III, and Shimony, A., The Lattice of Verifiable Propositions of the Spin-1 System, *J. Math. Phys.* **18**, 381 (1977).

34. Swift, A. R. and Wright, R., Generalized Stern-Gerlach Experiments and the Observability of Arbitrary Spin Operators, *J. Math. Phys.* **21**, 77 (1980).

35. Bennett, C. H., Dissipation, Information, Computational Complexity and the Definition of Organization, *in*: "Emerging Syntheses in Science," D. Pines, ed., Santa Fe Institute, Santa Fe (1985) pp. 297-313.

36. Turing, A. M., On Computable Numbers, with an Application to the Entschneidungsproblem, *Proc. London Math. Soc.* **42**, 230 (1936).

37. Hopcroft, J. E., Turing Machines, *Sci. Am.* **250**, 86 (May 1984).

38. Ford, J. and Lunsford, G. H., Stochastic Behavior of Resonant Nearly Linear Oscillator Systems in the Limit of Zero Nonlinear Coupling, *Phys. Rev. A* **1**, 59 (1970).

39. Woo, C. H., Chaos, Ineffectiveness, and the Contrast between Classical and Quantal Physics (preprint, 1987).

40. Feingold, M., Moiseyev, N. and Peres, A., Ergodicity and Mixing in Quantum Theory, *Phys. Rev. A* **30**, 509 (1984).

41. Eckhardt, B. and Jung, C., Regular and Irregular Potential Scattering, *J. Phys. A* **19**, L829 (1986).

42. Peres, A. and Zurek, W. H., Is Quantum Theory Universally Valid? *Am. J. Phys.* **50**, 807 (1982).

43. Primas, H., Contextual Quantum Objects and Their Ontic Interpretation (in these Proceedings).

44. Pomphrey, N., Numerical Identification of Regular and Irregular Spectra, *J. Phys. B* **7**, 1909 (1974).

45. Peres, A., Perturbation Theory for Chaotic Quantum Systems, *Bull. Am. Phys. Soc.* **31**, 47 (1986).

46. Feingold, M. and Peres, A., Distribution of Matrix Elements of Chaotic Systems, *Phys. Rev. A* **34**, 591 (1986).

47. Wilkinson, M., A Semiclassical Sum Rule for Matrix Elements of Classically Chaotic Systems, *J. Phys. A* **20**, 2415 (1987).

POSITION AND MOMENTUM AS SPECTRAL MEASURES ON LOCALLY COMPACT ABELIAN GROUPS

Kari Ylinen
Department of Mathematics-University of Turku
SF-20500 Turku
FINLAND

ABSTRACT

Physical quantities are treated from the point of view of harmonic analysis on locally compact abelian groups. The position and momentum of a free particle and of a particle in a box appear as examples of this approach. Commutation properties are discussed separately and in connection with uncertainty and complementarity.

1. INTRODUCTION

In the usual Hilbert space formulation of quantum mechanics physical quantities are taken to be self-adjoint operators in a complex Hilbert space H. There is a bijective correspondence $A \leftrightarrow E^A$ between such operators and Borel spectral measures on the real line, and it is customary also to talk about these spectral measures as physical quantities or observables. The purpose of this note is to point out some advantages of a general treatment of the position and linear momentum observables of a particle as (regular) spectral measures defined on the Borel σ-algebras of a locally compact abelian group G and its dual group Γ, respectively. This point of view is of course well established. For instance the Stone-von Neumann uniqueness theorem for (the Weyl form of) the canonical commutation relations is conveniently handled in the framework of abstract harmonic analysis, and then even nonabelian groups may be allowed (see e.g. Rieffel[9)] and earlier work referred to there).

The abstract approach - not an end in itself - has here its usual merits: it helps one to perceive the essential features of a given situation and yields a unified treatment of superficially different special cases. In particular, the position and momentum of a free particle and of a particle closed in a box can be studied in terms of a common structural setup. (We only deal with the case of one space dimension for simplicity.)

We are mainly concerned with the commutation properties of the position and momentum observables, and the connections between total noncommutativity, complementarity and the uncertainty relation. Our presentation is largely expository, surveying the main results in Lahti-Ylinen[7] and Ylinen[12,13], but §4 also contains some proofs.

2. NOTATION AND PRELIMINARIES

Unless specified, G will be an arbitrary locally compact abelian group and Γ its dual group. We take for granted the basics of abelian harmonic analysis as expounded e.g. in Rudin[10], whose notation we also use, though in matters pertaining to measure and integration we mainly follow Hewitt-Ross[5].

A fixed Haar measure λ (also denoted dx) for G is chosen, and the Haar measure $d\gamma$ of Γ is then so adjusted that the Plancherel theorem holds (see Rudin[10], p. 26). The complex function (class) spaces $L^2(G)$ and $L^\infty(G)$, and similarly for Γ, are defined with respect to dx and $d\gamma$ as in Hewitt-Ross[5]. We let $\mathcal{F}: L^2(G) \to L^2(\Gamma)$ denote the Fourier-Plancherel transformation; it is an isometric isomorphism. The Borel σ-algebra of e.g. G is denoted by $\mathcal{B}(G)$, and M(G) is the space of (automatically bounded) regular complex measures defined on $\mathcal{B}(G)$. The characteristic function of a set X is denoted by χ_X.

Unless specified, H is an arbitrary (complex) Hilbert space with inner product $\langle\cdot|\cdot\rangle$, linear in the second argument; often $H = L^2(G)$. The space of bounded linear operators on H is denoted by L(H), and P_H is the set of self-adjoint idempotents in L(H), i.e., the projection lattice of H.

3. POSITION AND MOMENTUM OBSERVABLES, GENERAL AND SPECIFIC

In keeping with the emphasis of this note we state:

<u>3.1. Definition</u>. A mapping $E: \mathcal{B}(G) \to L(H)$ is said to be an <u>observable</u>, if E is a regular spectral measure, i.e.,

(i) $E(X) \in P_H$ for all $X \in \mathcal{B}(G)$,

(ii) $E(G) = I$ $(= id_H)$, and

(iii) for all $\phi, \psi \in H$ the mapping $E_{\phi,\psi}: \mathcal{B}(G) \to \mathbb{C}$ defined by $E_{\phi,\psi}(X) = \langle\psi|E(X)\phi\rangle$ belongs to $M(G)$.

The requirement of regularity is a technical one which is automatically satisfied in the physical applications we have in mind. The following special cases are our main objects of study.

<u>3.2. Definition</u>. Let $H = L^2(G)$, and define $E^q(X)\phi = \chi_X\phi$ for all $X \in \mathcal{B}(G)$, $\phi \in H$. We call this $E^q: \mathcal{B}(G) \to L(H)$ the (generalized) <u>position observable</u> (related to G). We also define $E^p(Y)\phi = \mathcal{F}^{-1}(\chi_Y \mathcal{F}\phi)$ for all $Y \in \mathcal{B}(\Gamma)$, $\phi \in H$, and call this $E^p: \mathcal{B}(\Gamma) \to L(H)$ the (generalized linear) <u>momentum observable</u> (related to G).

<u>3.3. Example</u>. The usual quantum mechanical description of a free particle in one dimension (in suitable units) involves the Hilbert space $L^2(\mathbb{R})$ (with respect to the Lebesgue measure m), the position observable Q taken as the self-adjoint operator with domain $D_Q = \{\phi \in L^2(\mathbb{R}) |$ the function $x \mapsto x\phi(x)$ is in $L^2(\mathbb{R})\}$ and satisfying $Q\phi(x) = x\phi(x)$ for $\phi \in D_Q$, and the momentum observable P defined by $P\phi(x) = -i\frac{d}{dx}\phi(x)$ (a.e.) in its domain $D_P = \{\phi \in L^2(\mathbb{R}) | \phi$ is absolutely continuous on every compact interval, and $\phi' \in L^2(\mathbb{R})\}$. In this example take $G = \mathbb{R}$ with $\lambda = m$, so that Γ is $\mathbb{R}$, and its Haar measure is the Lebesgue measure divided by 2π. It is well known that $P = \mathcal{F}^{-1}Q\mathcal{F}$, and that E^q (resp. E^p) in the sense of Definition 3.2 is the spectral measure of the self-adjoint operator Q (resp. P).

<u>3.4. Example</u>. Let m be the Lebesgue measure on the interval $[0,\alpha]$ for a fixed $\alpha > 0$, and denote $H_o = L^2([0,\alpha],m)$. Define $Q: H_o \to H_o$ by $Q\phi(x) = x\phi(x)$, and $P: D_P \to H_o$ by $P\phi(x) = -i\frac{d}{dx}\phi(x)$ (a.e.), where $D_P = \{\phi \in H_o | \phi$ is absolutely continuous, $\phi' \in H_o$, and $\phi(0) = \phi(\alpha)\}$. It is

well known that Q and P are self-adjoint operators; Q is the position and P the momentum observable of a one-dimensional particle "in the box $[0,\alpha]$". As in the preceding example, the spectral measures of Q and P are naturally subsumed in Definition 3.2. Indeed, let G be quotient group $\mathbb{R}/\alpha\mathbb{Z}$, and choose for G the Haar measure with total mass α. If $\rho: [0,\alpha] \to \mathbb{R}/\alpha\mathbb{Z}$ is the quotient map restricted to $[0,\alpha]$, then $\theta: H \to H_o$ where $H = L^2(G)$ and $\theta(\phi) = \phi \circ \rho$ is an isometric isomorphism. For each $n \in \mathbb{Z}$ we define $\gamma_n: G \to \mathbb{C}$ by $\gamma_n(x + \alpha\mathbb{Z}) = \exp(i2\pi\alpha^{-1}nx)$, $x \in \mathbb{R}$. Then the dual group Γ of G is $\{\gamma_n \mid n \in \mathbb{Z}\}$; we identify $2\pi n$ with γ_n, so that $\Gamma = 2\pi\alpha^{-1}\mathbb{Z}$ (which is the annihilator of $\alpha\mathbb{Z}$ in the dual group of $\mathbb{R}$). In the notation of Definition 3.2, the spectral measure E^Q of Q satisfies $E^Q(X) = \theta E^q \rho(X \cap [0,\alpha])\theta^{-1}$ for $X \in \mathcal{B}(\mathbb{R})$, and the spectral measure E^P of P satisfies $E^P(X) = \theta E^p(X \cap 2\pi\alpha^{-1}\mathbb{Z})\theta^{-1}$, $X \in \mathcal{B}(\mathbb{R})$ (Ylinen[12]). In particular, we see at once the fact that the spectrum of P is $2\pi\mathbb{Z}$.

4. THE COMMUTATIVITY DOMAINS OF PAIRS OF OBSERVABLES AND THEIR INVERSE FOURIER TRANSFORMS

One measure of the degree of commutativity of two observables is their commutativity domain introduced in Hardegree[4]. We use the following more general definition.

4.1. Definition. Let Ω_1 and Ω_2 be sets and $\Phi_i: \Omega_i \to L(H)$ mappings. We denote $\mathrm{com}(\Phi_1,\Phi_2) = \{\phi \in H \mid \Phi_1(x)\Phi_2(y)\phi = \Phi_2(y)\Phi_1(x)\phi$ for all $x \in \Omega_1$, $y \in \Omega_2\}$, and call $\mathrm{com}(\Phi_1,\Phi_2)$ the commutativity domain of Φ_1 and Φ_2.

The following result has been observed in special cases (see Hardegree[4], p. 500, Ylinen[11], p. 692).

4.2. Lemma. In the situation of Definition 3.1, $\mathrm{com}(\Phi_1,\Phi_2)$ is a closed subspace of H. If the range of Φ_i is closed with respect to operator multiplication, then $\mathrm{com}(\Phi_1,\Phi_2)$ is invariant under each $\Phi_i(x)$, $x \in \Omega_i$.

Proof. Since $\mathrm{com}(\Phi_1,\Phi_2) = \bigcap_{x\in\Omega_1, y\in\Omega_2} \ker(\Phi_1(x)\Phi_2(y) - \Phi_2(y)\Phi_1(x))$, $\mathrm{com}(\Phi_1,\Phi_2)$ is a closed subspace of H. Suppose now that e.g. $\Phi_1(\Omega_1)$ is closed with respect to multiplication. If $x, y \in \Omega_1$, $z \in \Omega_2$, and $\phi \in \mathrm{com}(\Phi_1,\Phi_2)$, choose $u \in \Omega_1$ such that $\Phi_1(x)\Phi_1(y) = \Phi_1(u)$, so that

$\Phi_1(x)\Phi_2(z)\Phi_1(y)\phi = \Phi_1(x)\Phi_1(y)\Phi_2(z)\phi = \Phi_1(u)\Phi_2(z)\phi = \Phi_2(z)\Phi_1(u)\phi = \Phi_2(z)\Phi_1(x)\Phi_1(y)\phi$. Thus $\Phi_1(y) \in \mathrm{com}(\Phi_1,\Phi_2)$. A similar argument works for $i = 2$. □

Two important instances where the assumption in the second part of Lemma 4.2 holds are spectral measures and group representations. Recall the bijective (Stone-Naimark-Ambrose-Godement) correspondence between observables (i.e., regular spectral measures) $E: \mathcal{B}(\Gamma) \to L(H)$ and continuous unitary representations $U: G \to L(H)$ connected by the formula $U(x) = \int_\Gamma (x,\gamma)dE(\gamma)$; we call U the <u>inverse Fourier transform</u> of E. (Note that E is a spectral measure on the dualgroup Γ, and (x,γ) is without complex conjugation.)

<u>4.3. Theorem</u>. Let G_1 and G_2 be locally compact abelian groups with dual groups Γ_1 and Γ_2, respectively. Let $E_i: \mathcal{B}(\Gamma_i) \to L(H)$ be an observable and $U_i: G_i \to L(H)$ its inverse Fourier transform, $i = 1, 2$. Then $\mathrm{com}(E_1,E_2) = \mathrm{com}(U_1,U_2)$.

<u>Proof</u>. We denote $K_1 = \mathrm{com}(E_1,E_2)$ and $K_2 = \mathrm{com}(U_1,U_2)$. By Lemma 4.2 K_2 is a closed subspace which is invariant under each $U_1(x)$ and $U_2(y)$, $x \in G_1$, $y \in G_2$. Thus we get by restriction continuous unitary representations $U_i^{K_2}: G_i \to L(K_2)$. Similarly, since the orthogonal complement of an invariant subspace under a unitary representation is also invariant, we get continuous unitary representations $U_i^{K_2^\perp}: G_i \to L(K_2^\perp)$ such that $U_i = U_i^{K_2} \oplus U_i^{K_2^\perp}$. But the representations $U_i^{K_2}$ and $U_i^{K_2^\perp}$ may be obtained as inverse Fourier transforms of regular spectral measures $E_i^{K_2}: \mathcal{B}(\Gamma_i) \to L(K_2)$ and $E_i^{K_2^\perp}: \mathcal{B}(\Gamma_i) \to L(K_2^\perp)$. Obviously U_i is then the inverse Fourier transform of the regular spectral measure $E_i^{K_2} \oplus E_i^{K_2^\perp}$, and by the bijectivity of the Stone-Naimark-Ambrose-Godement correspondence we must have $E_i = E_i^{K_2} \oplus E_i^{K_2^\perp}$. In particular, $K_2^\perp$ is invariant under each $E_1(X)$ and $E_2(Y)$ for $X \in \mathcal{B}(\Gamma_1)$, $Y \in \mathcal{B}(\Gamma_2)$, and we have e.g. $E_1(X)\phi = E_1^{K_2}(X)\phi$, $\phi \in K_2$. Since $U_1^{K_2}(x)$ and $U_2^{K_2}(y)$ commute for all $x \in G_1$, $y \in G_2$, the projections $E_1^{K_2}(X)$ and $E_2^{K_2}(Y)$ commute for all $X \in \mathcal{B}(\Gamma_1)$, $Y \in \mathcal{B}(\Gamma_2)$. (It is well known that, in general, if $U: G \to L(H)$ is the inverse Fourier transform of a regular spectral measure $E: \mathcal{B}(\Gamma) \to L(H)$, an operator $T \in L(H)$ commutes with every $U(x)$, $x \in G$, if and only if T commutes with

every $E(X)$, $X \in \mathcal{B}(\Gamma)$.) This means that $K_2 \subset K_1$. An analogous argument could be used to show that $K_1 \subset K_2$, but in fact this direction is easier since each $U_1(x)$ (resp. $U_2(y)$) can be approximated in norm by linear combinations of some $E_1(X)$, $X \in \mathcal{B}(\Gamma_1)$ (resp. $E_2(Y)$, $Y \in \mathcal{B}(\Gamma_2)$). If $\phi \in K_1$, then the product of such linear combinations gives the same value at ϕ in either order, and the same is true in the limit. Thus $K_1 \subset K_2$. □

Denoting $H = L^2(G)$, the inverse Fourier transform of E^p: $\mathcal{B}(\Gamma) \to L(H)$ is known (and easily shown) to be U^p: $G \to L(H)$ where $[U^p(x)\phi](y) = \phi(x + y)$, and that of E^q: $\mathcal{B}(G) \to L(H)$ is U^q: $\Gamma \to L(H)$ where $U^q(\gamma)\phi = \gamma\phi$. (We regard G as the dual group of Γ in accordance with the Pontryagin duality theorem.) From the Weyl relation $U^p(x)U^q(\gamma) = (x,\gamma)U^q(\gamma)U^p(x)$, which immediately follows from these remarks, one sees that if G has at least two elements, so that $(x,\gamma) \neq 1$ for some $x \in G$ and $\gamma \in \Gamma$, then $\phi \in \mathrm{com}(U^q,U^p)$ implies that $[1 - (x,\gamma)]U^p(x)U^q(\gamma)\phi = 0$, i.e., $\phi = 0$ since $U^p(x)$ and $U^q(\gamma)$ are injective. Thus Theorem 4.3 implies the following fact (which in standard terminology is expressed by saying that E^p and E^q are <u>totally noncommutative</u>):

<u>4.4. Corollary</u>. If $G \neq \{0\}$, then $\mathrm{com}(E^q,E^p) = \{0\}$.

In view of Examples 3.3 and 3.4, (the spectral measures of) the position and momentum observables of a free particle and of one in a box are therefore totally noncommutative. The first case was already noted in Hardegree[4]; both also follow from different arguments in Lahti-Ylinen[7] (see §6 below).

5. COMMUTING SPECTRAL PROJECTIONS OF POSITION AND MOMENTUM

Even though for any G with more than one element E^q and E^p are totally noncommutative, some nontrivial projections $E^q(X)$ and $E^p(Y)$ may commute. In Busch-Lahti[2] a sufficient condition (in the case $G = \mathbb{R}$) was pointed out: $E^q(X)$ and $E^p(Y)$ commute, if X and Y are periodic sets with suitably related periods. In Busch-Schonbek-Schroeck[3] it was shown that in effect periodicity is also necessary; the following is a slight reformulation of a statement in that paper.

<u>5.1. Theorem</u> (Busch-Schonbek-Schroeck). Let $f, g \in L^\infty(\mathbb{R})$. Then $fF^{-1}[gF(\phi)] = F^{-1}[gF(f\phi)]$ for all $\phi \in L^2(\mathbb{R})$ if and only if one of the following conditions holds:

(a) f or g equals a constant a.e.;

(b) both f and g coincide a.e. with periodic functions with minimal periods α and β, respectively, such that $\frac{2\pi}{\alpha\beta}$ is an integer.

For the proof of the following generalization of this result to the case of any locally compact abelian group we refer to Ylinen[13].

<u>5.2. Theorem</u>. Suppose $f \in L^\infty(G)$ and $g \in L^\infty(\Gamma)$. The following conditions are equivalent:

(i) $fF^{-1}[gF(\phi)] = F^{-1}[gF(f\phi)]$ for all $\phi \in L^2(G)$;

(ii) there is a closed subgroup $K \subset G$ such that for every $y \in K$ the equality $f(x + y) = f(x)$ holds for locally almost every $x \in G$, and for every $\delta \in K^\perp$ (annihilator of K in Γ) $g(\gamma + \delta) = g(\gamma)$ for locally almost every $\gamma \in \Gamma$.

If $G = \mathbb{R}$, Theorem 5.2 is easily seen to imply Theorem 5.1. Applying Theorem 5.2 to the quotient group $\mathbb{R}/\alpha\mathbb{Z}$ one sees from Example 3.4 (in its notation) that if $f, g: \mathbb{R} \to \mathbb{C}$ are bounded Borel functions, their spectral integrals $\int_{\mathbb{R}} f dE^Q$ and $\int_{\mathbb{R}} g dE^P$ commute if, and only if, one of the following conditions holds:

(a) the restriction $f|[0,\alpha]$ equals a constant a.e. or $g|2\pi\alpha^{-1}\mathbb{Z}$ is constant;

(b) there is an integer $n \geq 2$ such that $2\pi\alpha^{-1}n$ is a period of $g|2\pi\alpha^{-1}\mathbb{Z}$, and whenever $j = 1,\dots,n-1$, the equality $f(x) = f(x + j\alpha n^{-1})$ holds for almost every $x \in [0,\alpha n^{-1}]$.

In particular, one has a characterization for the commuting spectral projections of the position and momentum of a particle in a box.

6. TOTAL NONCOMMUTATIVITY, COMPLEMENTARITY AND THE UNCERTAINTY RELATION

In this section we return to the setup of conventional quantum mechanics and regard physical quantities as self-adjoint operators on H (or, equivalently, as their spectral measures defined on $\mathcal{B}(\mathbb{R})$). However, the group theoretic approach is relevant here, too; we again consider the position and momentum of a free particle and of one in a box in the spirit of Examples 3.3 and 3.4. Some of the results we quote could easily be extended to more general groups than $\mathbb{R}$ or $\mathbb{R}/\alpha\mathbb{Z}$.

Let A and B be physical quantities, i.e., self-adjoint operators in H with spectral measures E^A and E^B, respectively. If T is a (mixed) state, i.e., a positive normalized trace class operator, we write e.g. $E_T^A(X) = \mathrm{tr}(TE^A(X))$ for $X \in \mathcal{B}(\mathbb{R})$, and denote by Var(A,T) the variance of the probability distribution E_T^A on the real line. If there is a constant $c > 0$ such that $\mathrm{Var}(A,T)\mathrm{Var}(B,T) \geq c$ for every state T, then A and B are said to satisfy the <u>uncertainty relation</u>.

If $E^A(X) \wedge E^B(Y) = 0$ for all bounded X, $Y \in \mathcal{B}(\mathbb{R})$ such that $E^A(X) \neq I$ and $E^B(Y) \neq I$, then A and B are said to be <u>complementary</u>. (References to background material can be found e.g. in Lahti[6].) Short proofs of the results in the following theorem are given in Lahti-Ylinen[7]. (For earlier parallel work in a more general setting, see Pulmannova-Dvurecenskij[8].)

<u>6.1. Theorem</u>. (a) If A and B satisfy the uncertainty relation, then $\mathrm{com}(E^A, E^B) = \{0\}$.
(b) If A and B are nonconstant (i.e., $E^A(\mathcal{B}(\mathbb{R})) \neq \{0,I\}$ and $E^B(\mathcal{B}(\mathbb{R})) \neq \{0,I\}$) and complementary, then $\mathrm{com}(E^A, E^B) = \{0\}$.

<u>6.2. Remark</u>. For a free particle (Example 3.3), P and Q are known to be complementary (see e.g. Busch-Lahti[1]), and for a particle in a box (Example 3.4) this is shown in Lahti-Ylinen[7], where it is also observed that as a consequence of Theorem 6.1, in both cases P and Q are totally noncommutative; recall Corollary 4.4, too.

REFERENCES

1. Busch, P. and Lahti, P., "A note on quantum theory, complementarity, and uncertainty", Philosophy of Science <u>52</u>, 64-77 (1985).
2. Busch, P. and Lahti, P., "To what extent do position and momentum commute?", Physics Letters A <u>115</u>, 259-264 (1986).
3. Busch, P., Schonbek, T.P., and Schroeck, F.E., Jr., "Quantum obser-

vables: compatibility versus commutativity and maximal information", manuscript.

4. Hardegree, G.M., "Relative compatibility in conventional quantum mechanics", Found. Phys. 7, 495-510 (1977).
5. Hewitt, E. and Ross, K.A., "Abstract harmonic analysis", Vol. I, Springer-Verlag, Berlin, 1963.
6. Lahti , P.J., "Uncertainty, complementarity, and commutativity", in: Recent Developments in Quantum Logic (eds. Mittelstaedt and Stachow), B.I.-Wissenschaftsverlag, Mannheim, 1985.
7. Lahti, P.J. and Ylinen, K., "On total noncommutativity in quantum mechanics", J. Math. Phys. (to appear).
8. Pulmannova, S. and Dvurecenskij, A., "Uncertainty principle and joint distributions of observables", Found. Phys. 10, 641-655 (1980).
9. Rieffel, M.A., "On the uniqueness of the Heisenberg commutation relations", Duke Math. J. 39, 745-752 (1972).
10. Rudin, W., "Fourier analysis on groups", Interscience, New York and London, 1962.
11. Ylinen, K., "On a theorem of Gudder on joint distributions of observables", In: Symposium on the Foundations of Modern Physics (eds. Lahti and Mittelstaedt), pp. 691-694, World Scientific Publishing Co., Singapore, 1985.
12. Ylinen, K., "Commutation properties of observables as spectral measures on locally compact abelian groups", University of California, Riverside, Department of Mathematics, Technical Report, March, 1987.
13. Ylinen, K., "Commuting functions of the position and momentum observables on locally compact abelian groups", J. Math. Anal. Appl., to appear.

POSTERS

QUANTUM MECHANICS AND CLASSICAL STATISTICAL MECHANICS

Alexander Bach

Institut für Theoretische Physik I, Westfälische Wilhelm Universität, D-4400 Münster, Federal Republik of Germany

TIME SYMMETRY AND THE INTERPRETATION OF QUANTUM MEASUREMENT

Michel Bitbol

Institut de Biologie Physico-Chimique, 13 Rue Pierre et Marie Curie, F-75005 Paris, France

PERSISTENCE OF INTERACTIONS IN QUANTUM THEORY

Philip B. Burt

Department of Physics and Astronomy, Clemson University Clemson, South Carolina 29634-1911, USA

BOHR, EINSTEIN AND REALISM

Wojciech Daniel

Institute of Physics, Nicholas Copernicus University, Grudziadzka 5/7, 87-100 Torun, Poland

THE ROLE OF THE GLEASON EXPANSION IN THE INTERPRETATION OF QUANTUM MECHANICS

J.E.G. Farina

Department of Mathematics, The University of Nottingham, University Park, Nottingham NG7 2RD, England

THE PHILOSOPHY OF BOHR AND WITTGENSTEIN

Christoffer Gefwert and Stig Stenholm

Research Institute for Theoretical Physics, Helsinki, Finland

A POSSIBLE EXPERIMENTAL DECISION BETWEEN THE COPENHAGEN AND REALISTIC INTERPETATION OF QUANTUM THEORY AND A PROPOSED EXPERIMENT OF HEISENBERG'S UNCERTAINTY RELATION OF SPACE AND MOMENTUM IN AN ORIGINAL EPR CONFIGURATION

Klaus Götsch, Thomas Koch, Günter Lüning, Jens Scheer, Michael Schmidt

University of Bremen, Physics Department, D-2800 Bremen 33, Federal Republic of Germany

PROPOSED EXPERIMENT OF HEISENBERG'S UNCERTAINTY RELATION OF SPACE AND MOMENTUM IN AN ORIGINAL EINSTEIN-PODOLSKY-CONFIGURATION

Klaus Götsch, Thomas Koch, Günter Lüning, Jens Scheer, Michael Schmidt

University of Bremen, Physics Department, D-2800 Bremen 33, Federal Republik of Germany

COMPATIBILITY OF OBSERVABLES REPRESENTED BY POSITIVE OPERATOR VALUED MEASURES

P. Kruszynski[1] and W.M. de Muynck[2]

[1]Department of Mathematics and Informatics, Delft University of Technology, Delft, The Netherlands, [2]Department of Physics, Eindhoven University of Technology, Eindhoven, The Netherlands

MEASUREMENT BY PHASE SEVERANCE

H. Pierre Noyes

Stanford Linear Accelerator Center, Stanford CA 94305, USA

QUANTUM MECHANICS DOES NOT FORCE THE EPR ON US IF WE DO NOT FORCE THE ACTION AT A DISTANCE ON QUANTUM MECHANICS

Oreste Piccioni and Werner Mehlhop

Department of Physics, University of California San Diego, La Jolla, California 92093, USA

MÖBIUS TRANSFORMATION - THE MISSING LINK

Reijo Piirainen

Wihuri Physical Laboratory, University of Turku, SF-20500 Turku, Finland

THOUGHT, SCIENCE AND THE WHOLENESS OF THE UNIVERSE - A LAYMAN'S VIEW ON QUANTUM PHYSICS

Klaus Rahikainen

Pitkämäki, SF-73100 Lapinlahti, Finland

QUANTUM CAUSAL STRUCTURE AND THE EPR EXPERIMENT

Laszlo Szabo

Institute for Theoretical Physics, Eötvös University Budapest, H-1088 Puskin u. 5-7, Hungary

THE REALITIES OF BOHR AND d'ESPAGNAT: A COMPARISON

Carol Westerlund

Physics Department, University of Helsinki, Helsinki, Finland

CAN QUANTUM UNPREDICTABILITY AND WAVE PACKET COLLAPSE BE EXPLAINED BY A DETERMINISTIC THEORY?

J.W.G. Wignall

School of Physics, University of Melbourne, Victoria, Australia

A DEMONSTRATION OF AN EXPERIMENTAL SET-UP OF A PHOTON SPLIT BEAM EXPERIMENT

Anssi Mälkki & Rauno Hämäläinen

Department of Physics, University of Joensuu, Joensuu, Finland

SYMPOSIUM ON THE FOUNDATIONS OF MODERN PHYSICS

The Copenhagen Interpretation 60 years after the Como Lecture

Joensuu, Finland, August 6-8, 1987

List of Participants

Alanen, P.
Institute of Dentistry
University of Turku
20500 Turku
Finland

Andersson, W.
Department of Physics
Iowa State University
Ames, Iowa 50011
USA

Bach, A.
Institut für Theoretische Physik I
Westfälische Wilhelms Universität
D-4400 Münster
Bundesrepublik Deutschland

Bitbol, M.
Institut de Biologie Physico-Chimique
13 Rue Pierre et Marie Curie
75005 Paris
France

Bleuler, K.
Institut für Theoretische Kernphysik
Universität Bonn
D-5300 Bonn 1
Bundesrepublik Deutschland

Bohm, D.
Department of Physics
Birkbeck College
London WCIE 7HX
England

Borg, F. G.
Palo 15
67400 Karleby
Finland

Burt, P.
Department of Physics and Astronomy
Clemson University
Clemson, South Carolina 29634-1911
USA

Busch, P.
Institut für Theoretische Physik
Universität zu Köln
D-5000 Köln 41
Bundesrepublik Deutschland

Carpentier, P.
French Embassy
Ruoholahdenkatu 23
00180 Helsinki
Finland

Cattaneo, G.
Department of Information Science
University of Milano
20133 Milano
Italy

Christiansen, P.
Institute of Mathematics and Physics
Roskilde University Center
DK-4000 Roskilde
Denmark

Daniel, W.
Institute of Physics
Nicholas Copernicus University
87-100 Torun
Poland

d'Espagnat, B.
Laboratoire de Physique
Théorique et Particules Elémentaires
Université de Paris XI
Centre d'Orsay Batiment 211
91405 Orsay Cedex
France

Enz, C.P.
Département de Physique Théorique
Université de Genève
CH-1211 Genève 4
Switzerland

Farina, J.E.G.
Department of Mathematics
The University of Nottingham
Nottingham NG7 2RD
England

Fennema, J.W.R.
Marterlaan 42
1216EX, Hilversum
The Netherlands

Finkelstein, D.
Georgia Institute of Technology
Atlanta Georgia 30332
USA

Folse, H.J.
Department of Philosophy
Loyola University New Orleans
Louisiana 70118
USA

Garola, C.
Dipartimento di Fisica
Universita de Lecce
73100 Lecce
Italy

Gebhard, G.
Abrahaminkatu 9 B 36
00180 Helsinki
Finland

Golshani, M.
Department of Physics
Sharif University of Technology
Teheran
Iran

Grabowski, M.
Institute of Physics
Nicholas Copernicus University
Torun
Poland

Gregorius, P.
The Delhi Orthodox Centre
2 Tughlagabad Institute Area
New Delhi 110062
India

Görnitz, T.
c/o Prof. C.F. von Weizsäcker
Bahnhofplatz 4
D-8130 Starnberg
Bundesrepublik Deutschland

Hokkanen, V.-M.
EMT 3-5 B 21
40740 Jyväskylä
Finland

Holmberg, K.
Swedish Herbal Insitute
41314 Göteborg
Sweden

Holmberg, L.
University of Stockholm
Institute of Physics
S-11346 Stockholm
Sweden

Hooton, D.J.
Department of Physics
The University of Auckland
Auckland
New Zealand

Huhmar, E.
Rantalaivurintie 15
00850 Helsinki
Finland

Hyvönen, T.
Töölönkatu 52 A 19
00250 Helsinki
Finland

Hämäläinen, R.
Department of Physics
University of Joensuu
80101 Joensuu
Finland

Jones, M.
Wave Propagation Laboratory
Environmental Research Laboratories
U.S. Department of Commerce
Boulder, Colorado 80303
USA

Just, K.
Department of Physics
University of Arizona
Tucson
Arizona 85721
USA

Jämsä, H.
Papinkuja 2
81700 Lieksa
Finland

Kouki, T.
Department of Physics
University of Joensuu
80101 Joensuu
Finland

Kruszynski, P.
Dept. of Mathematics and Informatics
Delft University of Technology
2600 AY Delft
The Netherlands

Kähkönen, K.
Yliopistokatu 13
80130 Joensuu
Finland

Lahti, P.
Department of Physical Sciences
University of Turku
SF-20500 Turku
Finland

Laurikainen, K.V.
Research Institute of Theoretical Physics
University of Helsinki
00170 Helsinki
Finland

Lensu, M.
Karstulantie 4 A 3
00550 Helsinki
Finland

Lipas, P.
Department of Physics
University of Jyväskylä
40100 Jyväskylä
Finland

Liukkonen, S.
Department of Chemistry
Laboratory of Physical Chemistry
Helsinki University of Technology
02150 Espoo
Finland

Mansikka, K.
Department of Physical Sciences
University of Turku
20500 Turku
Finland

Mehra, J.
7830 Candle Lane
Houston, Texas 77071
USA

Mittelstaedt, P.
Institut für Theoretische Physik
Universität zu Köln
D-5000 Köln 41
Bundesrepublik Deutschland

de Muynck, W.
Department of Theoretical Physics
Eindhoven University of Technology
Eindhoven
The Netherlands

Mälkki, A.
University of Helsinki
00101 Helsinki
Finland

Nagel, B.
Department of Theoretical Physics
Royal Institute of Technology
S-100 44 Stockholm
Sweden

Niiniluoto, I.
Department of Philosophy
University of Helsinki
00101 Helsinki
Finland

Noyes, P.
Stanford Linear accelerator Center
Stanford, CA 94305
USA

Ohba, I.
Department of Physics
Waseda University
Tokyo 160
Japan

Peres, A.
Department of Physics
Technion
3200 Haifa
Israel

Piccioni, O.
Department of Physics
University of California
La Jolla, California 92093
USA

Piirainen, R.
Linnank. 5 C 63
20110 Turku
Finland

Porra, J.
Servin-Maijan tie 10 C 31
02150 Espoo
Finland

Prieur, A.
Institut für Theoretische Physik
Universität zu Köln
5000 Köln 41
Bundesrepublik Deutschland

Primas, H.
Laboratory of Physical Chemistry
ETH
CH-8092 Zürich
Switzerland

Pylkkänen, P.
Kylänevantie 16 D 58
00320 Helsinki
Finland

Qvarnström, B.-O.
Hurtinkatu 11 D 28
20600 Turku
Finland

Raczynski, A.
Institute of Physics
Nicholas Copernicus University
87-100 Torun
Poland

Rahikainen, K.
Pitkämäki
73100 Lapinlahti
Finland

Rainio, H.
Finnish Meteorological Institute
00101 Helsinki
Finland

Rastas, J.
Department of Chemistry
Helsinki University of Technology
02150 Espoo
Finland

Rayski, J.
Institute of Physics
Jagellonian University
30-059 Krakow
Poland

Rugh, S.E.
Regensen
St. Kunnikestrade 2
1169 Kbhk
Denmark

Ryff, L.C.B.
Universidade Federal do Rio de Janeiro
Instituto de Fisica
Ilha do Fundao, Bloco A, 30 and. Centro de Technologia
CEP: 21.944-Rio de Janeiro
Brasil

Sandelin, S.
Department of Education
University of Helsinki
SF-00140 Helsinki
Finland

Scheer, J.A.
Department of Physics
University of Bremen
D-2800 Bremen 33
Bundesrepublik Deutschland

Seppälä, T.
Koskelantie 52 E 34
00610 Helsinki
Finland

Sibelius, P.
Department of Mathematics
Åbo Akademi
20500 Åbo
Finland

Smulter, L.
Department of Physics
Åbo Akademi
20500 Turku
Finland

Solombrino, L.
Dipartimento di Fisica
Universita de Lecce
73100 Lecce
Italy

Szabo, L.
Institute for Theoretical Physics
Eötvös University
Budabest
Hungary

Szamosi, G.
Department of Physics
University of Windsor
Windsor, Ontario N9B 3P4
Canada

Venkula, J.
University of Helsinki
00100 Helsinki
Finland

Viiri, J.
Teacher Training Institute
University of Joensuu
80101 Joensuu
Finland

von Weizsäcker, C.F.
Bahnhofplatz 4
D-8130 Starnberg
Bundesrepublik Deutschland

Westerlund, C.
Niemenmäentie 10 A 28
00350 Helsinki
Finland

Wignall, J.
School of Physics
University of Melbourne
Parkville, Victoria 3052
Australia

Wikman, G.
Swedish Herbal Institute
41314 Göteborg
Sweden

Ylinen, K.
Department of Mathematics
University of Turku
20500 Turku
Finland

Zeilinger, A.
Atominsitut der Österreichischen Universitäten
A-1020 Wien
Austria

AUTHOR INDEX